Lecture Notes in Mechanical Engineering

Lecture Notes in Mechanical Engineering (LNME) publishes the latest developments in Mechanical Engineering—quickly, informally and with high quality. Original research reported in proceedings and post-proceedings represents the core of LNME. Also considered for publication are monographs, contributed volumes and lecture notes of exceptionally high quality and interest. Volumes published in LNME embrace all aspects, subfields and new challenges of mechanical engineering. Topics in the series include:

- Engineering Design
- Machinery and Machine Elements
- Mechanical Structures and Stress Analysis
- Automotive Engineering
- Engine Technology
- Aerospace Technology and Astronautics
- Nanotechnology and Microengineering
- Control, Robotics, Mechatronics
- MEMS
- Theoretical and Applied Mechanics
- Dynamical Systems, Control
- Fluid Mechanics
- Engineering Thermodynamics, Heat and Mass Transfer
- Manufacturing
- Precision Engineering, Instrumentation, Measurement
- Materials Engineering
- Tribology and Surface Technology

More information about this series at http://www.springer.com/series/11236

Sanjay Singh · Pushkar Raj · Samir Tambe
Editors

Proceedings of the International Conference on Modern Research in Aerospace Engineering

MRAE-2016

Springer

Editors
Sanjay Singh
Amity Institute of Aerospace Engineering
Amity University
Noida, Uttar Pradesh
India

Pushkar Raj
DTC Secretariat
Ghaziabad, Uttar Pradesh
India

Samir Tambe
Department of Aerospace Engineering and Engineering Mechanics
College of Engineering and Applied Sciences, University of Cincinnati
Cincinnati, OH
USA

ISSN 2195-4356 ISSN 2195-4364 (electronic)
Lecture Notes in Mechanical Engineering
ISBN 978-981-10-5848-6 ISBN 978-981-10-5849-3 (eBook)
https://doi.org/10.1007/978-981-10-5849-3

Library of Congress Control Number: 2017946633

Printed on acid-free paper

This Springer imprint is published by Springer Nature
The registered company is Springer Nature Singapore Pte Ltd.
The registered company address is: 152 Beach Road, #21-01/04 Gateway East, Singapore 189721, Singapore

Preface

This volume comprises select papers from the International Conference on Modern Research in Aerospace Engineering (MRAE 2016) organized by Amity Institute of Aerospace Engineering, Amity University, Uttar Pradesh, India. The conference, with the core theme of "Promoting Innovation & Research in Aerospace Engineering for self-reliance," was held on September 22–23, 2016.

The era of productivity has seen the emergence of innovation research in aerospace engineering. Major organizations, private and government, numerous research groups from academia, educational institutions, focus on research in various areas of aerospace engineering all around the world.

MRAE 2016 brought together academic research communities and experts from the industry to exchange novel ideas and share new research and development in their respective fields of specialization. The main objective of organizing the conference was to exchange ideas and to bridge the gap between academia and the industry and to look closely at the future prospects of aerospace engineering.

We were fortunate to generate a high level of interest among the aerospace engineering community. MRAE 2016 received an appreciable participation of manuscripts from all around the globe. Some of these articles after due reviews are included in this volume.

Noida, India
Ghaziabad, India
Cincinnati, USA

Sanjay Singh
Pushkar Raj
Samir Tambe

Acknowledgements

We are thankful to the keynote and invited speakers:

Mr. Yogesh Kumar, Technical and Business Consultant of CSIR/NAL and Former Director of LCA-HAL.

Mr. Kanchan Biswas, Consultant of CSIR-NAL for aircraft development project and an Adjunct Faculty of IIAEM, Jain University, Bangalore.

Mr. Arun Chopra, Retd. Joint Director General of Civil Aviation.

Wg. Cdr. (Dr.) Anil Kumar (Retd.), Director, Amity School of Engineering and Technology Lucknow Campus.

Dr. Ajay Misra, Head, Aerospace Engineering, Defence Institute of Advanced Technology, Pune, for delivering valuable lectures for the conference.

It would not have been possible to organize a conference without the unwavering support and expertise of these outstanding researchers and leaders.

The organizing committee headed by Dr. Sanjay Singh take this opportunity to thank Patron-in-Chief Dr. Ashok K. Chauhan, Hon'ble Founder President, RBEF, & Chairman, AKC Group of companies, Patron Prof. (Dr.) Balvinder Shukla, Vice Chancellor, Amity University, Uttar Pradesh, and Co-patron Dr. K.M. Soni, Deputy Dean, Domain–Engineering & Technology, AUUP, Noida, India, for their support and guidance. We also extend our gratitude to the expert Prof. A.K. Verma, AISST AUUP chairing technical committee. Further, we also thank the members of review committee for their valuable contributions that improved the quality of papers and made MRAE 2016 a resounding success. We would like to express deep gratitude to our sponsors namely Tata Technologies Ltd., DRDO, and Springer.

No conference would be successful without excellent papers and inspiring presentations. We thank all authors, presenters, and delegates for their contribution in making MRAE 2016 a grand success.

Contents

About the Editors

Dr. Sanjay Singh completed his B.E. (Aero) in 1998 at PEC Chandigarh. He completed his M.Tech. (Aerospace) and Ph.D. (Aerospace) at the Indian Institute of Technology Kanpur (IIT Kanpur) in 2002 and 2008, respectively. He served as the Project Associate at IIT Kanpur from March to May 2002 and July to December 2006. He has been working as an academic since 2007, and he was appointed as the Director of the Aerospace Engineering Department, Amity University, UP, in 2012. He has authored several articles published in journals and conferences. He has worked in the fields of flight mechanics, stability and control, and neural network. He has been a member of the Institution of Engineering and Technology (IET, UK) since 2013 and a life member of the International Association for Continuing Engineering Education (IACEE), USA, since January 2014. Under his leadership, the Aerospace Engineering Department, Amity University, received accreditation from National Assessment and Accreditation Council (NAAC), IET, UK.

Dr. Pushkar Raj received his M.Tech. from the Indian Institute of Technology Bombay (IIT Bombay) in 1991 and his doctorate from the National Institute of Technology (NIT) Kurukshetra in 2013. He is a dynamic professional with 33 years of experience of managing scientific research and development projects at the Defence Research and Development Organization (DRDO), focusing on long-term planning, policymaking, organizational restructuring and change management and coordination with other government departments. He has published several research papers at national and international levels. He is a life member of the Computer Society of India and a fellow of the Institution of Electronics and Telecommunication Engineers (IETE), Delhi. He has advised on matters related to budget allocation, projects, and induction of DRDO products in the armed forces, paramilitary forces, and police organizations. He has been involved in about 200 DRDO projects to analyze the reasons for cost and program delivery control (PDC) overruns and suggested solutions. He retired from his position of Director-Defense Technology Commission at the DRDO in 2015.

Dr. Samir Tambe received his B.Tech. in Aerospace Engineering from the Indian Institute of Technology Bombay (IIT Bombay) in 2002, M.S., Ph.D., and post-doctorate fellowship from the Department of Aerospace Engineering and Engineering Mechanics, University of Cincinnati, in 2004, 2010, and 2012, respectively. He has been working as a Research Assistant Professor in the Department of Aerospace Engineering and Engineering Mechanics, College of Engineering and Applied Sciences, University of Cincinnati, since May 2012. He is a senior member of the prestigious American Institute of Aeronautics and Astronautics (AIAA) and a member of the Institute for Liquid Atomization and Spray Systems (ILASS). His work in the fields of propulsion, combustion, and heat transfer has been pioneering.

Thermal Decomposition and Stabilization Studies of ADN with AN Impurities in the Presence of Catalysts

Pratim Kumar, Rajiv Kumar and Puran Chandra Joshi

1 Introduction

With increasing frequency of rocket launch every year for putting navigational satellite in low earth orbits, fascination for space tourism, space exploration, and for studying zero-gravitational combustion processes at International Space Station (ISS), it's the demand of the time to replace conventional solid propellants with green propellants for providing clean and green environment. Presently, Ammonium perchlorate (AP) based solid propellants are the most frequently used in booster stages during rocket launch. Solid propellants are generally used in initial booster stages for producing high thrust at the time of take-off due to its higher accelerating capabilities. AP is a workhorse oxidizer from last several decades. It is because of its better burning characteristic, burning rate tailoring capacity, less shock and friction sensitivity, non-hygroscopic, easy availability and cost effectiveness [1]. Inspite of having these advantages, AP also shows certain undesirable properties too. The major problems with AP propellants are its chlorinated exhaust which includes; Hydrochloric acid (HCl), Chlorine oxides (ClO_x), and Chlorine (Cl) gas [2]. All these gaseous entities are highly corrosive in nature, detrimental for the earth environment, and also for the upper ozone layer.

Renewed interest in green-oxidizers for rocket propulsion led to the exhaustive research in green energetic materials (GEM's) area. Ammonium nitrate (AN), Hydrazinium nitroformate (HNF), and Ammonium dinitramide (ADN) are some of the available green oxidizers as possible candidates to replace AP [3]. Among these, ADN is the most promising candidate because of its high specific impulse, thrust, and burning rate [4, 5].

P. Kumar (✉) · R. Kumar · P.C. Joshi
Birla Institute of Technology, Mesra, Ranchi 835215, Jharkhand, India
e-mail: pratim.kumar.86@gmail.com

S. Singh et al. (eds.), *Proceedings of the International Conference on Modern Research in Aerospace Engineering*, Lecture Notes in Mechanical Engineering,
https://doi.org/10.1007/978-981-10-5849-3_1

ADN is the member of dinitramide anion [DA, $N(NO_2)_2]^-$ family with molecular formula of $[NH_4^+]$ $[N(NO_2)_2^-]$. The presence of Dinitramide anion imparts high heat of formation, density, oxygen balance, and oxygen content. There are numerous other synthesized DA based energetic compounds too with many possible applications areas [6]. Although, there are some disadvantages associated with ADN too, i.e. (a) ADN is much hygroscopic in nature, (b) having needled shaped crystal structure, and (c) quiet unstable [7]. However, hygroscopicity problem can be resolved by coating ADN crystals with a non-hygroscopic chemical i.e. *encapsulating method* [8], needle shaped crystals can be reshaped into spherical crystals by using *prilling technique* [8], while stabilization of ADN can be done by adding different chemicals with requisite properties [9].

In the present study, three catalysts viz. Cupric Oxide, Ammonium Metavandate, and Copper Chromite are used for stabilizing ADN. Evaporative crystallization method was used to prepare ADN + Catalyst samples. Simultaneous Thermal Analyzer (STA) was used to study the thermal characteristics of prepared samples by using DSC/TG-DTG techniques. All the three catalysts were found to be effective in stabilization of ADN. This study helps us to stabilizes ADN, to increase its energy content, and also to use it more efficiently as a green oxidizer in solid propellants.

2 Experimental Methodology

2.1 Preparation of ADN

ADN was synthesized using the lab facility available in Department of Applied Chemistry, and Department of Space Engineering and Rocketry, BIT Mesra, and the procedures adopted are as described in ref. [10]. All the starting materials for synthesizing ADN are of ACS/analytical reagent grade. The list of chemicals used are, Sulfamic acid (Sigma-Aldrich, 99.3%, A.C.S Reagent), Potassium hydroxide (Sigma-Aldrich, 90%, Reagent grade), Ethanol (Merck, analytical reagent grade), Fuming nitric acid (Merck, 100% pure), Sulfuric acid (Acros Organics, 95%, A.C.S. Reagent), Acetone (Merck, for synthesis, reagent grade), Isopropyl alcohol (Sigma-Aldrich, USP, Analytical reagent), Ammonium sulfate (Sigma-Aldrich, Bioxtra, 99% pure, reagent grade), and Petroleum ether (Sigma-Aldrich, A.C.S. Reagaent, 40–60 °C b.p.). Preparation of ADN was completed in five steps. The steps are.

(a) Preparation of potassium sulfamate (K^+ $[NH_2SO_3^-]$) by using sulfamic acid (H_2NSO_3H), potassium hydroxide (KOH) and ethanol.
(b) Nitration of potassium sulfamate in nitrating mixture of nitric and sulfuric acid at −40 °C. This nitrating mixture was then neutralized by using potassium hydroxide solution to give mixture of three salts namely, potassium dinitramide $[KN(NO_2)_2]$, potassium bisulfate ($KHSO_4$) and potassium nitrate (KNO_3).

(c) Extraction of potassium dinitramide using acetone, as only potassium dinitramide is soluble in acetone.
(d) Separate solution of KDN and Ammonium sulfate [$(NH_4)_2SO_4$] were prepared in distilled water. Both solutions were mixed in another beaker. As both chemicals interact with each other, cation exchange reaction starts and NH_4^+ of ammonium sulfate replaced with K^+ of potassium dinitramide, and Ammonium dinitramide [$NH_4N(NO_2)_2$] formed.
(e) Formed ADN is in dissolved state with water and the solid precipitate formed was of potassium sulfate (K_2SO_4). Isopropyl alcohol was added to the solution and potassium sulfate was filtered off. ADN precipitates as a needle like crystals with yellow to white in color.

2.2 *Characterization of ADN*

Synthesized ADN was analytically characterized using following analytical instruments:

(a) Differential Scanning Calorimetry/Thermo-gravimetry (DSC/TG)
(b) Ultra-Violet (UV) Spectrophotometer
(c) Carbon Hydrogen Nitrogen Sulphur (CHNS) analyzer.

Simultaneous Thermal Analyzer (STA Model 409 PG luxx) of NETZSCH, Germany was used to obtained the differential scanning calorimetry (DSC), and thermo gravimetry (TG) of ADN at the heating rate of 10 °C/min. Nitrogen at a flow rate of 60 ml/min was used as a purge gas in all the analysis. Alumina crucible with lid having a small perforation in it was used to place the samples in STA sample carrier. The sample weight taken for each analysis was 6 ± 0.5 mg. The heat flow and temperature calibrations of the instrument were done using indium standard.

DSC trace of ADN (Fig. 1) shows one minor endothermic melting peak (m.p.) at around 50 °C, and one major exothermic decomposition peak at 190 °C. In general, m.p. of pure ADN is at around 92 °C [11], while in present case, m.p. of ADN is at 50 °C. This may be due to the presence of Ammonium nitrate (AN) in ADN crystals. Since ADN is not much thermally stable, and hence AN formed during the storage of ADN. This AN and ADN together forms an eutectic mixture, and the m. p. of AN/ADN eutectic mixture get dropped to 50 °C. TG curve (Fig. 2) of ADN show two decompositions in the temperature range of 170–270 °C. First decomposition temperature range was 170–210 °C, which corresponds to ADN decomposition, while second decomposition temperature range was 210–270 °C, which corresponds to AN decomposition. Reported decomposition temperature of ADN is around 180 °C, while that of AN is about 270 °C [12, 13].

UV-spectrophotometer (Make: Perkin Elmer, Model No. Lambda 25, USA) was used to obtain UV-spectrum of ADN. The observed UV spectrum of synthesized ADN is shown in Fig. 3. UV-spectrum of ADN shows two peaks at 213 and

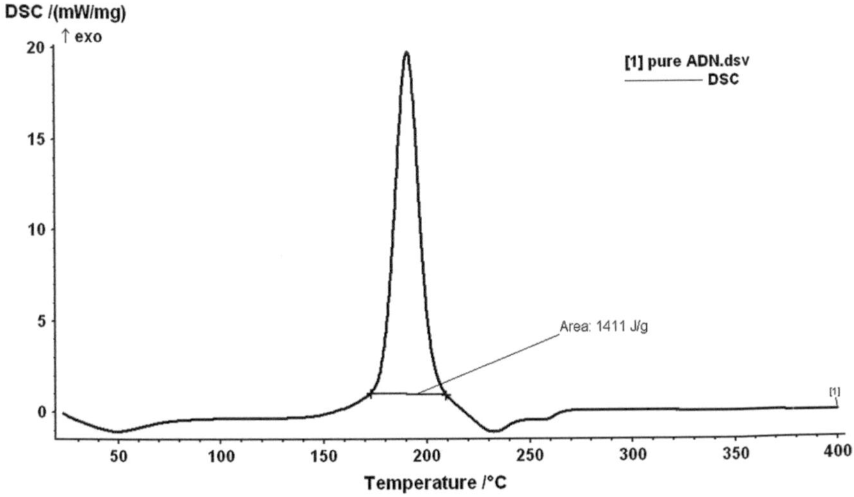

Fig. 1 DSC of ADN

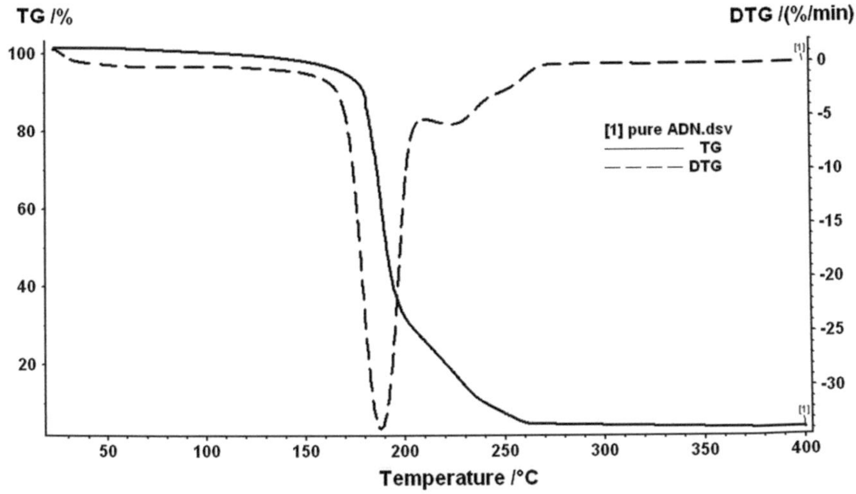

Fig. 2 TG-DTG of ADN

284 nm. Absorbance at 284 nm is the characteristics of $–N(NO_2)^-$ ion due to the low energy n-π* transition and the absorption maximum at 213 nm is due to high energy σ-σ* transition.

CHNS analyzer [Make: Elementar, Model No. Vario EL III, Germany] was used to obtain the percentage content of the various elements present in ADN. Table 1 shows the percentage content of C, H, N and S in ADN. The calculated values of N and H in ADN is 45.16% and 3.25% respectively. However, measured value of N

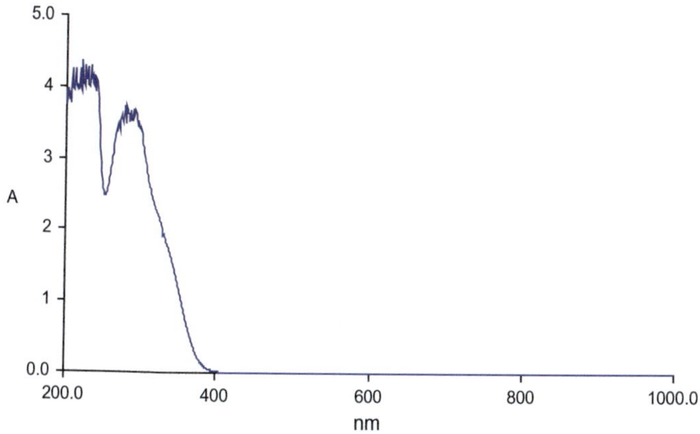

Fig. 3 UV-spectrum of ADN

Table 1 CHNS data of ADN

Nitrogen (N)	Carbon (C)	Hydrogen (H)	Sulphur (S)
31.01	0.110	5.356	1.283

and H is 31.01% and 5.35% respectively. This result also confirmed the presence of AN in ADN crystals.

2.3 Sample Preparation

2% by mass of cupric oxide, ammonium metavanadate, and copper chromite were mixed thoroughly in the ADN crystals for studying their effectiveness on the thermal stabilization and on the thermal decomposition of these three prepared oxidizers. All the samples were prepared by ***evaporative crystallization method*** [14]. For preparing sample, like ADN + CuO, 100 mg of ADN was first dissolved in 2 ml of distilled water with continuous stirring. After complete dissolution, 2% by weight of CuO was added into the ADN solution. After 10 min of stirring, the ADN + CuO solution was kept at 50 °C over hot plate with continuous stirring till dry crystals obtained. The obtained crystals were further dried in vacuum desiccators. It was further crashed and used for thermal analysis.

2.4 Instrumentation

Simultaneous Thermal Analyzer (STA Model 409 PG luxx) of NETZSCH, Germany was used to obtain the Differential Scanning Calorimetry (DSC), Thermo

Gravimetric (TG), and Differential Thermo Gravimetric (DTG) thermograms for each sample. Nitrogen at a flow rate of 60 ml/min was used as a purge gas in all the analysis. Alumina crucible with lid having a small perforation in it was used to place the samples in STA sample carrier. The sample weight taken for each analysis was 6 ± 0.5 mg. The heat flow and temperature calibrations of the instrument were done using indium standard. All experiments were carried out at heating rate of 10 °C/min.

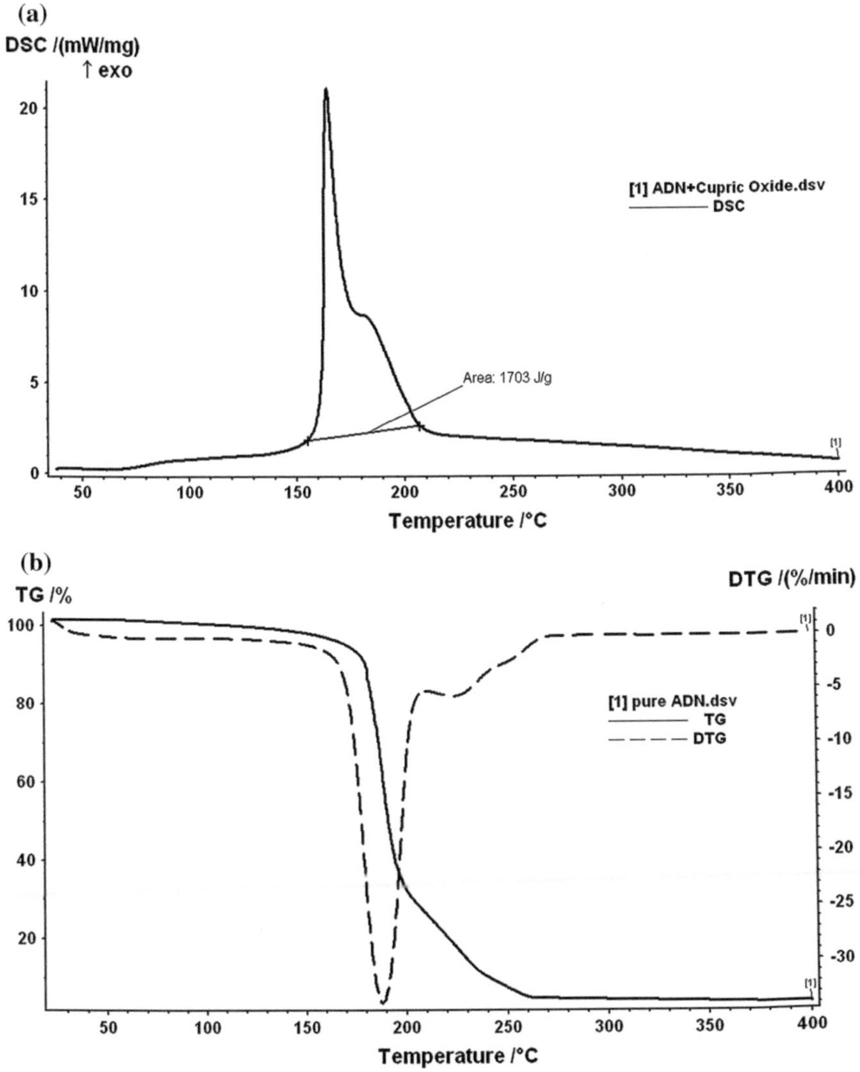

Fig. 4 **a** DSC of ADN + CuO; **b** TG-DTG of ADN + CuO

Table 2 DSC/TG-DTG data of ADN and ADN + Catalysts

Oxidizer	Melting peak	Decomposition peak		DTG peak	TG range
	°C	°C	Enthalpy (J/g)	°C	°C
ADN	50	190 and 230	1411	190	150–270
ADN + CuO	70	165 and 185	1703	165	160–210
ADN + NH_4VO_3	95	165 and 185	1923	180	160–220
ADN + $Cu_2Cr_2O_5$	80	165 and 190	1921	180	150–200

3 Results and Discussion

3.1 *DSC/TG-DTG of ADN + Cupric (II) Oxide (CuO)*

Figure 4a, b shows the DSC and TG-DTG curves of AN + CuO respectively. From Fig. 4a, it was observed that there are two exothermic decomposition peaks in ADN + CuO in the temperature range of 160–210 °C. First peak at 170 °C was for ADN decomposition, while second peak at 190 °C corresponds to AN decomposition. One minor endothermic melting peak was observed at around 70 °C. This melting peak at 70 °C confirms that CuO helps in stabilization of ADN. From Fig. 4b, it was observed that, 98% decomposition completed in the range of 160–210 °C with DTG max peak at around 165 °C. The summary of DSC/TG-DTG peaks is also provided in Table 2.

3.2 *DSC/TG-DTG of ADN + Ammonium Metavanadate (NH_4VO_3)*

Figure 5a, b shows the DSC and TG-DTG curves of AN + NH_4VO_3 respectively. From Fig. 5a, it was observed that there are two major exothermic decomposition peaks in ADN + NH_4VO_3 in the temperature range of 150–220 °C. First peak was at 160 °C, while second peak was at 190 °C. First peak corresponds to ADN decomposition while second peak represents AN decomposition respectively. One minor endothermic melting peak was observed at around 95 °C. This melting peak at 95 °C confirms that NH_4VO_3 is better stabilizer of ADN. From Fig. 5b, it was observed that, 80% decomposition completed in the range of 160–220 °C with DTG max peak at around 180 °C. The summary of DSC/TG-DTG peaks is provided in Table 2.

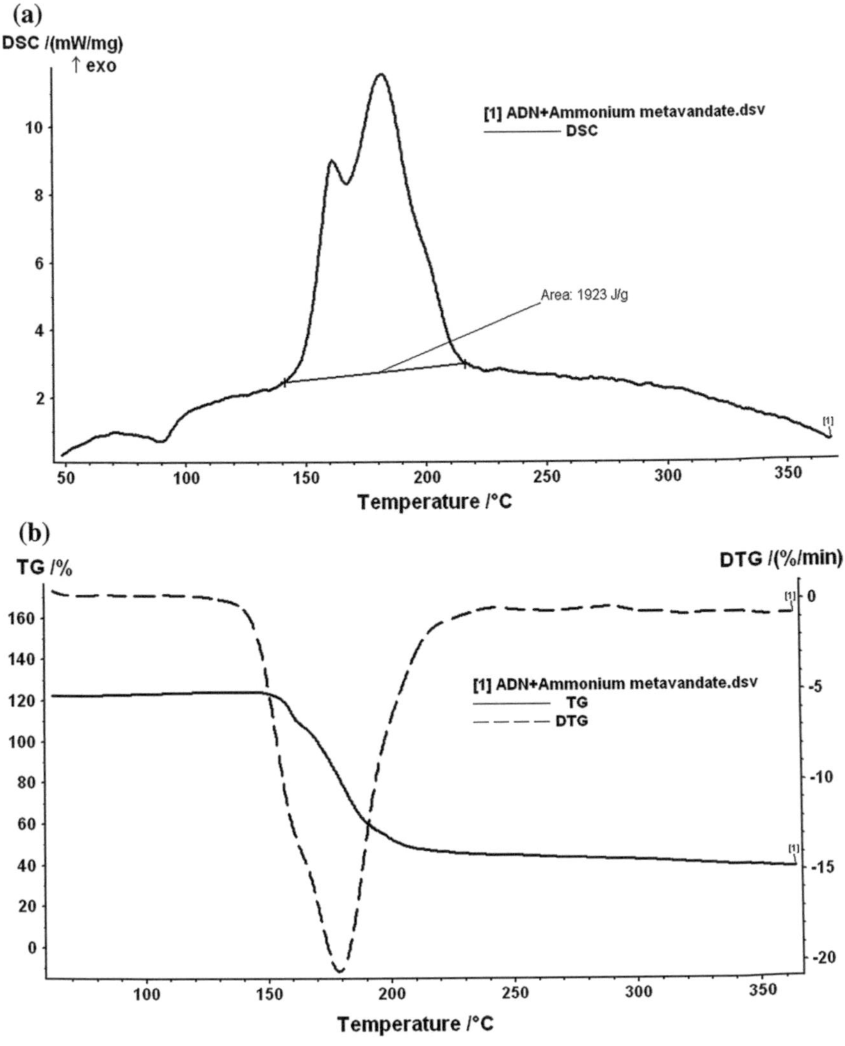

Fig. 5 **a** DSC of ADN + NH_4VO_3; **b** TG-DTG of ADN + NH_4VO_3

3.3 DSC/TG-DTG of ADN + Copper Chromite ($Cu_2Cr_2O_5$)

Figure 6a, b shows the DSC and TG-DTG curves of ADN + $Cu_2Cr_2O_5$ respectively. From Fig. 6a, it was observed that there are two exothermic decomposition peaks in ADN + $Cu_2Cr_2O_5$ in the temperature range of 150–210 °C. First peak at 160 °C corresponds to ADN decomposition, while second peak at 190 °C corresponds to AN decomposition respectively. One minor endothermic melting peak was observed at around 80 °C. This melting peak at 80 °C confirms that $Cu_2Cr_2O_5$

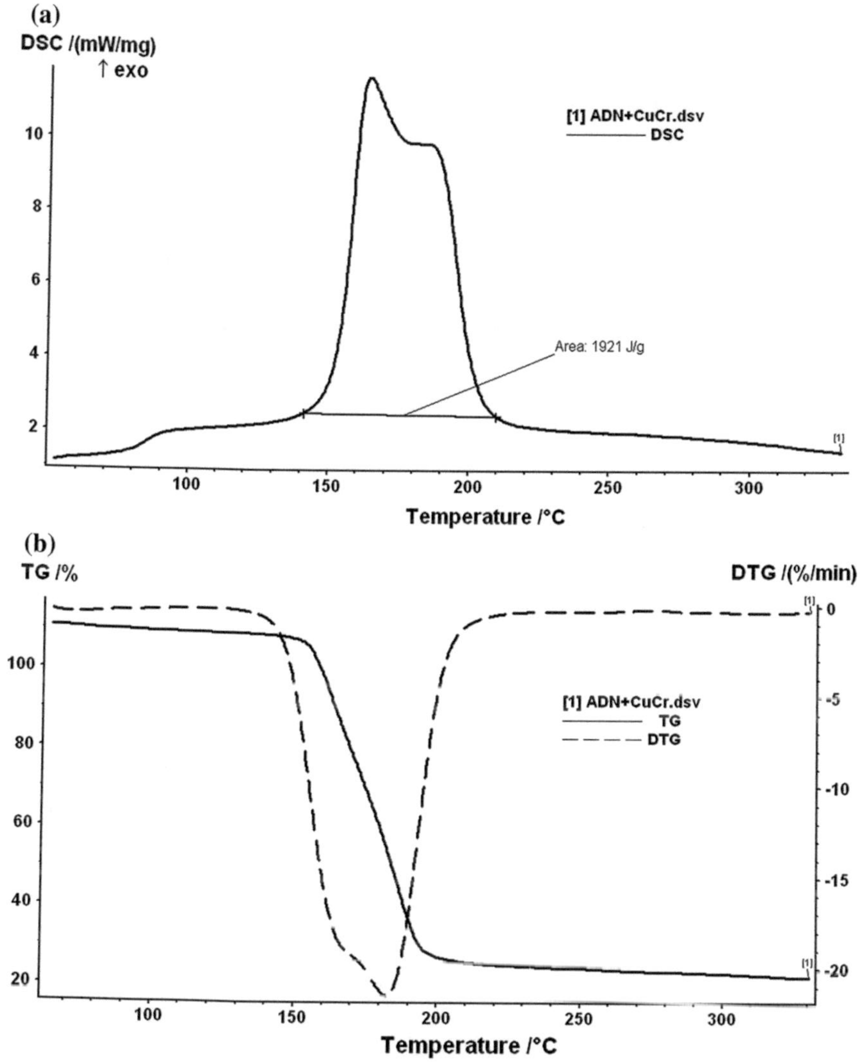

Fig. 6 **a** DSC of ADN + $Cu_2Cr_2O_5$; **b** TG-DTG of $Cu_2Cr_2O_5$

is better stabilizer of ADN than CuO but less than NH_4VO_3. From Fig. 6b, it was observed that, 90% decomposition completed in the range of 150–200 °C with DTG max peak at around 180 °C. The summary of DSC/TG-DTG peaks is given in Table 2.

4 Conclusion

ADN has great scope in coming future as an alternative oxidizer that has the capabilities of replacing the Ammonium Perchlorate. Present work deals with the thermal stabilization of ADN with the incorporation of three catalysts namely cupric (II) oxide, ammonium metavanadate, and copper chromite. The following conclusions are drawn based on the present study:

(a) Being ADN a thermally unstable compound, during storage AN was formed in the ADN crystals. This AN and ADN together forms an eutectic mixture which reduces the melting point of ADN from 92 to 50 °C.
(b) In pure ADN, decomposition completed in two steps. First exothermic decomposition was for ADN, while and second endothermic decomposition was for AN.
(c) In catalyzed ADN, both ADN and AN decompositions are exothermic in nature. In the presence of catalyst, early decomposition was observed with net increase in heat release too.
(d) Among all the three catalysts used, ammonium metavandate was found to be most effective in thermal stabilization of ADN, since melting point was observed at 95 °C.
(e) Maximum exothermic decomposition heat release was nearly equal in the presence of both ammonium metavanadate and copper chromite.
(f) In presence of CuO, early decomposition of ADN was observed with DTG peak at 165 °C. In all other cases, DTG peaks were at 180 °C or above.

Acknowledgements Authors are thankful to the Dptt of Space engg. and Rocketry, Dptt of Applied Chemistry, and Central Instrumentation Facility (CIF), BIT Mesra for providing all the necessary equipments for carrying out present experimentations.

References

1. Sutton GP (2001) Rocket propulsion elements seventh edition. Wiley New York, Chapter 12, pp. 474–517
2. Jacobs PWM, Whitehead HM (1969) Decomposition and combustion of ammonium perchlorate. Chem Rev 69(4):551–590
3. Talawar MB, Sivabalan R, Mukundan T, Muthurajan H, Sikder AK, Gandhe BR, Rao AS (2009) Environmentally compatible next generation green energetic materials (GEMs). J Hazard Mater 161:589–607
4. Sinditskii VP, Egorshev VY, Levshenkov AI, Serushkin VV (2006) Combustion of ammonium dinitramide, Part 1: burning behavior. J Propuls Pow 22(4):769–776
5. Sinditskii VP, Egorshev VY, Levshenkov AI, Serushkin VV (2006) Combustion of ammonium dinitramide, Part 2: combustion mechanism. J Propuls Pow 22(4):P777–784
6. Vandel AP, Lobanova AA, Loginova VS (2009) Application of dinitramide salts (Review). Russ J Appl Chem 89(10):1763–1768

7. Ostmark H, Bemm U, Langlet A, Sanden R, Wingborg N (2000) The properties of ammonium dinitramide (ADN): part 1, basic properties, and spectroscopic data. J Energ Mater 18(2):123–138
8. Heintz T, Pontius H, Aniol J, Birke C, Leisinger K, Reinhard W (2009) Ammonium dinitramide (ADN)-prilling, coating, and characterization. Propellants Explos Pyrotech 34:231–238
9. Mishra IB, Russell TP (2002) Thermal stability of ammonium dinitramide. Thermochim Acta 384:47–56
10. Golofit T, Maksimowski P, Biernacki A (2013) Optimization of potassium dinitramide preparation. Propellants Explos Pyrotech 38:261–265
11. Wingborg N, de Flon J, Johnson C, Whitelow W (2008) Green propellants based on ADN, space propulsion, 5–8 May', Heraklion, Crete, Greece. ESA, 3AF, SNPE
12. Larsson A, Wingborg W, Green propellants based on ammonium dinitramide, Swedish Defence Research Academy, Sweden, Advances in Spacecraft Technology
13. Oommen C, Jain SR (1999) Ammonium nitrate: a promising rocket propellant oxidizer. J Hazard Mater 67:253–281
14. Sudhakar AOR, Mathew S (2006) Thermal behavior of CuO doped phase stabilized ammonium nitrate. Thermochim Acta 451:5–9

Experimental Investigation of Flame Speed of Fuel–Air Mixture for Varying Air/Fuel Ratio

Ankur Vats, Vivek Kumar and Saquib Reza

1 Introduction

Combustion, a process consisting of several chemical and physical phenomena, is widely used in industries, scientific research, various professions, and households. Various areas like physics, chemistry, and mechanics hold the knowledge of these phenomena, where their inter-relationship holds a prime importance to deal with flame propagation and associated processes. Therefore, it becomes necessary to discuss these phenomena at an elementary level and study the co-relation of the underlying processes.

A flame is caused by a highly exothermic self-propagating reaction accompanied by a luminous reaction zone. It is the visible, gaseous part of a fire. The emission of light is one of the most important characteristic properties of flame and is useful in locating the position and form of the flame front. Two theories, namely, 'Heat conduction' and 'Diffusion' are used to explain the phenomena of flame propagation. In the first theory of heat conduction, flow of heat occurs from the flame front, which refers to the area where combustion takes place, to the inner cone where unburned mixture of air and fuel is present. On heating this unburned air and fuel mixture, as it reaches the ignition temperature, the mixture starts to burn in the flame front, and now, as a result of this reaction, heat again moves to the inner cone to establish a self-propagating cycle. As per the second theory of diffusion, the reactive molecules which are generated in the flame front diffuse into the inner cone

A. Vats (✉) · V. Kumar · S. Reza
Amity University, Noida, Uttar Pradesh, India
e-mail: avats1@amity.edu

V. Kumar
e-mail: vkumar13@amity.edu

S. Reza
e-mail: sreza@amity.edu

S. Singh et al. (eds.), *Proceedings of the International Conference on Modern Research in Aerospace Engineering*, Lecture Notes in Mechanical Engineering,
https://doi.org/10.1007/978-981-10-5849-3_2

which initiates the combustion of the air and fuel mixture. To sustain a flame, a certain range of fuel percentage within the mixture is required. The minimum percentage is known as lower limit of flammability and the maximum percentage is known as upper limit of flammability of fuel gas.

The linear velocity of flame front with respect to itself and relative to unburned gas is known as burning velocity (S_u). In general terms, it is equal to the volume of the unburned fuel and air mixture consumed per unit time per unit flame front area in which the flame front is consumed. This burning velocity varies with the air/gas mixture ratio and the chemical composition of the fuel gas. Burning velocity could be measured by observing the flame front using following methods:

1. Stationery Flames:
   - Flat Flame Burner
   - Cylindrical Burner
   - Slot Burner
2. Non Stationery Flames:
   - Cylindrical Tube Method
   - Constant Volume Bomb Method
   - Constant Pressure Bomb Method
   - Double Flame Karnel Methods

Cylindrical Tube Method: In this present investigation, cylindrical tube method is used to determine the burning velocity. In cylindrical tube method, a horizontal tube opened at one end contains the fuel–air mixture which is ignited at the open end of the tube. The rate of progress of the flame into the unburned gas is known as the flame speed. During the propagation of the combustion wave, the column of unburned gas remains stationery with respect to the tube.

The cylindrical tube method to measure the flame speed was first used by Mallard and Le Chatelier [1]. The fuel and air mixture was filled in a tube, closed at one end, and was ignited at the open end. The propagation of flame was observed and photographed. Coward and Hartwell [2] used explosive gas filled in a tube, closed and one end and the gas ignited at open end. During the propagation of combustion wave, the column of unburned gas remains stationery with respect to the tube and the burned gas flows from the wave toward the open end. Guenoche et al. [3] measured the speed of gas flow in the unburned zone by measuring the rate of growth of a soap bubble through the orifice at the unburned end. It was found that the gas flow at the unburned end varies directly with the flame speed. It was also found that flame instabilities increased with increase in pressure. Xian Feng et al. [4] explored the flame structure and propagation behavior of premixed fuel and air, in the transition from laminar to turbulent combustion using high speed camera and Schilieren photography for recording the photograph of flame propagation process in a semi vented pipe. Tripathi et al. [5] utilizes a few standard methods of flame speed measurements from LPG-air-diluent mixtures as a function of air fuel ratio and diluent concentrations. The variation in burning velocity obtained by using two different experimental procedures, such as orifice burner and cylindrical tube method at varying air fuel ratios is quantified.

2 Experimentation

Fuel

In the present investigation, liquified petroleum gas (LPG) is used as the fuel to burn with air. LPG is a mixture of light gas phase hydrocarbons (60% volume of propane, 40% volume of butane) as shown in Fig. 1 which are gaseous at normal, ambient temperature, and atmospheric pressure conditions, but turned liquid when moderate pressure is applied. It has an explosive range from 2.1 to 9.5% volume of gas in air which is considerably narrower than other common gaseous fuels.

Experimental Setup

The experimental setup used in the present work was P.A. Hilton's C551 Flame Propagation and Stability unit, supplied by P.A. Hilton Ltd., Horsebridge Mill, King's Somborne, Stockbridge, Hampshire, England. It consists of a bench top panel housing the single stage centrifugal air blower of 1.7 kW operating at 220/240, single phase, 50 Hz electric power, control valves, and flow meters for fixing the air/fuel ratio in the premix, an air–fuel mixing unit, PVC plastic tube of varying sizes and Meker-Fisher Burner fitted with spark plug and flame arrestor. The flow meters work in the range of 0–0.6 l/s for air flow and 0–0.2 l/s for gas flow, respectively. The line diagram of this apparatus is shown in Fig. 2.

There are two igniters, one is piezo igniter, and is used to ignite the unburned gases on the burner top, the other one is spark igniter, and is used to ignite the unburned gases trapped inside the tube for measuring the flame speed. Flame speed tube is made of PVC, of varying diameter of certain length. A mixing block is been provided for premixing the air and fuel before they are sent through the pipe. In

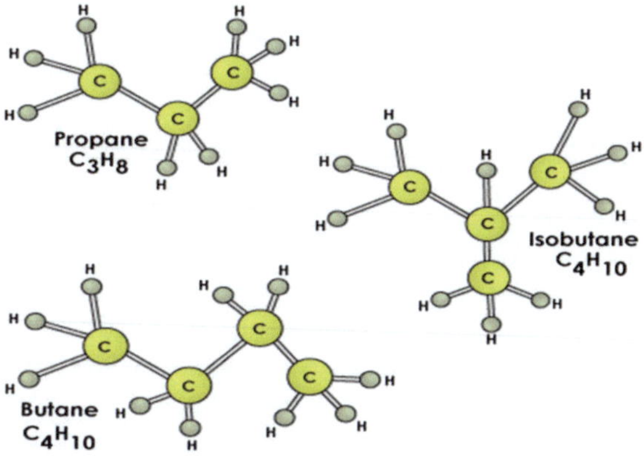

Fig. 1 Constituents of LPG

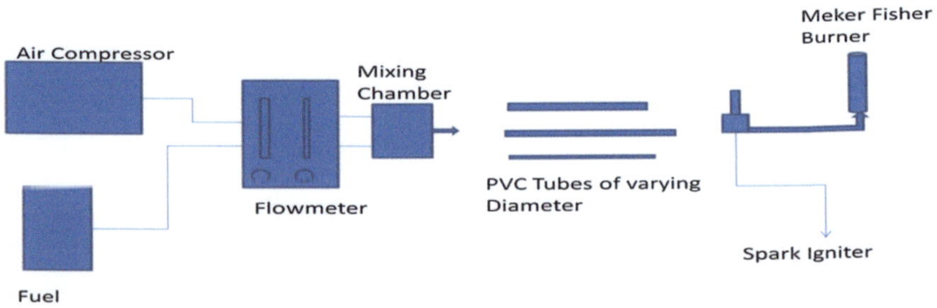

Fig. 2 Line diagram of experimental setup

addition to this, there were two burner adaptors, one is connected over the mixing block, and the other is connected at the burner top. For safety precautions, it was made sure that all the components are connected to common earth/ground points.

The flame tube in the P A Hilton apparatus is initially marked at a meter distance as stations using a permanent marker. After switching on the apparatus, it is ensured that the LPG is passing through the apparatus. The air is collected by the compressor from the atmosphere. The regulated amount of air and fuel, after mixing in the provided chamber flow through the flame tube toward the burner top, where they are ignited using a peizo igniter. After the flame is stabilized, air and fuel valve were shut off, with simultaneously switching on the spark igniter to ignite the mixture in tube. The time taken by the moving flame through each station is noted down, and averaged, hence used to calculate the flame speed at various Air/fuel ratio for varying tube diameter.

3 Results and Discussion

In the first set of experiments, the tube used was of 1.25 in. of diameter. The flame speed was calculated for varying air/fuel ratio 20/1 to 54/1 using method mentioned above. The time travelled per station was noted down and averaged. The results shows that below 22/1 A/F ratio, due to insufficient premixed oxidizer, the combustion at burner rim occurred due to the diffusion of atmospheric oxygen into the lame front and hence, a diffusion flame was obtained. The flame speed was calculated for varying A/F ratio. It was observed that from A/F ratio 22/1 to 34/1, the flame speed increased with increase in air–fuel ratio and after 34/1, the flame speed decreases till the flame blows off at 50/1. The flame speed variation with varying A/F ratio for 1.25 in. tube diameter has been shown in Fig. 3.

In the second set of experiments, 1 in. diameter tube was used. The flame speed was calculated for varying air/fuel ratio 20/1 to 54/1 using the same method as above. The time travelled per station was noted down and averaged. The results shows that below 22/1 A/F ratio, due to insufficient premixed oxidizer, the combustible mixture couldn't burn in the tube; however, the combustion at burner rim

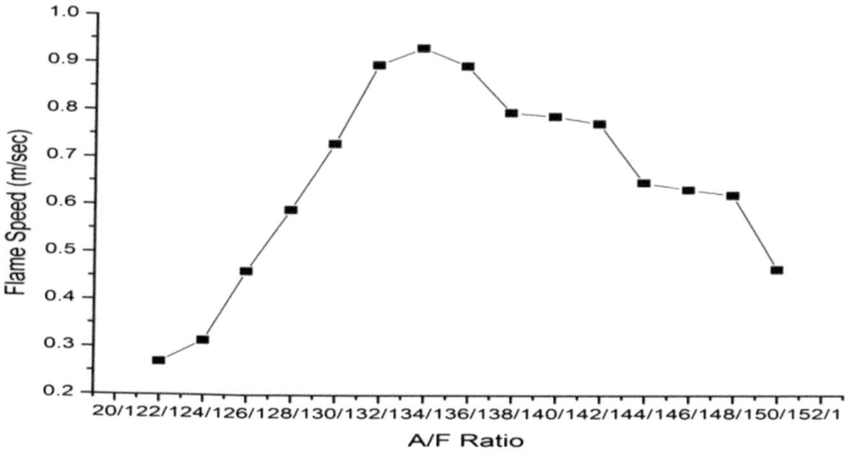

Fig. 3 Flame speed versus A/F ratio: 1.25 in. tube

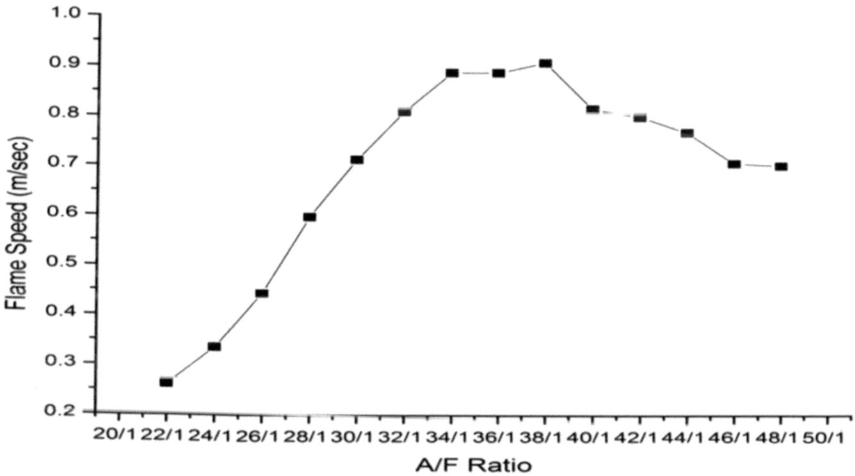

Fig. 4 Flame speed versus A/F ratio: 1 in. tube

occurred due to the diffusion of atmospheric oxygen into the flame front and hence, a diffusion flame was obtained. The flame speed was calculated for varying A/F ratio. It was observed that from A/F ratio 22/1 to 34/1, the flame speed increased with increase in air–fuel ratio and after 34/1, the flame speed decreases till the flame blows off at 52/1. The flame speed variation with varying A/F ratio for 1 in. tube diameter has been shown in Fig. 4.

In the third set of experiments, the tube was 0.75 in. in diameter of same length. The flame speed was calculated for varying air/fuel ratio using the same method as described above. The time travelled per station was noted down and averaged. It was

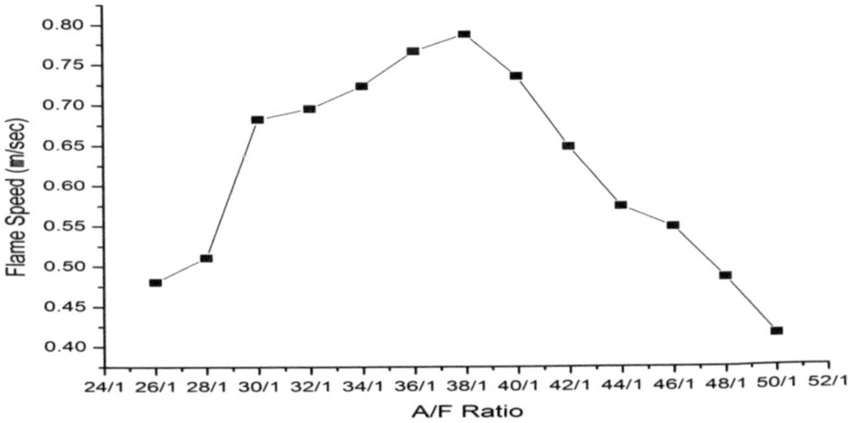

Fig. 5 Flame speed versus A/F ratio: 0.75 in. tube

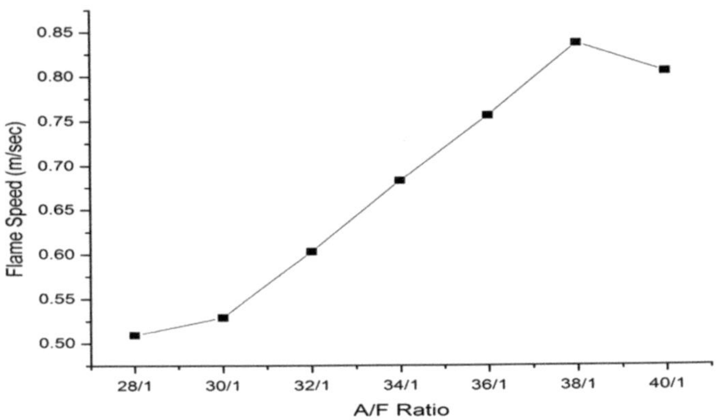

Fig. 6 Flame Speed versus A/F ratio: 0.5 in. tube

found that at ratio 24/1, on igniting the combustible mixture the flame propagated for about 1 m distance and then extinguished. At 28/1, the flame propagated for 3 m distance and after that it extinguished. Above this ratio, the flame propagates throughout the length of the flame tube till ratio 50/1, after which it lifts off and extinguishes. The flame speed for ratio 24/1 to 38/1 increased with increasing air fuel ratio and after 38/1 it started decreasing till the flame blow off. The flame speed variation with varying A/F ratio for 0.75 in. tube diameter has been shown in Fig. 5.

In the fourth set of experiments, the tube was replaced by 0.5 in. of diameter and of same length. The flame speed was calculated for varying air fuel ratio using the same method as above. The time travelled per station was noted down and averaged. The results show that till 28/1, no flame propagation occurred in the tube. For

28/1 and 30/1, the flame propagated for some distance and then quenched. From 32/1 to 40/1, the flame propagated in the tube and an increasing flame speed was obtained till 38/1. After 40/1 ratio, the gas mixture ceased to flow in the tube and no combustion could occur. The flame speed variation with varying A/F ratio for 0.5 in. tube diameter has been shown in Fig. 6.

In the fifth set of experiments, an 8 mm (0.314961 in.) diameter tube was used. The results show that no flame propagation occurred for this diameter of tube at any air fuel ratio. Hence, flame speed could not be calculated.

4 Conclusion

Based on the work carried out in the present investigation, the following conclusions are drawn:

- From the comparison of flame speed at different air fuel ratio in 0.5 in., 0.75 in., 1 in. and 1.25 in., respectively, it is observed that the flame speed increased as the flame tube diameter increased and vice versa. This may be due to quenching effect on the wall, which decreases with increase in tube diameter.
- The highest flame speed for all the tubes was obtained around 34/1-38/1, till which the flame speed increases with increase in air fuel ratio and then decreases. This may be because in lower air/fuel ratios, due to insufficient oxygen incomplete combustion occurs and the flame temperature is therefore low. And for higher than stoichiometric air/fuel ratios, due the dilution effect of excessive air the flame temperature is low which in turn decreases the flame speed.

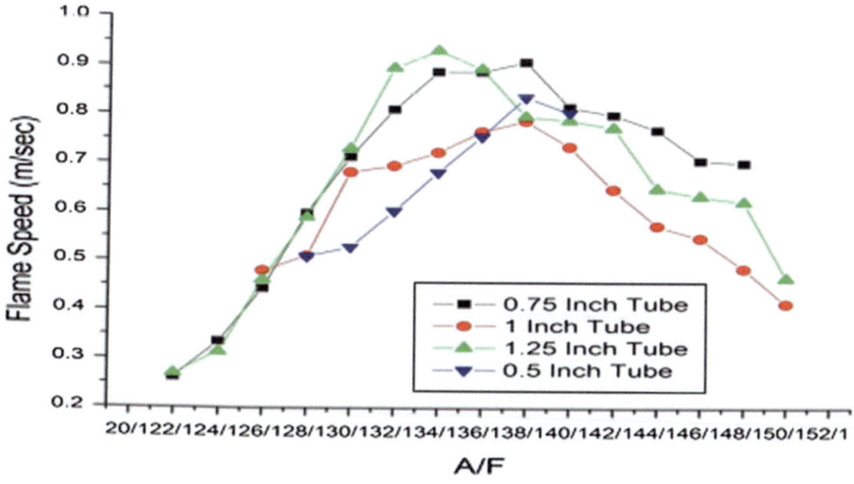

Fig. 7 Comparison of flame speed versus A/F Ratio for varying tube diameter

- The flame speed variation with varying A/F ratio comparison for all tube diameters has been shown in Fig. 7. It is observed that lower is the tube diameter, higher is the air/fuel ratio to obtain higher flame speed. Flame speed tends to increases with increase in tube diameter for varying A/F ratio and vice versa. Flame speed increases with varying A/F ratio till stoichiometric ratio is reached, after which it starts decreasing again.

References

1. Mallard E, Le Chatelier HL (1883) Annals of Mines 4:379
2. Coward HF, Hartwell FJ, Studies in the mechanism of flame movement. Part I. The uniform movement of flame in mixtures of methane and air, in relation to tube diameter. J Chem Soc Part 2 1996–2004
3. Bradley D (2008) Flame propagation in a tube: the legacy of henri guenoche. Combust Sci Technol 158(1)
4. Xianfeng C, Jinhua S, Yi L, Xuanya L, Sining C, Shouxiang L, Premixed propane/air flame microstructure in the transition from laminar to turbulent combustion
5. Tripathi A, Chandra H, Agrawal M, Effect of mixture constituents on the laminar burning velocity Of Lpg-co2-air mixtures. Asian Research Publishing Network (ARPN)

Medium Scale Jet Engine Design and Design Validation Through Simulation

Vamsikrishna Undavalli, Anmol Kumar, Yash Gupta, S. Ramanan and Md. Saquib Reza

Nomenclature

A	Area
T	Temperature
P	Pressure
TIT	Turbine Inlet Temperature
$\dot{m}$	Mass flow rate
M	Mach number
ρ	Density
V	Velocity
η	Efficiency
c	Specific heat
γ	Specific heat ratio
π	Pressure ratio
τ	Temperature ratio

Subscripts

abs	Absolute
Tan	Tangential
rad	Radial

V. Undavalli · A. Kumar · Y. Gupta · S. Ramanan · Md. S. Reza (✉)
Amity Institute of Aerospace Engineering, Amity University,
Noida 201303, Uttar Pradesh, India
e-mail: sreza@amity.edu

V. Undavalli
e-mail: u.vamsikrishna@yahoo.com

A. Kumar
e-mail: narutoanmol@gmail.com

Y. Gupta
e-mail: 2.11.1994yash@gmail.com

S. Ramanan
e-mail: ramanan.kkn@gmail.com

S. Singh et al. (eds.), *Proceedings of the International Conference on Modern Research in Aerospace Engineering*, Lecture Notes in Mechanical Engineering,
https://doi.org/10.1007/978-981-10-5849-3_3

o	Stagnation
ov	Overall
p	Propulsive
t	Thermal
C	Cold Section
H	Hot Section
1	Diffuser Inlet
2	Compressor Inlet
3	Combustion Chamber Inlet
4	Turbine Inlet
5	Nozzle Inlet
6	Nozzle Exit
V1	Absolute stage exit velocity
U1	Tangential stage exit velocity
α	Gas angles
αin	Absolute inlet gas angles
αexit	Absolute exit gas angles
α1	Absolute Inlet gas angle
α2	Absolute outlet gas angle
cx	Stator chord
cr	Rotor chord
s	Spacing
Zs	Tangential force coefficient of stator
ϒ	Blade angle
ϒ1	Inlet blade angle
ϒ2	Outlet blade angle
Λ	Stagger angle
б	Solidity
D	D + hx
d	Disc diameter
hx	Blade position above disc
Cl	Lift coefficient for rotor

1 Introduction

The development of gas turbine technology has being very crucial in every country due to increase in demands of civil and military requirements. As this is the one of the confidential technologies that no country shares with others, even its allies. A country like India, which had been trying to develop indigenous gas turbine, KAVERI has been a milestone for Indian aerospace history. With this, it is quite necessary to develop indigenous technology for aviation gas turbines i.e. Jet engine. A jet engine has many variants in its classification based on component

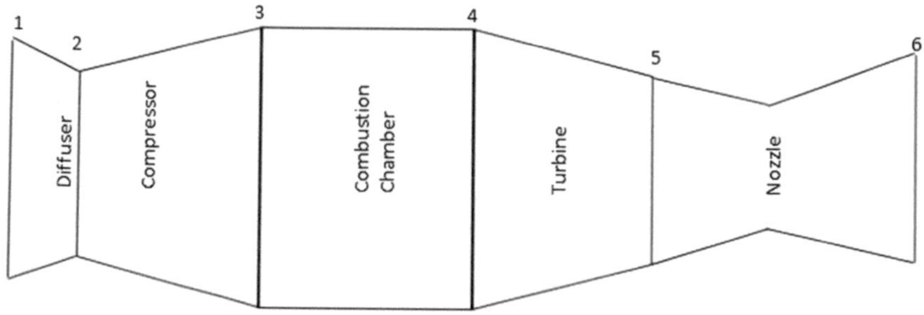

Fig. 1 Schematic sketch of engine

composition of the engine. So it is mandated to study existing design methods and the latest improvements in the design process to meet up to industry standards [1]. In this paper, we aimed at developing turbojet designs for upcoming technology of the drone industry which can generate 500 N thrust. As we must also be aware optimization of core components, they play a vital role in engine performance [2]. Medium scale jet engine requirements are predominantly increasing with unmanned aerial vehicle usage due to their thrust to weight ratio [3]. The development of medium scale turbojets is much more complex and challenging due to their lower mass flow rates [4]. However, design of smaller engines is different from conventional jet engine designs because of flow residence time and operation at higher temperatures and pressure profiles. The preliminary design was conventionally adopted from [5, 6]. MATLAB and C++ codes are developed to obtain the design parameters for the flow regime of Mach 0.0–3.0 thereafter the design was optimized to make a feasible system with the maximum possible efficiency (Fig. 1).

2 Methodology

2.1 Design Specifications

M_1	0.1–3
T_{o1}	300 K
γ_C	1.4
γ_H	1.3
c_C	1.004 kJ/kg.K
c_H	1.239 kJ/kg.K
hpr	42,800 (KJ/kg)
$\pi_{1-2\ max}$	1
π_{2-3}	6
π_{3-4}	0.9
π_{5-6}	0.96
$\eta_{\text{polytrophic compressor}}$	0.9

$\eta_{\text{polytrophic turbine}}$	0.9
$\eta_{\text{mechanical}}$	0.9
η_{burner}	0.98
TIT	1800 K
Overall pressure ratio P1/P6	0.75

The preliminary design was carried out with the following initially considered parameters on the Brayton cycle for having the operating envelope of the engine with its inlet flow regime from Mach 0–3. With this we are able to obtain cycle parameters with different flow parameters. With this we decided to make an engine max efficiency, operating Mach no at 2.2 as shown in Figs. 2 and 3. With the engine operating envelope the geometric design was preceded. By the obtained parameters, it was evident that it can generate a specific thrust of 483 N where the engine should have mass flow rates of 1 kg/s at compressor pressure ratio 6.

2.2 *Design Procedure*

From the operating envelope and max operating conditions, the geometric design was carried out and the design procedure for each component varies which would be discussed further, but at every section it was verified whether the component design obey flow field governing equations. As the engine operating Mach is at supersonic regime, the inlet diffuser is designed in such a way that the downstream of diffuser is a subsonic flow by inserting a series of oblique shocks in the diffuser. Diffuser design is purely based on the continuity equation and shock relations [5]. As the required $\pi_{2-3} = 6$ instead of choosing a centrifugal compressor we opted for four stage axial compressor where it requires less frontal area for same pressure ration which is essential for medium scale jet engines. As of compressor, the axial compressor was chosen because with increase of pressure ratio the diameter of the impeller also increases, which would be complicated for compact engine design and High RPM. Centrifugal compressor has small feasible operating range and high work input requirement from the turbine. It also elevates pressure ratio requirement

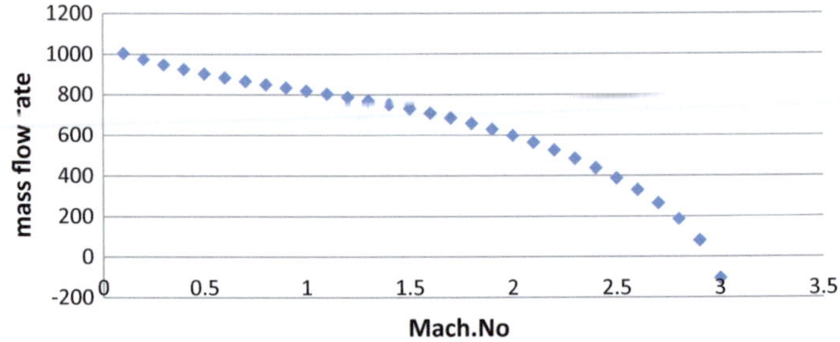

Fig. 2 Thrust mass flow rate with Mach no

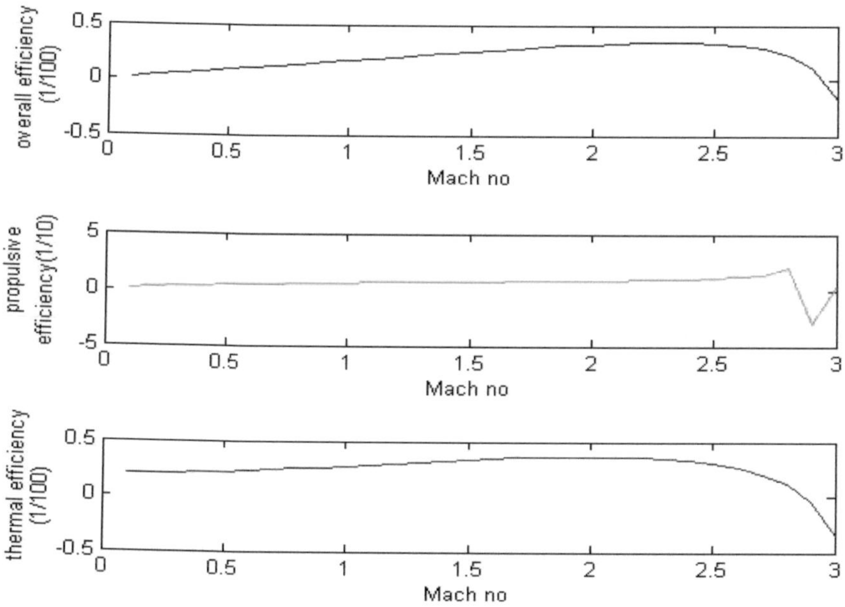

Fig. 3 Various efficiencies of engine with Mach no

with low exit velocity. In contrast with an increase in speeds and mass flow rates, the efficiency of axial compressor increases. With all the above considerations, we opted to go for axial compressor in which we had designed the compressor for a pressure ratio of 6 i.e. By theoretical analysis it had been obtained as 427 kPa. The geometrical profile of compressor is obtained in the iterative procedure of the following Eqs. (1–4) and by constant mean diameter design process. The cascade design of blades was performed on solidity ratio 1 where the stagger angle for rotor is 19.54° and Stator is 19.02° and circumferential blade arrangement of 28 rotors and stator.

$$\textit{Workdone by compressor} = \sigma \Psi \mathrm{V}_{\tan}^2 = \mathrm{c_p}(\mathrm{T_{02}} - \mathrm{T_{01}}) \tag{1}$$

$$r_{rms} = \left[\frac{r_{hub}^2 + r_{tip}^2}{2} \right]^{1/2} \tag{2}$$

$$r_{tip} = \left[\frac{Area}{\pi \left(1 - \left\{ \frac{r_{hub}}{r_{tip}} \right\}^2 \right)} \right]^{1/2} \tag{3}$$

$$U_{mean} = \left[\frac{2\pi r_{rms} N}{60}\right] \quad (4)$$

The combustion chamber is another crucial component in an engine which is a heat source and the performance, durability and efficiency of the chamber decides the engine performance and reliability. Design of combustion chamber is mainly oriented on the annular combustor as if to make the component compact and smaller in size. Combustor design is one of the most crucial and critical exercises because of which it requires a better optimization for turbine inlet by operating compressor output. In combustor design there is ambiguity for industry to industry design as the individual design standards they set up are from years of research and experimental data that have excelled. By following standard format from [7] the Eqs. (5–8) are adapted to obtain geometrical parameters and dilution zones design Table 1 are completely based on continuity and momentum equations rather than following the conventional Carnfield and NASA methods [7].

$$\text{Reference length } t_{res} = \frac{L}{V_{avg}} \approx \frac{L}{V_{ref}} = \frac{\rho_{t3} A_{ref} L}{\dot{m}_3} \quad (5)$$

$$\text{Reference area } A_{ref} = \left[\frac{R}{2}\left(\frac{\dot{m}_3 T_3^{0.5}}{P_3}\right)^2 \frac{\Delta P_{3-4}}{q_{ref}}\left(\frac{\Delta P_{3-4}}{P_3}\right)^{-1}\right]^{0.5} \quad (6)$$

$$\text{Profile factor } P_f \equiv \frac{T_{t\,\max av} - T_{t\,in}}{T_{t\,av} - T_{t\,in}} \quad (7)$$

$$\text{Pattern factor PF} \equiv \frac{T_{t\,\max av} - T_{t\,in}}{T_{t\,av} - T_{t\,in}} \quad (8)$$

$V_{ref} = 50$ m/s; $A_{ref} = 9.569\text{e-}03$ m^2

Length of the combustion chamber = 55 cm.

The turbine is one of the most sensitive parts in the engine where it has to operate within the specified temperature envelope. From the evolution of jet engines, material science has seen many improvements and better material compositions have been developing. But in this paper, we restricted to design a medium scale axial turbine. From parametric cylce analysis, it was clear that we require a single stage axial turbine that can generate sufficient power of 248 KW required by the compressor. The cascade blade design was considered from conventional Turbomachinery design [8] with Eqs. (9–16). Rotor blades staggering angles of 19° and stator blades with staggering angles of 19° as shown in Table 2 were obtained.

Table 1 Combustor dilution zone design

Section	No of holes	Diameter of hole (mm)
Fuel mixture	16	1.8
Primary	24	7
Secondary	36	10
Tertiary	54	8.2

Table 2 Turbine design for TIT 1800 and 2200 K

TIT (K)	V1	U1	φ	α_2	Degree of reaction	P_{05}	β_3	M_3
1800	254.839	254.835	1.7043	0.5025	0.2069	672.717	0.5679	0.8886
2200	281.735	281.736	1.7043	0.4303	0.1614	504.205	0.524	0.9186

Stator Design

$$(cx/s)m = [(2(\cos(\alpha 2m))^2) * (\tan(\alpha 1m) + (u2/u1) * \tan(\alpha 2m)) * (u1/u2)^2]/Zs \quad (9)$$

$$\Upsilon 2m = (\Upsilon 1m + 8 * \alpha 2m * (\sigma)^{0.5})/(8 * (\sigma)^{0.5-1}) \quad (10)$$

$$\lambda m = (\Upsilon 2m - \Upsilon 1m)/2 \quad (11)$$

$$\sigma m = (cx/s)/\cos(\lambda m) \quad (12)$$

Rotor Deisgn

$$s = \pi D/N \quad (13)$$

$$(b/s) = |(2/Cl) * \cos(\alpha exit) * (\tan(\alpha in) - \tan(\alpha exit))| \quad (14)$$

$$|\alpha exit| 0 < Mexit < 0.5 = [(7/6) * [|\cos - 1(o/s)| - 100] + 4o(s/e)] \quad (15)$$

$$\mathrm{Cos}\lambda = (b/cr) \quad (16)$$

where

$0.25 < (s/e) < 0.625$

$0.015c < te < 0.05c$

$C_l = 1$

As of our case, the engine must be able to fly at supersonic speeds and at the same time it should be able to generate thrust requirement as of mission require ment. In this case we require to design convergent, divergent nozzle that can generate nozzle lip Mach 2.3 and above to make self-sustainable flight. The nozzle design as shown in Table 3 was carried out with gas dynamics consideration and the isentropic relations with adiabatic consideration for real cycle. The nozzle length was obtained from boundary layer consideration and flow governing principles and Throat diameter was obtained from the Eq. (17) as 34 cm. Obtained nozzle geometric parameters were tabulated in Table 3.

$$A_t = \omega_t / P_t \sqrt{RT} / \gamma g_c \quad (17)$$

Table 3 Nozzle design parameters

Parameters	Inlet	Outlet
Diameter (cm)	38	58
Temperature (k)	1725	1725
Pressure (pa)	525422	101325
MACH no	0.9	2.3

2.3 *Simulation Modelling*

The aim of this simulation was to verify the design rather than knowing the behavior in real time due to limited computation power availability, but despite that, the considerations were same as real time. The simulations were carried out in CFX and CFD of ANSYS and geometries were developed in Solid works and ANSYS workbench, Gambit, Blade Gen. Programs. Diffuser section simulation was carried out on 2D gambit geometry in CFD to visualize the boundary layer with the help of SST model with Y+ model. Instead of simulating the compressor and turbine, simulation was carried on a set of stator and rotor blades to verify the same amount pressure rise and expansion possible in theoretical calculations [9]. The 2D grid was generated over each set of stator and rotor blades of compressor and turbine blades which provides the compression and expansion ratio at the end of the blade tip. The combustion chamber was carried in CFX with 3D where the species were considered of 63 in the methane with mole specifications [10] and with turbulent non-premixed combustion. Nozzle simulations were carried out in back pressure boundary conditions through which the flow adjusts to atmospheric conditions. For combustion chamber and nozzle, volumetric meshes were generated by using ANSYS Workbench.

3 Results and Discussions

As chosen from the inlet conditions of the engine to be Mach 2.2 it would be a huge problem for compressor to generate a required compressor ratio due to shock formation at the tip of the blades. Due to this, the normal shock existence was designed in the diffuser section through which the flow becomes subsonic, a feasible condition for operating the engine. The boundary layer thickness across the diffuser cross section which had been validated from the simulation results is shown across the diffuser section.

Rather validating compressor design at 3D scale, we opted to validate in stages where the stage pressure ratio theoretically across 4 stages were 1.81, 1.62, 1.50, and 1.45 respectively. The 3D simulation would be a task of complexity due to turbo machinery. So due to inadequate resources we validated first and last stages of

Table 4 Comparison of theoretical and simulated results

Sections		Parameters	Theoretical	Simulation
1	Diffuser outlet	Boundary layer thickness	1 mm	0.3 mm
2	Compressor outlet	Total pressure	426.7178 kPa	425–430 kPa
		Absolute velocity	184.64 m/s	176.05–245 m/s
3	Combustion chamber outlet	Total temperature	1800 K	2160–2250 K
		Static pressure	972.720 kPa	897.320–919.560 kPa
4	Turbine outlet	Static pressure	414.124 kPa	403.084–413.831 kPa
		Absolute velocity	704.74 m/s	690–712 m/s
5	Nozzle outlet	Thrust mass flow rate	483 N/kg/s	478 N/kg/s

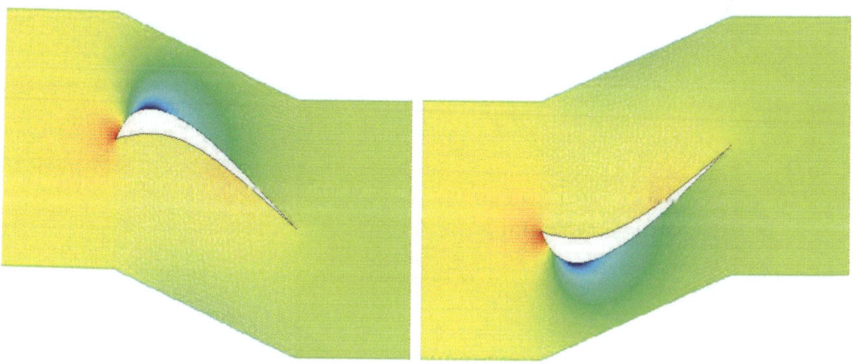

Fig. 4 Compressor last stage rotor and stator static pressure contour

compressor which are having pressure ratio across stages 1.81 and 1.45 respectively. The simulation results showed up to be satisfied as tabulated in Table 4 and the pressure contour shown in Fig. 4. The Combustion chamber 3D model was simulated as shown in Fig. 5, in which the temperature of turbine inlet was 2200 K from Fig. 6, which was inappropriate according to parametric cycle as it has to be 1800 K. But due to absence of heat exchanging devices which is predominantly used in industries, it requires for further reduction of heat to 1800 K but moving to 2200 K provides an edge in overall efficiency and operating range as shown in Fig. 9. The above TIT 2200 K results which better suits the turbine configuration as well, which is clear from Table 2. The turbine was a crucial and most sensitive part of the engine and its design is most complex where one should have a better idea of the aero-thermodynamics before dealing with it.

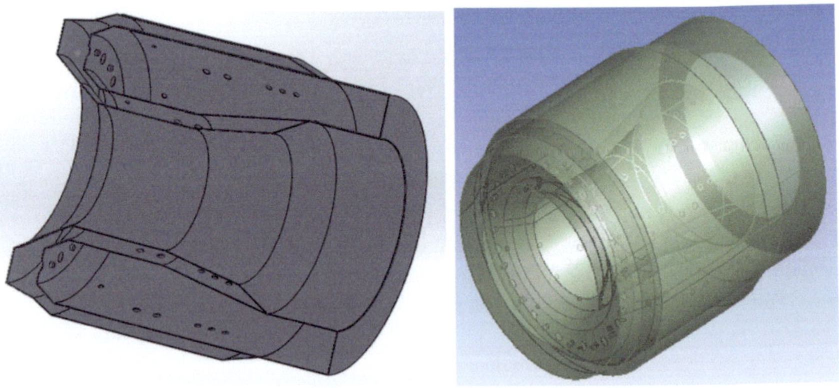

Fig. 5 Combustor 3D model

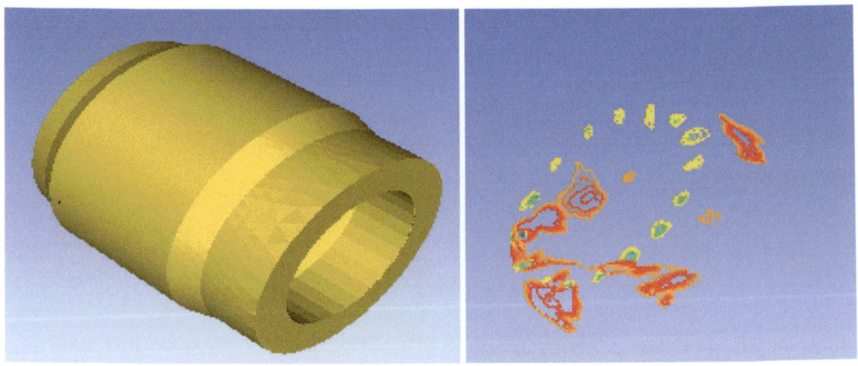

Fig. 6 Combustor temperature profile and flame temperature

As shown in Fig. 7, the simulation results showed an approximate expansion ratio of 0.691 from pressure profile. The static pressure can be compared from Table 3 by theoretical and simulation which was differing 0.2% of actual value. Because of the high expansion ratio, it is a high pressure turbine, which is providing the required output in a single stage expansion and its temperature profile can vary from 1800–2200 K as shown in Table 2. The nozzle was designed to be C-D as the engine operating at supersonic speeds, and to sustain the same we opted for C-D nozzle which could generate a thrust of 500 N but as of parametric cycle analysis it was clear that it would generate 483 N thrust. So the design considerations were made for the same and design parameters were shown in Table 3. The same was simulated in an under altitude condition of 25,000 m and it was observed that the exit Mach 2.3 from Fig. 8, was providing thrust of 478 N. In general, the nozzle was able to provide the thrust of 478 N and was able to maintain the engine to fly at

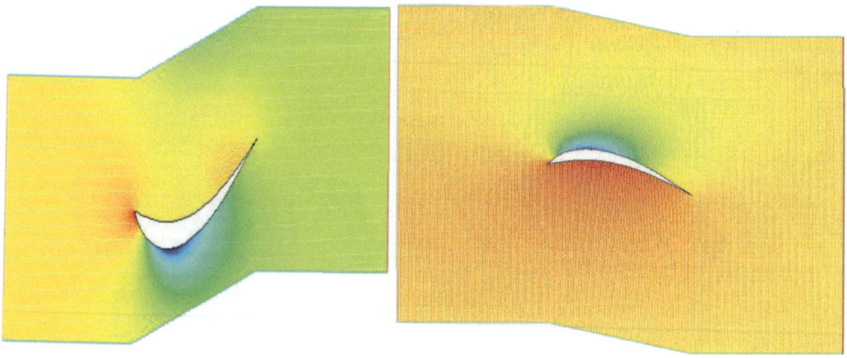

Fig. 7 Turbine rotor and stator static pressure contour

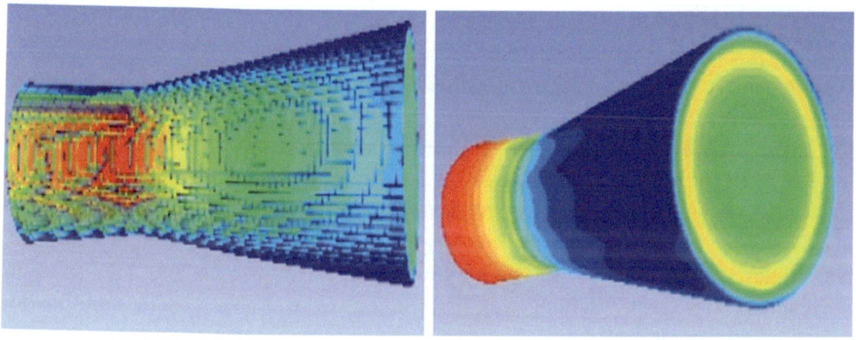

Fig. 8 Nozzle velocity vector and static pressure contour

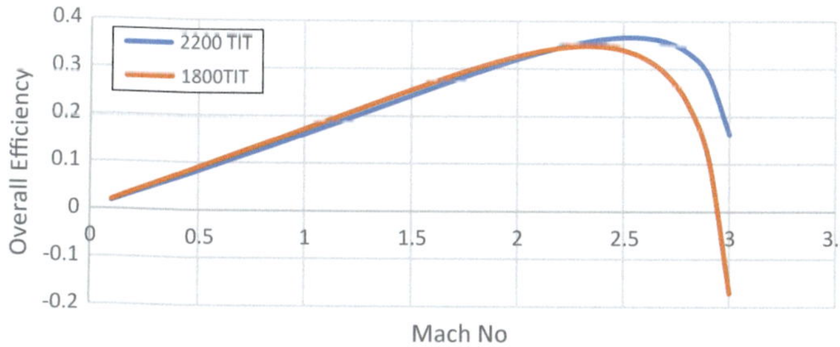

Fig. 9 Overall efficiency comparison with TIT

the same velocity as inlet velocity. From this it is clear that this design of jet engine was feasible and self-sustaining with additional generation of thrust of 478 N.

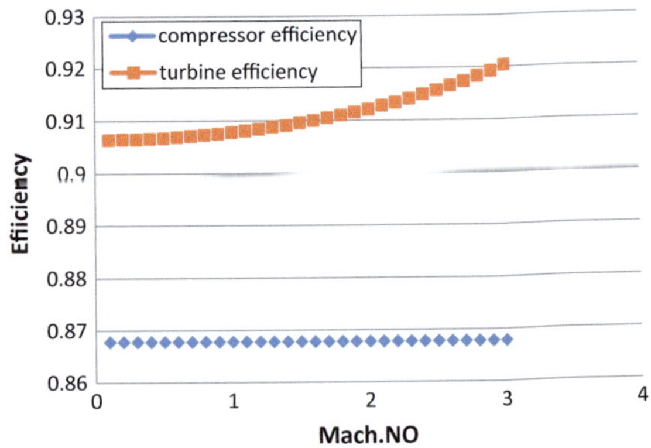

Fig. 10 Compressor and turbine efficiency comparison with Mach

4 Conclusions

With the above obtained results we validated our design, but there are many factors that need to be considered to improve its overall performance off the engine. As of now we are able to achieve an overall efficiency of about 38% from the TIT of 1800 K with an engine operating at Mach 2.2–2.5 (Fig. 9). Following our compressor design, a pressure ratio 6 was designed across 4 stages with pressure ratios of 1.81, 1.62, 1.50, and 1.45 respectively in stages. With the same simulation results had shown appropriate 1.8 pressure ratio over the compressor and 1.47 pressure ratio across last stage, which was interpreted from Table 4. The combustion chamber had shown underestimated performance in terms of dilution as of required TIT was 1800 K according to cycle analysis, but the simulation shown it to be 2200 K. With this we moved forward to the parametric cycle analysis to verify the feasibility of temperature and interestingly, it showed that overall efficiency increased by 2% and the operating range of the engine increased from Mach 2.5 to 3 (Fig. 10). Turbine designed showed the appropriate expansion ratio 0.691 pressure with high efficiency, this might be because of our underestimation of efficiencies during our parametric cycle analysis. The expansion ratio shown by turbine simulation is nearly double of expansion ratio in the parametric cycle analysis. The design of turbine is even suitable for TIT 2200 K. The Nozzle designed was made on C-D in order to provide thrust mass flow rate of 483 N accordingly to parametric cycle. The same can be obtained in simulation results with a velocity of 478 m/s but when the flow expansion was not as perfect as outlet pressure, it was not equal to back pressure of the atmosphere.

5 Future Work

Initially, a full pledge 3D simulation can be conducted on the individual components and whole engine assembly depending upon the simulation power. Moreover, the complex turbo machinery part can be better understood in order analyze the aero-thermodynamics and aero elasticity problems related to turbine and compressor. Once the complete simulation is performed, we can design some of the auxiliary components of an engine, like heat exchange and even a turbine blade cooling system to compete with industry standards. Even an afterburner device to produce wet thrust at time requirements can be developed. With these design templates, we can proceed to manufacturing stage where the whole engine manufacturing can take place, but before proceeding to production, we need to develop an idea of materials and manufacturing processes. Furthermore, integrating the propulsion system for a suitable medium drone can be done.

Acknowledgements Asst. Prof. Ankur Vats, Asst. Prof Jayanth Sinha and Dr. Konark Arora for their valuable guidance

References

1. Becker RG, Reeitenbach S, Klein C, Otten T, Nauroz M, Sigel M (2015) An Integrated method for propulsion system conceptual design. In: GT2015-43251 turbine technical conference and exposition ASME
2. Tyapin I, Sandberg M, Kokkolaras M, Lundbladh A, Isakson O (2012) Jet engine design optimization using a knowledge-based master model. In: GT2012-69309 turbo expo ASME
3. Benini E, Giacometti S (2007) Design, manufacturing and operations of a small turbojet-engine for research purposes. Appl Energy 84 (Elsevier)
4. Rodgers C (2003) Some effects on size on the performance of small gas turbines. In: GT2003-38027 ASME
5. Mattingly JD (1992) Elements of propulsion: gas turbines and rockets. Addison-Wesley
6. Oates GC (1997) Aircraft propulsion and rocket propulsion, 3rd edn. AIAA publications
7. Lefebvre AH (1999) Gas turbine combustion chamber, 2nd edn. Taylor and Francis
8. Horlock JH (1966) Axial-flow turbines. Butterworths, London
9. Calabria A, Capata R, Di Veroli M, Pepe G (2013) Development of an axial micro turbine for a portable gas turbine generator. Sci Res
10. Hashim SA (2013) Design and fabrication of an annular combustion chamber for the micro gas turbine engine applications. IJERT 2(8)

Eco-Friendly Propellant for Hybrid Rocket Motor

Aaditya Shrivastava

1 Introduction of Hybrid Rocket Motor

Hybrid Rocket Motor is such a class of rocket motor which is having the advantages of solid motor as well as liquid engines, in which mostly solid fuel is a polymeric material and oxidizer is a liquid or gaseous material. Earlier the application of Hybrid Rocket Motor has been limited which further increases because of its high safety and lower environmental impact. Initial Hybrid Rocket Motor contains one major disadvantage of having low fuel-burning rate and hence low thrust. Over the year when the research has been carried out to find out new types of fuel which can produce high regression rate as compared to conventional hybrid fuel, it was discovered that members of normal alkaline class of hydrocarbons are solid at room temperature, having low surface tension and viscosity, which include Paraffin waxes and polyethylene waxes (Fig. 1) [6].

2 Advantages of Hybrid Rocket Motor

(1) Easy thrust control because only the flow of liquid oxidizer is required to regulate.
(2) Chemical deterioration of fuel and oxidizer does not occur because both are kept separate.
(3) These are light in weight as compared to the liquid propellant because it required single propellant feed system.

A. Shrivastava (✉)
International School of Professional Studies, Navi Mumbai, India
e-mail: aaditya_shrivastava@yahoo.com

S. Singh et al. (eds.), *Proceedings of the International Conference on Modern Research in Aerospace Engineering*, Lecture Notes in Mechanical Engineering,
https://doi.org/10.1007/978-981-10-5849-3_4

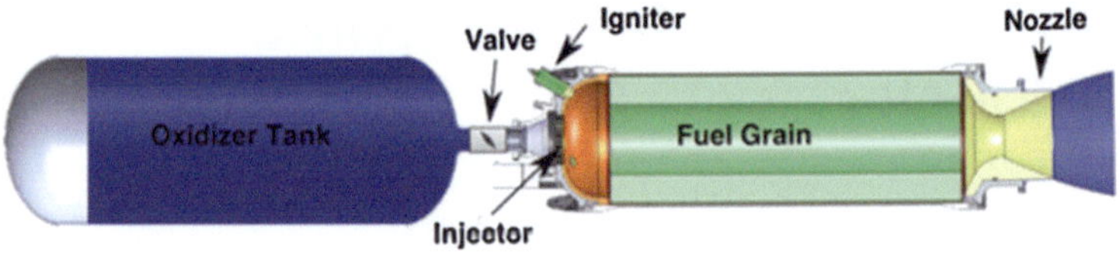

Fig. 1 Typical Hybrid Rocket Motor

(4) For fuel grain configurations, it is having higher number of options as compared to solid propellant rocket motor.
(5) In case any explosion occurred due to some failure, it is less destructive as compared to liquid propellant rocket engines [5].

3 Advantages of Paraffin Wax

(1) Regression rate of the Paraffin wax is 3–5 times high as compared to the polymeric fuels (including HTPB), which enables more efficient design.
(2) It is non-toxic, non-carcinogenic, non-hazardous and eco-friendly.
(3) By-products of combustion are carbon dioxide and water, while in conventional solid propellant with ammonium perchlorate (AP) oxidizer, by-products are acid forming gases such as hydrogen chloride. Apart from this, ammonium perchlorate (AP) also contaminated water during propellant manufacturing which is a significant environmental concern.
(4) Paraffin-based fuels are inexpensive.
(5) Processing of the fuel grain is simple. It does not require polymerization reaction. No curing agents are involved [1].
(6) Scrap pieces of fuel can be re-melted and reused.
(7) Paraffin waxes are hydrophobic, which makes them an ideal binder for metal, metal hydride and dense organic additives [2].
(8) These are inert; hence, Paraffin-based fuel is having infinite storage life.

4 Advantages of Nitrous Oxide

(1) It is an oxidizer with a vapour pressure of 736 psi at 68 °F which makes Nitrous Oxide a self-pressurizing propellant, which eliminates the requirement of complex structure and reduces propulsion system mass.
(2) Easy to handle, low toxicity, and high vapour pressure.
(3) Powerful solvent and very readily dissolve hydrocarbon materials [5].

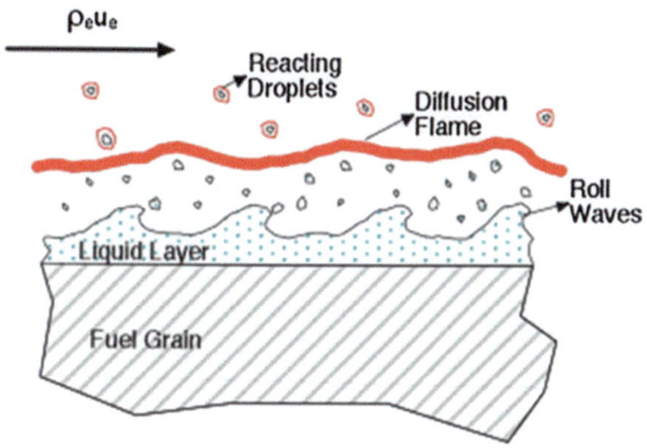

Fig. 2 Entrainment combustion mechanism of Paraffin wax

5 Combustion Process

The combustion system in the Paraffin wax follows the Entrainment theory in which heat is transferred through convection towards the fuel. Since Paraffin is having low melting point, it forms a liquid layer on the surface of the fuel layer which becomes unstable due to high velocity and turbulent gas flow in the port. The shear cause ripples on the liquid layer and due to viscosity, droplets are pulled up from the liquid layer and moved up through the boundary layer by means of Entrainment. When these droplets pass from the combustion flame or reach there, they themselves react in the oxidizer-rich environment and burn much faster. Finally, ripples which are formed in the liquid layer increase the surface area as well as roughness which enhance heat transfer (Fig. 2).

Regression rate equation is given by the following:

$$\dot{r} = \mathbf{a G_o^n},$$

where

a Empirical constant
n Pressure exponent
G_0 Oxidizer mass velocity.

6 Rocket Propulsion Principle

Thrust is a propulsive force which is produced by imparting change of moments to the exhaust combustion gases ejecting at the divergence section at convergent-divergent nozzle. Propellant is burnt into the combustion chamber of the rocket motor, and the resulting burn gases are allowed to eject through the convergent-divergent nozzle and increase the velocity and ambient pressure which gives change in momentum. According to the change in moment force called as thrust, this thrust has propelled the rocket in a forward direction [3].

The thrust equation is as follows:

$$T = \dot{m}V_e + (P_e - P_0)A_e,$$

where

$\dot{m}$ Mass flow rate
V_e Exit velocity
P_e Exit pressure
A_e Exit area
P_o Ambient pressure.

7 Cold Flow Test

A test to check and verify propulsion subsystem and flow of propellants, without firing the rocket. A cold flow experimental investigation is performed for Hybrid Rocket Engine test apparatus in order to gain a further understanding of transient phenomena affecting the engine's hot test firing results. The test allows for the evaluation of the fluid–structure interaction behaviour of the rocket engine's combustion chamber and oxidizer feed line/injection apparatus. Cold flow analysis and experiments enable us to see flow characteristics, velocity profile within the combustion system and rockets. To understand the loss caused by design is very important for the designer. Result obtained from this test is used to identify and analyse the faults of the design [4].

8 Static Testing

It is used to accurately determine the motor thrust, chamber pressure, burn time and other data which help in propellant evaluation, engine design and engineering. In any case, it is better to have a new motor malfunction in a safe test bay compared to flying high overhead out of control. By static testing, rocket engines can be tested before flying them to understand and learn how to rectify the errors and improve for further testing. For testing smaller rocket motors, it is required to measure and record the thrust performance. Thrust usually measured in ounces or grams or newton by various mechanisms. From these data, thrust curve is created which shows the engine performance. It shows the rise and fall of engine thrust over the period of propellant burning. This curve is one of the most important charts to evaluate engine and propellant performance. Chamber pressure is the next most important thing to measure along with burn time (duration of thrust). How much

Fig. 3 Static testing of Hybrid Rocket Motor

pressure is being generated in the motor case? Chamber pressure for most rockets is a few to thousands of pounds per square inch. Many rocket engines are designed to operate between 300 and 600 psi chamber pressure, higher the chamber pressure, higher the thrust (Fig. 3) [4].

9 Results and Discussion

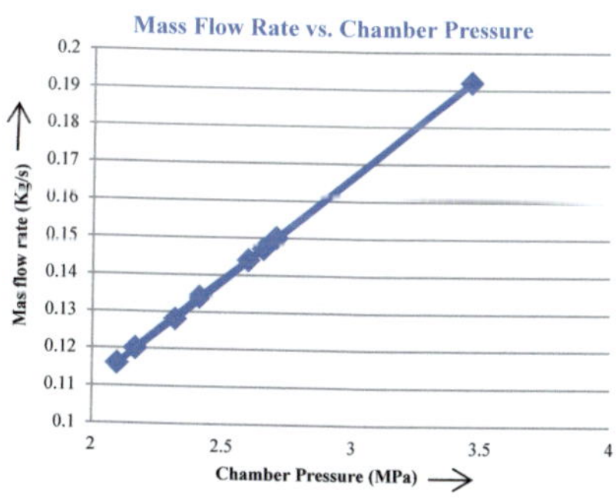

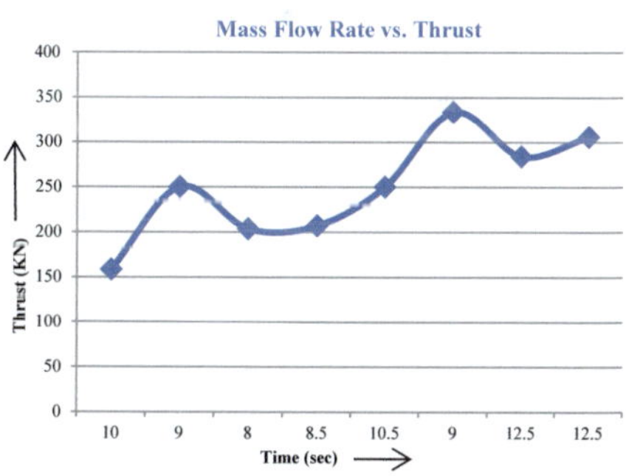

First graph shows the gradual increase in chamber pressure according to the mass flow rate of the Hybrid Rocket Motor. It indicates that with the increase in mass flow rate of the oxidizer, chamber pressure also increases.

In the case of Hybrid Rocket Motor, the thrust is a function of mass flow rate. This second graph shows the variation of thrust with respect to mass flow rate.

9.1 Chamber Pressure Verses Time Comparison

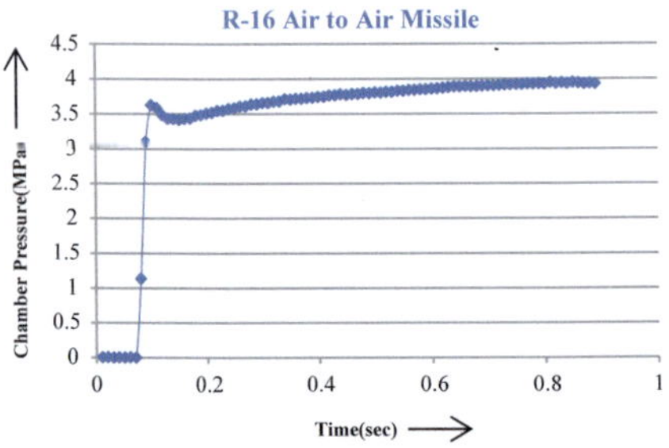

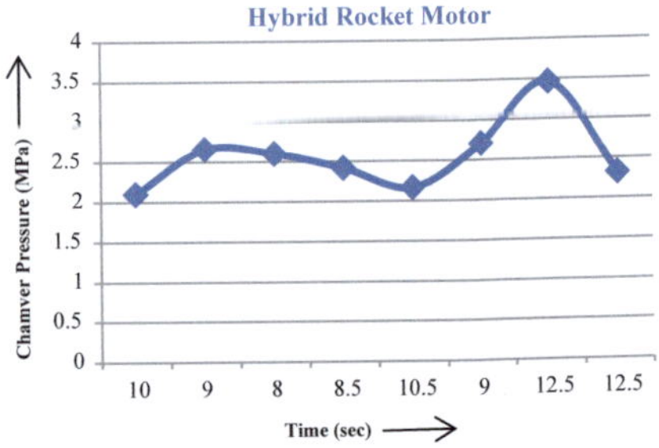

R-16 is a Solid Rocket air-to-air missile which uses cordite as propellant. Here, the comparison shows the performance of Hybrid Rocket Motor with Solid Rocket Motor. In the case of Solid Rocket Motor, at initially there is no increase in the chamber pressure, and after some microseconds, the graph shows rapid increase in the chamber pressure. After reaching the higher point, the graph shows the linear increase in the chamber pressure with respect to time.

While in Hybrid Rocket Motor, since beginning it shows the increase in chamber pressure, which also indicates no fluctuation in the graph, unlike R-16 where chamber pressure is increasing, but it also contains the fluctuations.

9.2 Thrust Verses Time Comparison

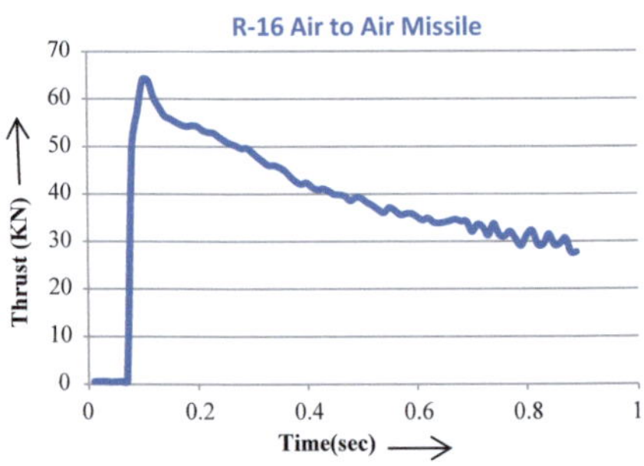

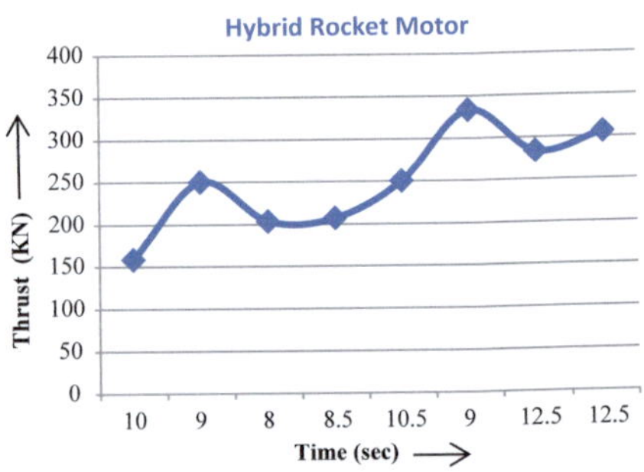

The graph shows the comparison between R-16 Solid Rocket air-to-air missile and Hybrid Rocket Motor. In the first graph of R-16, initially thrust increases rapidly which further decreases with respect to time. This decrease is not linear in fact having some fluctuations.

In the second graph, thrust is increasing with respect to time. This increases without any fluctuation which is much better than R-16 missile.

10 Conclusion

1. As compared to conventional hybrid propellant, Paraffin wax is a better option which eliminates the environmental hazards, easy to handle and provides better performance.
2. Unlike solid propellant, it never shows any fluctuation in its operation.
3. Regression rate can be enhanced by using additive material like charcoal and aluminium powder.

Acknowledgements Apart from my efforts, the success of this project is largely depended on the encouragement and guidance of many others. I express my thanks to my parents because since starting they support me when firstly I share my idea of doing project in this area, and they encouraged me whenever I lost my hope. I also thank to my near and dear ones for their encouragement.

References

1. Boros C, Konecny P (2012) Development of wax fuel grain for hybrid rocket motor. Adv Military Technol 4(2):5–11
2. Edwards T (2013) Liquid fuels and propellants for aerospace propulsion 1903–2003. J Propuls Pow 19(6)
3. Hasan T, Phillips J (2011) Potassium nitrate based rocket propulsion. NC Sch Sci Math 45(3)
4. McCormick A, Hultgren E, Lichtman M, Smith J, Azimi S (2012) Design Optimization and Launch of a 3' Diameter N2O/Aluminized Paraffin Rocket, vol 21, Iss 4. Stanford University, Stanford, CA
5. Patnala S, Chattaraj T, Joshi PC (2012) Combustion studies on paraffin wax in gaseous oxygen and nitrous oxide 7(7)
6. Stevens J, Karabeyoglu A et al (2011) High performance hybrid upper stage motor. In: 47th AIAA/ASME/SAE/ASEE Joint Propulsion Conference and Exibit, vol 32

Fluidic Thrust Vectoring of Engine Nozzle

R.B. Rakesh and Sijo Varghese

1 Introduction

Thrust vector control is the ability of aircraft to maneuver the direction of the thrust from its engine in order to control the angular velocity of the vehicle or its direction of motion. Thrust vectoring is used to control aircraft's or rocket's motions in roll, pitch, and yaw [1]. For fighter aircraft, the benefits of thrust vector control include control augmentation and improved survivability. Thrust vectoring has found increasing applications in recent years on aircraft, mostly as an augmentation to conventional control surfaces. Apart from post-stall maneuvering, advantages of thrust vectoring include reduced tail area and enhanced stealth, improved low-speed engine-out recoverability, and increased departure resistance. Several thrust vectoring methods are available for use on a jet engine. Some of these options include: rotation of the entire engine relative to the vehicle; rotation of the engine nozzle in order to direct the exhaust gases in a different direction; insertion of movable vanes or paddles into the exhaust flow; secondary fluidic injection; diversion of exhaust gases prior to the nozzle exit. Fluid thrust vectoring is a technology that deflects the main flow of an engine jet from the centerline in order to transfer some force to the targeted axis through primary and/or secondary fluid interaction [2]. By that imbalance momentum is created, and it changes the direction of motion. These techniques can be used to vector the exhaust flow in the pitch direction with the

R.B. Rakesh (✉) · S. Varghese
Department of Aeronautical Engineering, Sathyabama University, Chennai 600119, TN, India
e-mail: rbrakesh21@outlook.com

S. Varghese
e-mail: write2sijovarghese@gmail.com

S. Singh et al. (eds.), *Proceedings of the International Conference on Modern Research in Aerospace Engineering*, Lecture Notes in Mechanical Engineering, https://doi.org/10.1007/978-981-10-5849-3_5

magnitude measured by pitch vector angle or in the yaw direction with the magnitude measured by yaw vector angle. A means of comparing the thrust vector angles of different fluidic injection techniques or configurations is thrust vectoring efficiency, which is defined as the thrust vector angle in degrees, normalized by the percent of secondary fluidic injection. Use of fluid thrust vectoring is implying less complexity and faster dynamic response compared to other thrust vector control systems. Another aspect of secondary injection thrust vector control is that the moment arm of the resultant force is bigger than the mechanical thrust vector control techniques enabling to have lesser side forces since the ratio of the side force to the axial force allowed by this technique is limited. Using fluidic techniques for thrust vectoring can lead to weight reductions of as much as 80% over their mechanical counterparts.

Based on the conceptual design and module, there was a challenge of selection in papers as this topic hasn't been implemented into practice. Brendan A. Blake [3] titled 'Numerical Investigation of Fluidic Injection as a means of thrust control' helped in gaining an insight to the designs. A variety of different methods of fluidic vectoring have been developed since initial investigations of fluidic vectoring techniques began in the 1950s. Each method has been investigated both experimentally and numerically with different levels of success. All methods are fairly unique in the manner in which secondary flow is used to vector thrust.

The different types of fluidic thrust vectoring methods are classified as,

1. co-flow method
2. counterflow method
3. shock vector method
4. throat shifting method
5. recessed cavity method

Co-flow method is used when the fluid flow from the secondary injection is either directed toward the main flow. Due to low pressure flow, the main flow tends to deflect giving thrust angles. When a faster moving fluid flows along a slower moving fluid, the slower moving fluid tends to bend toward the faster moving fluid. The Coanda effect is the tendency of a flow to adhere to a curved surface due to the low pressure generated at the surface as the flow velocity over it increases [3]. The counterflow method uses suction between the trailing edge of the nozzle and an aft collar. The suction creates reversed flow along the collar. The end result is a pressure drop and an increase in velocity near the reversed flow area which provides a thrust differential. Shock vector control uses the injection of secondary flow downstream of the throat in the diverging section of the nozzle. Injection of flow here induces the formation of a shock wave at some angle. Maximum vector angle is achieved just prior to the shock impinging on the opposite side of the nozzle. Throat shifting uses the injection of secondary flow at or just upstream to the throat of a converging diverging nozzle [4]. The throat shifting method is considered to be the most efficient method of all combined, and this method is one of the main sources of the idea to provide secondary injection ducts for thrust vectoring in this project, but in this case, the secondary flow injection is made in the throat area itself before the flow reaches supersonic by skewing which forms a barrier, thereby deflecting the main flow at desired angles. The fluidic injections can be oriented desirably in order to get the required vectoring of the primary flow. Although the presence of recessed cavities does cause thrust vectoring, it is the combination of injected secondary flow at the throat in combination with the recessed cavities that cause vectoring. Although achieving significant results, combined methods are not widely investigated due to their inefficiency and difficulty to incorporate [5]. Recessed cavity method is a combination of throat shifting and formation of cavity, thus giving the project a base principle. A secondary flow is injected through a small injection port which is intended to travel into the cavity in the nozzle geometry such that the secondary fluid forms a vortex and acts as a barrier, and the main flow takes the shape of the rest of the exit nozzle geometry where vortex is not generated and the flow gets deflected.

2 Modeling Methodology

2.1 Computational Approach

A standard 2D—CD nozzle was designed in ANSYS 15(fluent). The nozzle boundary conditions consist of a main inlet, throat, outlet, wall, and three secondary inlets on either sides of the horizontal central axis (Table 1). Secondary inlet ducts are positioned at certain angle with respect to the main flow, but the ducts are perpendicular to the diverging section. The inlet ducts are equally spaced so as to provide a continuous minor obstructions to the primary flows which would provide efficient deflection and the deflection rate can be decided by controlling the flow through the secondary inlets and the flow velocities through each one of them (Fig. 1).

Table 1 Geometrical parameters of the nozzle design

Parameters	Values (mm)
Inlet diameter	25
Outlet diameter	35
Throat	10
Length of the diverging section	50
Length of the CD nozzle	75
Secondary inlet diameter	3
Secondary inlet ducts spacing	8
Secondary inlet ducts space from the throat	5.0562

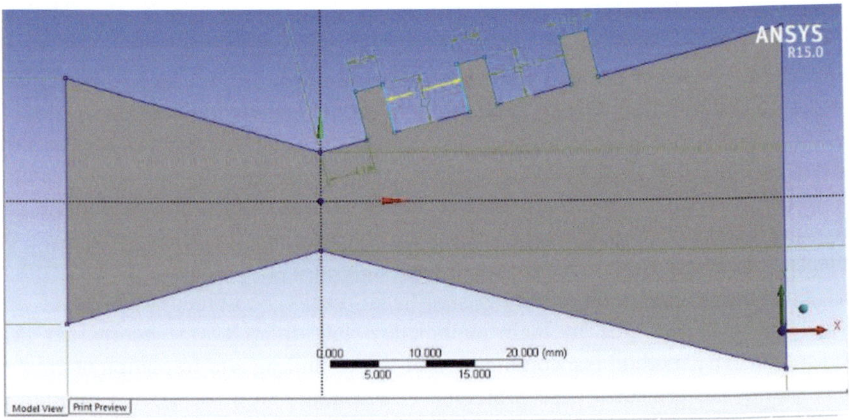

Fig. 1 CD nozzle

2.2 *Meshing of the Geometry*

Finite element analysis is dividing the geometry into various small numbers of elements. These elements are connected to each other at points called nodes. Each node may have two or more than two elements connected to it. A collection of these elements is called mesh. In ANSYS workbench, there are many tools and options available to create an effective mesh. An effective mesh is the one that requires less computational time and gives maximum accuracy. The meshing of the given CD nozzle geometry is performed to divide the geometry into number of cells so as to calculate the fluid flow at each and every point in the domain. By using unstructured quadrilateral mesh, there are about 23,000 cells. The refinement of the mesh is given specifically at the secondary inlets, and the outlet is made to identify any abrupt temperature change at the region where the secondary and the main flow interact, and to study whether there is any shock wave that is produced (Fig. 2).

Fig. 2 Mesh of the CD nozzle

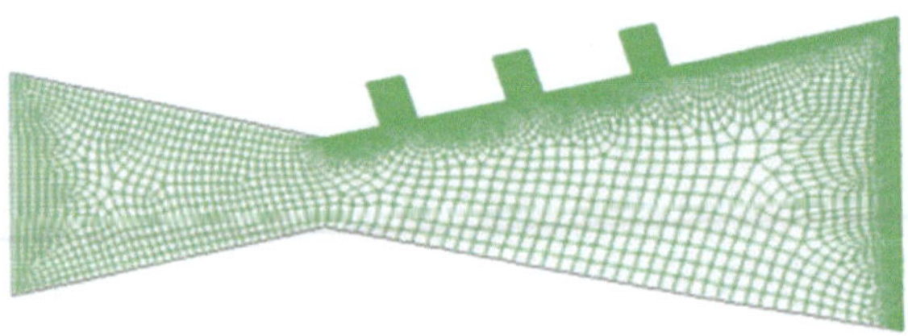

Table 2 Boundary conditions of the CD nozzle

Condition	Values (Pa)
Inlet pressure	3E5
Outlet pressure	0
Secondary inlet pressure	1E5
Ambient pressure (14000 m)	0.11E5

2.3 *Boundary Conditions*

The boundary conditions pertaining to the CD nozzle are the inlet, secondary inlet, outlet, and the wall (Table 2). There must be a pressure difference in the nozzle such that the flow goes from high pressure to low pressure (vacuum). The outlet pressure is maintained low because the flow needs to accelerate at supersonic speeds.

3 Performance Analysis

3.1 *Deflection Angle Value*

From the simulation, the fluid flow is vectored downward at a deflection angle of 23.8°.

Deflection angle of the vectored flow can be found using the formula and not by the computational approach as the software does not provide the angle at which the main flow turns while exiting the nozzle.

$$\delta = \tan^{-1}\left[F_y / F_x \right] \quad (1)$$

δ Deflection angle, $F_{x,}$ F_y = force in x, y directions
δ $\tan^{-1}$ [3505.233/7941]
δ 23.8

The deflection angle obtained is 23.8°.

3.2 *Results*

The computational method result indicates the thrust vectoring action with the maximum flow velocity of 1200 m/s (M > 3) without any abrupt heating/ temperature change in the secondary inlet and the main nozzle junction. The indicated velocity contour is after the application of the boundary condition to the nozzle with the thrust vectoring capability (Fig. 3).

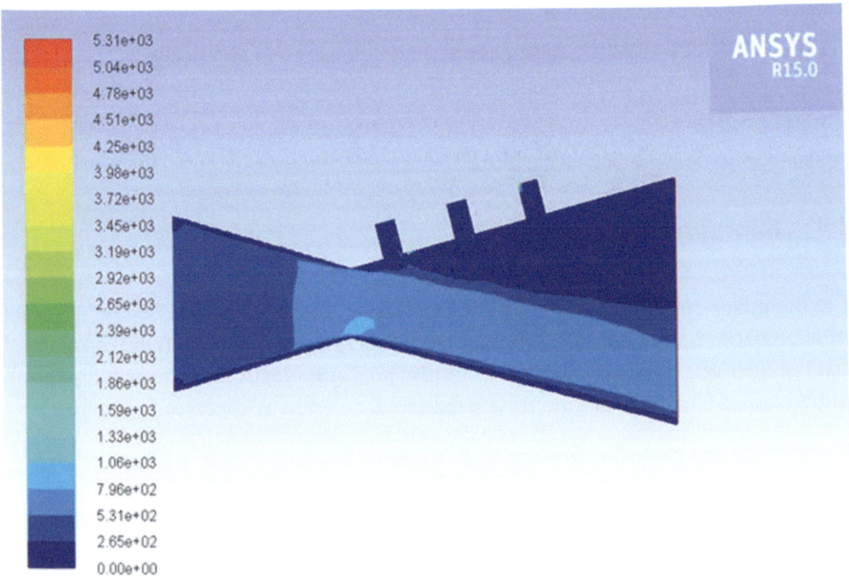

Fig. 3 Velocity contours

The table indicates the directions provided in which the thrust or the force values is intended to find, i.e., in the 'x' or 'y' direction whether positive or negative direction (Table 3).

The plots indicate the length or the distance of the nozzle's secondary inlet 1 in the 'x' direction while the velocity value in the secondary inlet 1 which has a diameter of 3 mm is indicated in the graph which shows certain fluctuations in the velocity (Fig. 4). The flow where the secondary injection and the main flow intersects experiences stagnation point where the velocity almost becomes zero, then the flow combines with the main flow and accelerates to a higher value. The intersection of the secondary flow with the main flow initially experiences stagnation, i.e., almost zero velocity but gradually combining with the main flow the flow velocity attains higher flow value.

The plots indicate the length or the distance of the nozzle's secondary inlet 1 in the 'x' direction while the pressure value in the secondary inlet 1 which has a diameter of 3 mm is indicated in the graph which shows certain fluctuations in the pressure at the secondary inlet 1 and the main flow junction (Fig. 5). In the pressure plot the point at which the stagnation point is experienced gives a higher static

Table 3 Values of vectored CD nozzle

No	X direction	Y direction	Force (N)
1.	1	0	7941
2.	1	1	3136
3.	1	−1	8093

Fig. 4 Plot between position and velocity

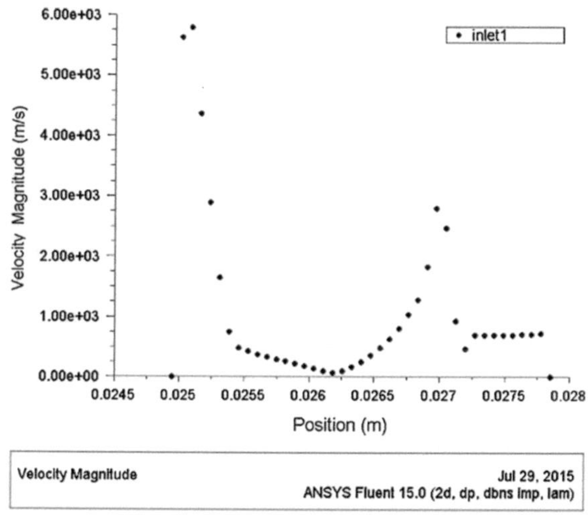

pressure value in the graph and then as the flow reaches the end if the secondary inlet results in increasing the flow velocity and decreasing the static pressure value.

The reports clearly indicate that the secondary inlets were kept at the upper part of the diverging section where the secondary injection deflects the main flow to the lower right-hand-side direction (1, −1) such that the resultant moment is obtained by the resultant force, i.e., pitching down tendency in the attitude of the aircraft.

Fig. 5 Plot between position and static pressure

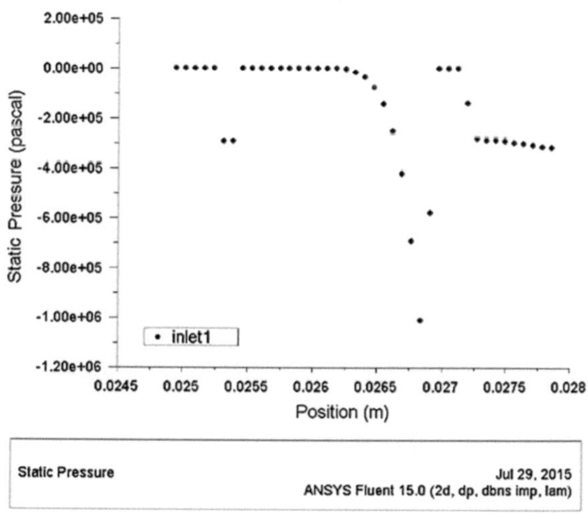

Compared to the main flow thrust (1, 0), the vectored thrust is greater than the main flow indicating an additional thrust by the secondary injection, and the vectored thrust in the lower direction is higher than that of the vectored thrust in the upper direction, this would give desired moment intended of changing the attitude of the aircraft and provide lower reaction time.

In order to evaluate thrust vectoring and nozzle performance, relevant efficiency parameters are defined to compare nozzle performance under different circumstances. Hence, specific impulse amplification coefficient C_{AL} is defined as the ratio between the global specific impulse in vectoring and non-vectoring mode operations, forces of such will give the thrust vectoring efficiency.

3.3 Thrust Vectoring Efficiency

$$\eta = C_{AL} = \text{Specific Impulse Amplification coefficient} = I^*_{SP}/I_{SP} \qquad (2)$$

$$C_{AL} = \{(F_X^2 + F_Y^2)^{1/2}/(\mathring{m}_i + \mathring{m}_j)\}/(F_a/m_i)$$
$$\{(7941^2 + 3505.233^2)^{1/2}/(0.3123 + 1.4378)\}/(1918.29/0.3123)$$
$$C_{AL} = 4959.83/6142.45$$
$$0.80 = 80\% = \eta = \text{efficiency}$$

The obtained thrust vectoring efficiency indicates the percentage of efficiency which is a parameter which can be related to the amount of flow being vectored at the exhaust.

4 Conclusion

The thrust produced by the vectoring CD nozzle is higher as the secondary flow gets added up with the main flow. The secondary injection when interacts with the main working fluid creates initial disturbance in the flow; hence, that interaction point acts as stagnation point leading to reduction in the flow velocity and drastic increase in the static pressure at the secondary inlet 1, where the obstruction in the flow is first experienced. The thrust values thus obtained having the 'x' and 'y' component provides the resultant moment in the desired direction.

Very high Mach number can be achieved in this type of vectoring where it can find its application in space vehicles like rockets and space shuttles.

References

1. Schaefermeyer MR (2006) Fluidic thrust vector control as a secondary injection. In: A Thesis Submitted in Partial Fulfilment of the Requirements for the Degree of MASTER OF SCIENCE in Mechanical Engineering, pp 15–45, Sep
2. Sutton GP, Biblarz O (2001) Aerodynamic thrust vectoring for attitude control of a vertically thrusting jet engine. In: "Thrust Vector Control," Rocket Propulsion Elements, 7th edn. Wiley, New York, NY, pp 11–52
3. Blake BA (2009) Numerical investigation of fluidic injection as a means of thrust control. In: Initial Thesis Report, ACME, UNSW@ADFA, pp 2–10
4. Yagle PJ, Miller DN, Ginn KB, Hamstra JW (2012) Demonstration of fluidic throat skewing for thrust vectoring in structurally fixed nozzles, Lockheed Martin Aeronautics Company, pp 4–6
5. Erdem E (2005) Thrust vector control by secondary injection. In: Thesis Submitted to Middle East Technical University, p 103, Sep 2006

2D-PIV Measurements in a Novel Swirl Burner Under Isothermal Conditions

Mourya Voleti, Naga Venkata Sai Aditya Burle, Raghu Jarpala and Rajesh Sadanandan

1 Introduction

One of the major design criteria in combustor design is pollutant reduction, especially the oxides of nitrogen (NO_x) and carbon monoxide (CO). To reduce these pollutants, global gas turbine (GT) manufacturers pursue different approaches. Some of the methods in use include fuel flexibility, staged combustion, moderate and intense low-oxygen dilution (MILD) and flameless combustion [1–3].

For GT engines, swirl is the preferred flame-stabilizing mechanism. A number of investigations elucidate the various parameters influencing swirl combustion [4]. Swirling flows result in strong shear regions, high turbulence, enhanced mixing and rapid mixing rates of hot products with fresh unburned gas mixture resulting in wider stability limits. A major feature in swirling flows is the presence of recirculation zone above a critical swirl number [5]. The hot combustion products in the recirculation zone constantly mix and ignite the fresh mixture and thereby ensure high stability of the flame.

With an aim to develop new burner designs to reduce pollutant formation, a novel gaseous non premixed model GT burner [6] is developed at IIST. The burner design is aimed at promoting intense mixing of the fuel and air with the recirculating product gases. Previous researches in MILD [2] or flameless combustion (FC) [3] has

M. Voleti · N. Venkata Sai Aditya Burle · R. Jarpala · R. Sadanandan (✉)
Indian Institute of Space Science and Technology (IIST), Thiruvananthapuram 695547, India
e-mail: rajeshsadanandan@iist.ac.in

M. Voleti
e-mail: mourya1995@gmail.com

N. Venkata Sai Aditya Burle
e-mail: bnvsaditya@gmail.com

R. Jarpala
e-mail: raghu4youall@gmail.com

S. Singh et al. (eds.), *Proceedings of the International Conference on Modern Research in Aerospace Engineering*, Lecture Notes in Mechanical Engineering,
https://doi.org/10.1007/978-981-10-5849-3_6

shown that intense mixing promotes uniform temperature distribution at lean operating conditions, resulting in low adiabatic flame temperature and thereby enabling ultra-low emissions.

In this regard, the goal of the current studies is to investigate the turbulence characteristics in the newly developed burner under isothermal conditions. The absolute flow velocities are measured using two-component particle image velocimetry (2D-PIV) to capture the temporal development of the flow field. The information is vital in understanding the development of large-scale coherent structures and their impact on the heat release by chemical reactions.

2 Burner Description

The model GT swirl burner (IIST-GS1) [6] used in the present work is shown schematically in Fig. 1. The swirling motion is imposed on the main airflow by means of a radial swirler of swirl angle 60° and consisting of 12 vanes. Air at room temperature and pressure passes through the radial swirler and enters into the flame zone through an annular nozzle. The annular space at the burner exit has outer and inner diameters of 42 and 36 mm, respectively, resulting in a hydraulic diameter (D) of 6 mm at the nozzle exit. The fuel is fed through a non-swirling annual gap of 0.5 mm between the nozzle body and the central bluff body. For the current isothermal studies, the fuel channel is closed. The central region of the bluff body is recessed in the shape of an inverted truncated cone; the recess height is approximately 11 mm. The airflow rate was regulated by digital mass flow meter (Make: ALICAT M-100 SLPM-D/10M) and metering valve and is fixed at 100 lpm. The nozzle Reynolds number based on the cold inflow bulk velocity and hydraulic diameter is Re = 1680. The flowmeter has an accuracy of typically 0.8% of the reading and +0.2% of full scale. Swirl number S of the burner, defined as the ratio of axial flux of swirl momentum to axial flux of axial momentum, is 1.01 [7].

3 Experimental Set-up

The PIV experimental set-up consists of two configurations, one for axial flow field and other for tangential flow field measurements. A Nd: YAG double-pulsed laser (532 nm, 10 Hz, 15 mJ/pulse energy) is used as the light source and LaVision Imager Pro X 4M CCD camera (2048 × 2048 pixels) is attached with a Nikkor lens (50 mm, f/1.8D) and bandpass filter (532 ± 10 nm) is used as a recording medium. The time delay (dt) between laser pulses is 55 and 45 μs for axial and tangential flow field measurements, respectively. LaVision Programmable Timing Unit (PTU) was used to synchronize the laser pulse with camera shutter timing. Light sheet optics is used to create a diverging laser sheet of 1 mm thickness for the tangential plane measurements, and a vertical parallel sheet (50 mm height) was used for the axial plane

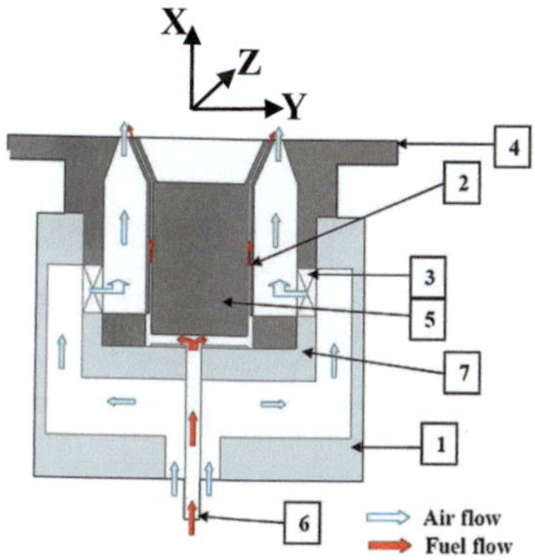

PART NAME	REFERENCE NUMERALS
1	Air Plenum
2	Fuel Duct
3	Swirler
4	Nozzle body
5	Inner Cone
6	Fuel injection tube
7	Thread sleeve

Fig. 1 Schematic of IIST-GS1 swirl burner

measurements. The flow was seeded with olive oil droplets (particle size ~1–10 μm) using a seeding generator to illuminate the flow. Approximately 300 pairs of instantaneous images are acquired for determining the average velocity vectors and RMS velocity fluctuations.

For the tangential flow field measurements, the camera is mounted at an angle to the burner axis. The image defocussing due to this angular configuration is corrected by mounting the camera on a Scheimpflug adapter and by rotating the camera body with respect to the lens as stipulated by the Scheimpflug principle [8]. Commercial PIV software (LaVision Davis 8.2) is used for the vector realizations by spatial cross-correlation of the particle images. An adaptive multi-pass vector evaluation technique is employed with interrogation window sizes decreasing from 64 × 64 to 16 × 16 pixels with 75–50% overlap. This results in a spatial resolution and vector spacing of approximately 3.15 mm and 1.57 mm, respectively. Considering the uncertainty of 0.1 pixel in the peak detection algorithm, the measurement uncertainty of the instantaneous velocities is estimated as 0.1 m/s. Spurious vectors are detected based on deviation from the local median and are iteratively removed.

4 Results and Discussion

4.1 Axial Flow Field

The instantaneous axial velocity flow field and the streamlines are shown in Fig. 2a, b, respectively. The fine-scale vortices seen in these images highlight the level of turbulence present. On the other hand, the time-averaged images in Fig. 2c, d show a well-structured flow field.

The flow field comprises three regions namely (i) bluff body-stabilized region close to the burner exit, (ii) swirl-stabilized region further downstream and

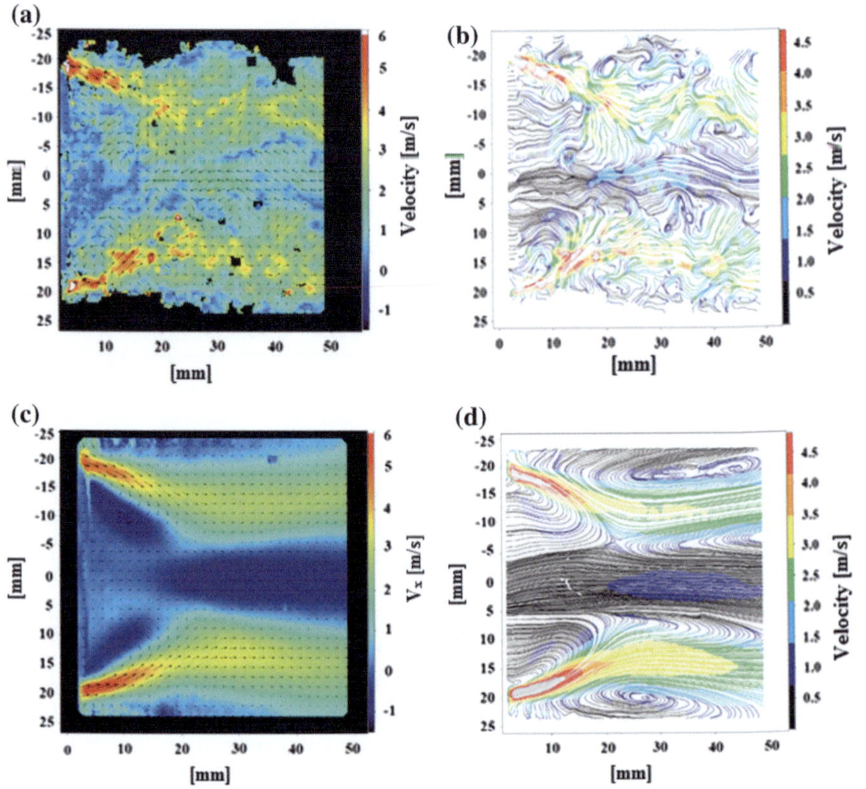

Fig. 2 **a** Instantaneous axial velocity flow field **b** Instantaneous streamlines **c** Time-averaged axial velocity flow field **d** Time-averaged streamlines

(iii) transition region in between the bluff body-stabilized region and swirl-stabilized region. Two inner recirculation zones are formed at the burner exit due to the bluff-body effect, and these zones extend up to 20 mm axially from the burner exit. This recirculating flow is enclosed by the high-velocity incoming flow with the streamlines converging up to downstream boundary of the bluff body-stabilized region.

Intense mixing between the incoming airflow, fuel and the burned gases is expected at the shear layer between the high-velocity inflow and the inner recirculation zone. Further downstream, the streamlines diverge and a swirling flow field with a central reverse flow region is observed from approximately 25 mm from the burner exit. The transition region from bluff body-stabilized to swirl-stabilized regions is observed from approximately 20–25 mm downstream of the burner exit. It can also observed that the central reverse flow region is extending from the swirl-stabilized region up to the burner exit in the bluff body-stabilized region.

For understanding the turbulence chemistry interactions, one of the important parameters desired is the lengthscales of the turbulent motion as they provide a measure of the energy-containing eddies. The turbulent lengthscales are estimated based on the space correlation coefficient [9] given by Eqs. (1) and (2)

$$R_{ii}^{(j)} = \frac{\overline{u_i(x,t)u_i(x+re_j,t)}}{\sqrt{\overline{u_i'^2(x,t)}}\sqrt{\overline{u_i'^2(x+re_j,t)}}} \tag{1}$$

$$L_{ii}^{(j)}(x) = \int_0^\infty R_{ii}^{(j)}(r,x)dr \tag{2}$$

where $R_{ii}^{(j)}$ is the correlation coefficient, $L_{ii}^{(j)}(x)$ is the integral lengthscale. The double script i refers to the velocity component, and the subscript j stands for the direction of moving probe displacement.

The variation of the correlation coefficient distribution along the radial direction for particular axial locations is shown in Fig. 3a. In case of homogeneous turbulence, the correlation coefficient ($R_{ii}^{(j)}$) distribution should be symmetric [9]. It is clearly seen that the correlation coefficient distribution is symmetric and the departure from symmetry remains small far from the axis. The variation of lengthscales estimated from axial velocities along the axis using the Eq. (2) is shown in Fig. 3b. Due to the symmetry of the correlation coefficients, the lengthscales are calculated in positive radial direction only.

It is evident that the turbulence characteristics in the three stabilization regions are dissimilar. From the burner exit, the integral lengthscale value gradually increases up to X/D = 2.65 in the bluffbody-stabilized region due to the presence of the recirculation zone. In transition region, the turbulence intensity increases and the lengthscale values decrease. From X/D = 3.4, the lengthscale value fluctuates and then reaches a peak value at an axial distance of X/D = 5.85, which means that there is low level of turbulence at this location.

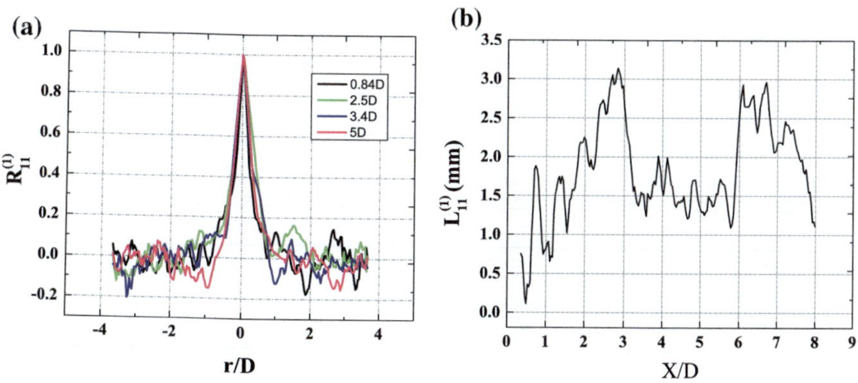

Fig. 3 **a** Correlation coefficient distribution along the radial direction at different axial directions **b** Integral lengthscale profile along axial distance

4.2 Tangential Flow Field

The instantaneous velocity field, time-averaged mean velocity field and the corresponding streamlines at an axial distance of 5 mm are shown in Fig. 4a, b, respectively. The anticlockwise rotation of the bulk flow about the centre due to the swirling motion is clearly seen in these images. The high inflow velocities (red-coloured regions), which are formed near to the burner edge, are the resultant of both tangential and axial velocity components. The corresponding time-averaged tangential velocity vector field and streamlines are shown in Fig. 4c, d, respectively. It can be seen that at the interface between the inflow region and the central recirculation region, the direction of the resultant inflow velocity vector is opposite to that of the circulating flow velocity vector due to the strong influence of both swirling and the bluff body effect.

The resolved vorticity field (ω) in the horizontal plane at different axial distances is shown in Fig. 5a. The positive and negative vortices at the shear layer indicate

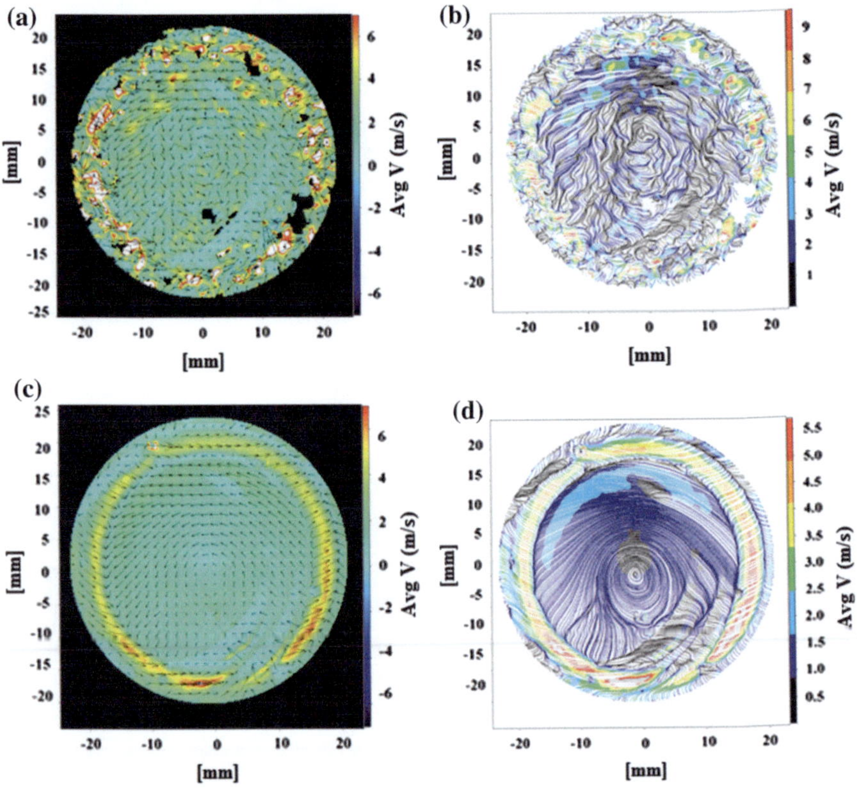

Fig. 4 **a** Instantaneous velocity flow field **b** Instantaneous streamlines **c** Time-averaged velocity flow field **d** Time-averaged streamlines

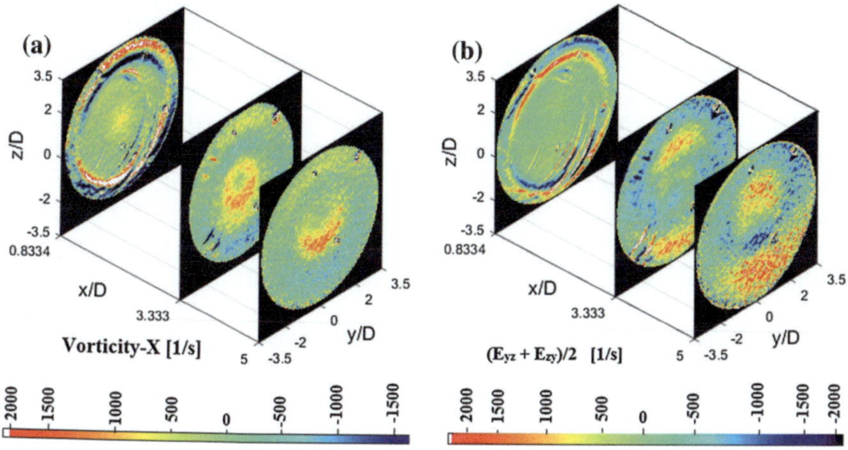

Fig. 5 Variation of **a** vorticity field and **b** extensive strain field along axial direction

the existence of counter-rotating vortex pairs at the shear layer in these locations. At the transition region (X/D = 3.3), where the flow converges, the intense amount of turbulence generated is evident from the relatively high levels vorticity in a narrow region. Further downstream (X/D > 3.33), the vorticity levels gradually decrease and spread out due to the diverging flow field in the swirl-stabilized region.

The two-dimensional fluid dynamic extensive strain rate field, defined as $S_{yz} = (\frac{\partial v}{\partial z} + \frac{\partial w}{\partial y})/2$ at different axial locations from the exit of the burner, is shown in Fig. 5b. The high shear area at the shear layer between the incoming flow and the recirculation zones is clearly reflected in the tangential strain field plots. The strain field plays a critical role in the flame stabilization as high local strain rates can lead to flame extinction in these locations. Previous researches in non-premixed CH_4/air flames have shown that localized flame extinction occurs when the strain rates are close to the critical strain rate of 400 1/s. So from the plots, flame stabilization at the shear layer closer to the burner exit is highly unlikely. The flame will most probably stabilize at low strain regions downstream, provided that the local mixture fraction is also favourable.

5 Conclusion

Non-intrusive laser-based measurement techniques (2D-PIV) were employed to understand the isothermal, turbulent flow field characteristics of a novel non-premixed model GT burner at atmospheric conditions.

The axial flow field measurements revealed the existence of two stabilization regions due to bluffbody effect and swirl effect and a transition region in between. The flow streamlines along the high inflow region converges in the bluffbody-stabilized region and then diverges in the swirl-stabilized region. The transition region in between is characterized by a turbulent region of finer length-scales.

The vorticity plots in the axial and tangential show intense turbulence being generated at the shear layers by the combined bluffbody and swirl effect. This will aid in improving the mixing of fuel and air with the burned gases which is critical for reducing the pollutant emissions. Also, the increased turbulence level at the transition regions along with the relatively low extensive strain at these locations will be helpful in attaining flame stabilization over a wide range of operating conditions.

Acknowledgements The authors would like to acknowledge the support of Vinil Kumar and Prakash from APLD Lab, IIST.

References

1. Fleck JM, Griebel P, Steinberg AM, Stohr M, Aigner M, Ciani A (2010) Experimental investigation of a generic, fuel flexible reheat combustor at gas turbine relevant operating conditions. In: ASME turbo expo 2010: power for land, sea, and air. American Society of Mechanical Engineers, pp 583–592
2. Cavaliere A, de Joannon M (2004) Mild combustion. In: Progress in energy and combustion science, vol 30. Elsevier, pp 329–366
3. Sadanandan R, Lckerath R, Meier W, Wahl C (2011) Flame characteristics and emissions in flameless combustion under gas turbine relevant conditions. J Propul Power 27:970–980
4. Weigand P, Meier W, Duan XR, Stricker W, Aigner M (2006) Investigations of swirl flames in a gas turbine model combustor: I. Flow field, structures, temperature, and species distributions. Combust Flame 144:205–224
5. Lefebvre AH (1998) Gas turbine combustion. CRC press
6. Sadanandan R (2015) Ultra lean non-premixed gaseous fuel burner. Indian Institute of Space Science and Technology applicant, India Patent 6632/CHE/2015
7. Gupta AK, Lilley DG, Syred N (1984) Swirl flows. Abacus Press, Tunbridge Wells, Kent, England
8. Douglas A, Kerr PE (2006) The Scheimpflug Principles, 70th birthday series, pp 1–18
9. Belmabrouk H, Michard M (2001) Analysis of the swirl effect on turbulent length scales in an ICE cylinder by two-point LDV. Int J Heat Fluid Flow 22:417–423

Design of 3D Guidance Law for Tactical Missiles

Sikander Kumar Pandit, Bhavnesh Panchal and S.E. Talole

1 Introduction

Traditionally, guidance laws for tactical missiles are designed by considering decoupled missile-target engagement dynamics. The dynamics is assumed to be decoupled into two mutually orthogonal planes assuming that the cross-coupling between the channels is negligible and guidance laws are then designed for both the planes separately. This approach can be implemented effectively if the elevation angle between the missile and target is considerably small [1]. However, in engaging highly maneuvering targets especially in terminal phase, yaw and pitch angles may not be small, and, therefore, the cross-coupling effects may show up resulting in degraded performance for the planar guidance laws. This fact has motivated to design a three-dimensional (3D) guidance law so that the cross-coupling effects are catered for. The 3D guidance law design based on proportional navigation is a natural choice as the PNG has been shown to be most effective strategy for guiding tactical missiles. However, while the classical PN offers desired performance in nonmaneuvering or slowly maneuvering targets cases, its performance in the presence of highly maneuverable targets may not be satisfactory. To deal with the highly maneuverable target scenario, various guidance law formulations based on different nonlinear control methods have appeared in literature [1–5]. In this work, the 3D PN guidance law is derived by using the input-output linearization (IOL) [6, 7] theory. Simulation by considering highly maneuvering targets is carried out, and the results are presented to showcase the effectiveness of the design. Further, performance of the proposed formulation is compared with augmented proportional navigation guidance laws and the related results are presented. The remaining article is organized

S.K. Pandit (✉) · B. Panchal · S.E. Talole
Department of Aerospace Engineering, Defence Institute of Advance Technology, Girinagar 411025, Pune, India
e-mail: sikander007.pandit@gmail.com

S. Singh et al. (eds.), *Proceedings of the International Conference on Modern Research in Aerospace Engineering*, Lecture Notes in Mechanical Engineering,
https://doi.org/10.1007/978-981-10-5849-3_7

as follows. The missile-target engagement model is presented is Sect. 2, and the formulation of feedback linearization-based 3D guidance law is presented in Sect. 3. In Sect. 4, simulation results are presented whereas Sect. 5 concludes this work.

2 Missile-Target Engagement Model

The 3D pursuit situation is as shown in Fig. 1 [8–10].

Where (X_I, Y_I, Z_I), (X_L, Y_L, Z_L), (X_M, Y_M, Z_M), and (X_T, Y_T, Z_T) represent reference, LOS, missile body, and target body coordinate systems, respectively, whereas i_L, j_L, k_L, i_M, j_M, k_M, and i_T, j_T, k_T are the unit vectors in the respective frames of reference. The other quantities $V_M(V_T)$ are the missile(target) velocity vectors, $\rho = V_T/V_M$, θ_L, ψ_L, θ_{IL}, ψ_{IL} and θ_m, ψ_m are the pitch and azimuth angles of missile and target, respectively, in the corresponding frame of reference; θ_t, ψ_t are the pitch and azimuth angle of target, respectively; N is the navigation constant; r is the missile to target range; $\dot{\lambda}_x, \dot{\lambda}_y, \dot{\lambda}_z$ are the components of angular velocity vector in LOS frame; A_M^C is the commanded acceleration; A_{ym}^c, A_{zm}^c are the commanded accelerations in yaw and pitch plane, respectively. A_{ym}, A_{zm} are the yaw and pitch accelerations of missile, and A_{yt}, A_{zt} are the yaw and pitch accelerations of target. L is the LOS vector, and A_{YL}, A_{ZL} and Y_L, Z_L are the components of A_L. θ_{ij}, ψ_{ij} are the Euler angles, and ω_L is the angular velocity vector of LOS.

To make the problem tractable, the following assumptions are made, missile and target are considered as point masses, and seeker dynamics are neglected. Autopilot is much faster than guidance loop, i.e., $A_{ym} = A_{ym}^c$ and $A_{zm} = A_{zm}^c$. Further, the velocity of target is assumed as constant and angle of attack of missile is considered small enough to be neglected [11]. Under these assumptions, the nonlinear differential equations which describe the pursuit situation of missile and target in 3D space are given as

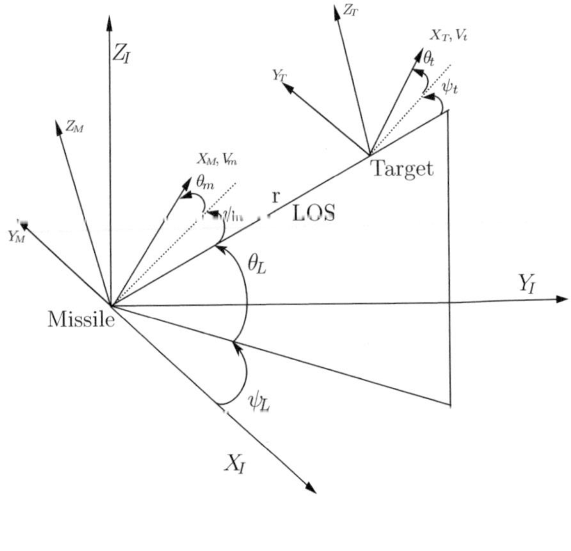

Fig. 1 3D missile-target engagement geometry

$$\dot{r} = (\rho c\theta_t c\psi_t - c\theta_m c\psi_m)V_m$$
$$r\dot{\lambda}_y = (s\theta_m - \rho s\theta_t)V_m$$
$$r\dot{\lambda}_z = (\rho c\theta_t s\psi_t - c\theta_m s\psi_m)V_m$$
$$\dot{\theta}_m = \frac{A_{zm}}{V_m} + \frac{V_m \tan\lambda_y s\psi_m(\rho c\theta_t s\psi_t - c\theta_m s\psi_m)}{r} - \frac{V_m c\psi_m(\rho s\theta_t - s\theta_m)}{r}$$
$$\dot{\psi}_m = \frac{A_{ym}}{V_m c\theta_m} - V_m s\theta_m c\psi_m \tan\lambda_y(\rho c\theta_t s\psi_t - c\theta_m s\psi_m)/(rc\theta_m)$$
$$-V_m s\theta_m s\psi_m(\rho s\theta_t - s\theta_m)/rc\theta_m - V_m(\rho c\theta_t s\psi_t - c\theta_m s\psi_m)/r$$
$$\dot{\theta}_t = \frac{A_{zt}}{\rho V_m} + V_m s\psi_t \tan\lambda_y(\rho c\theta_t s\psi_t - c\theta_m s\psi_m)/r - V_m c\psi_t(\rho s\theta_t - s\theta_m)/r$$
$$\dot{\psi}_t = \frac{A_{yt}}{\rho V_m c\theta_t} - V_m s\theta_t c\psi_t \tan\lambda_y(\rho c\theta_t s\psi_t - c\theta_m s\psi_m)rc\theta_t$$
$$-V_m s\theta_t s\psi_t(\rho s\theta_t - s\theta_m)/rc\theta_t - V_m(\rho c\theta_t s\psi_t - c\theta_m s\psi_m)/r \quad (1)$$

From the engagement geometry, it can be shown that the angular LOS rate is given by

$$\omega_L = \dot{\psi}_L s\theta_L i_L - \dot{\theta}_L j_L + \dot{\psi}_L c\theta_L k_L = \dot{\lambda}_x i_L + \dot{\lambda}_y j_L + \dot{\lambda}_z k_L \quad (2)$$

The conventional pure proportional navigation law generates missile acceleration command proportional to LOS rate; however, in three-dimensional case the commanded acceleration is given by $A_M^C = A_{ym}^c j_m + A_{zm}^c k_m$. The three-dimensional APNG law is practically given by [11]

$$A_{yL} = A_{yt}c\psi_t - A_{zt}s\theta_t s\psi_t, \quad A_{zL} = A_{zt}c\theta_t$$
$$A_{ym}^c = -NV_m\dot{\lambda}_y s\theta_m s\psi_m + NV_m\dot{\lambda}_z c\theta_m + \frac{1}{2}N(A_{yL}c\psi_m - A_{zL}s\theta_m s\psi_m) \quad (3)$$
$$A_{zm}^c = -NV_m\dot{\lambda}_y c\psi_m + \frac{1}{2}NA_{zL}c\theta_m \quad (4)$$

The augmented proportional navigation guidance laws assuming target information is known will be used for comparison with the designed guidance law.

3 Guidance Law Design

In this work, a guidance law using feedback linearization methodology is proposed. To this end, the dynamics of λ_y and λ_z relating to the control input, i.e., lateral accelerations in pitch and yaw plane are derived using (1)

$$\ddot{\lambda}_y = \frac{-2\dot{r}}{r}\dot{\lambda}_y + \frac{A_{zm}}{r} + d_1 \quad (5)$$
$$\ddot{\lambda}_z = \frac{-2\dot{r}}{r}\dot{\lambda}_z - \frac{A_{ym}}{r} + \frac{A_{zm}}{r} + d_2 \quad (6)$$

where

$$d_1 = \frac{1}{r}[A_{zm}(c\theta_m - 1) - A_{zt}c\theta_t + V_m^2 c\theta_m \tan\lambda_y s\psi_m(\frac{p}{r})$$
$$- \rho V_m^2 c\theta_t s\psi_t \tan\lambda_y(\frac{p}{r}) + \rho V_m^2 c\theta_t c\psi_t(\frac{q}{r}) - \rho V_m c\theta_t c\psi_t(\frac{q}{r})]$$
$$d_2 = \frac{1}{r}[A_{yt}c\psi_t - \rho V_m^2 c^2\psi_t s\theta_t \tan\lambda_y(\frac{p}{r}) - A_{zt}s\theta_t s\psi_t$$
$$- \rho V_m^2 s^2\psi_t s\theta_t \tan\lambda_y(\frac{p}{r}) + A_{ym}(1 - c\psi_m) + A_{zm}(s\theta_m s\psi_m - 1)$$
$$+ V_m^2 c^2\psi_m s\theta_m \tan\lambda_y(\frac{p}{r}) + V_m^2 s^2\psi_m s\theta_m \tan\lambda_y(\frac{p}{r}))]$$

It can be noted from (5) and (6) that the LOS rate dynamics are highly coupled in elevation and azimuth planes, although from (5) and (6) it can also be observed that the lateral acceleration, A_{zm}, has an effect not only on elevation channel but also on azimuth channel. Naturally, if guidance law is designed by considering decoupled dynamics, the performance may not be satisfactory.

3.1 Guidance Law Design by Feedback Linearization

Rewriting equation (5) and (6) as

$$\begin{bmatrix} \ddot{\lambda}_y \\ \ddot{\lambda}_z \end{bmatrix} = \begin{bmatrix} -\frac{2\dot{r}}{r} & 0 \\ 0 & -\frac{2\dot{r}}{r} \end{bmatrix} \begin{bmatrix} \dot{\lambda}_y \\ \dot{\lambda}_z \end{bmatrix} + \begin{bmatrix} \frac{1}{r} & 0 \\ \frac{1}{r} & -\frac{1}{r} \end{bmatrix} \begin{bmatrix} A_{zm} \\ A_{ym} \end{bmatrix} + \begin{bmatrix} d_1 \\ d_2 \end{bmatrix} \tag{7}$$

$$Y = A + Bu \tag{8}$$

where $A = \begin{bmatrix} -\frac{2\dot{r}}{r} & 0 \\ 0 & -\frac{2\dot{r}}{r} \end{bmatrix}$, $B = \begin{bmatrix} \frac{1}{r} & 0 \\ \frac{1}{r} & -\frac{1}{r} \end{bmatrix}$, $u = \begin{bmatrix} A_{zm} \\ A_{ym} \end{bmatrix}$ and $d = \begin{bmatrix} d_1 \\ d_2 \end{bmatrix}$.

The LOS rates are driven to zero by using the control inputs A_{zm} and A_{ym}, and thus it represents an output regulation problem. From (5) and (6), it can be verified that the relative degree [6, 7] for both the outputs is 2. To this end, (7) can be written as

$$Y^r = A + Bu + d \tag{9}$$

where r is the relative degree. Applying IOL theory, one can get the control law as

$$u = B^{-1}[-A - d + \nu] \tag{10}$$

where ν is the external imposed dynamics and is given by

$$\nu = \begin{bmatrix} -k_1\dot{\lambda}_y \\ -k_2\dot{\lambda}_z \end{bmatrix} \tag{11}$$

From (7), if the LOS rates $\dot{\lambda}_y$, $\dot{\lambda}_z$, and d_1 and d_2 are available, the dynamics of (7) will be completely linearized as

$$\begin{bmatrix} \ddot{\lambda}_y \\ \ddot{\lambda}_z \end{bmatrix} = \begin{bmatrix} -k_1 & 0 \\ 0 & -k_2 \end{bmatrix} \begin{bmatrix} \dot{\lambda}_y \\ \dot{\lambda}_z \end{bmatrix} \tag{12}$$

Thus for $k_1 > 0$ and $k_2 > 0$, the dynamics are closed loop stable will be there, and by choosing the values of k_1 and k_2, desired transient performance for LOS rate can be obtained. The proposed guidance (10) will be referred to as the IOL-GL law and is implementable if knowledge of target maneuver and nonlinearities of d_1 and d_2 are precisely known apart from the measured quantities $\dot{\lambda}_y$ and $\dot{\lambda}_z$.

4 Simulation Results

In this section, simulation results are presented to demonstrate the efficacy of the proposed guidance law. As stated earlier, following the assumptions that the autopilot dynamics are much faster than guidance loop and so are neglected, one has $A_{ym} = A^c_{ym}$ and $A_{zm} = A^c_{zm}$. Due to physical limitation of dynamic actuators, commanded acceleration is always constrained in practice and for this purpose, the guidance law (10) is implemented with a saturation function assuming that the maximum lateral acceleration in magnitude is restricted to 30 g. The saturation is implemented as

$$A^c_m = \begin{cases} sat_N(A^c_{zm}) \\ sat_N(A^c_{ym}) \end{cases} \tag{13}$$

$$sat_N(x) = \begin{cases} N, & x > N \\ x, & |x| \leq N \\ -N, & x < -N \end{cases}$$

where $N > 0$ The simulation is carried out for maneuvering target with constant velocity. The missile-target parameters are taken similar as in [11] and are given in Table 1. Two target maneuvering cases as shown in Figs. 2a and 3a are considered in the simulation. To cater for the effect of autopilot, a second-order system with

Table 1 Simulation data

Parameter	Value	Parameter	Value
r(0)	2000 m	$\theta_m(0)$	10°
$\psi_m(0)$	10°	$\theta_T(0)$	20°
$\psi_T(0)$	135°	V_m	450 m/s
V_T	350 m/s	$\theta_L(0)$	30°
ψ_L	45°		

7

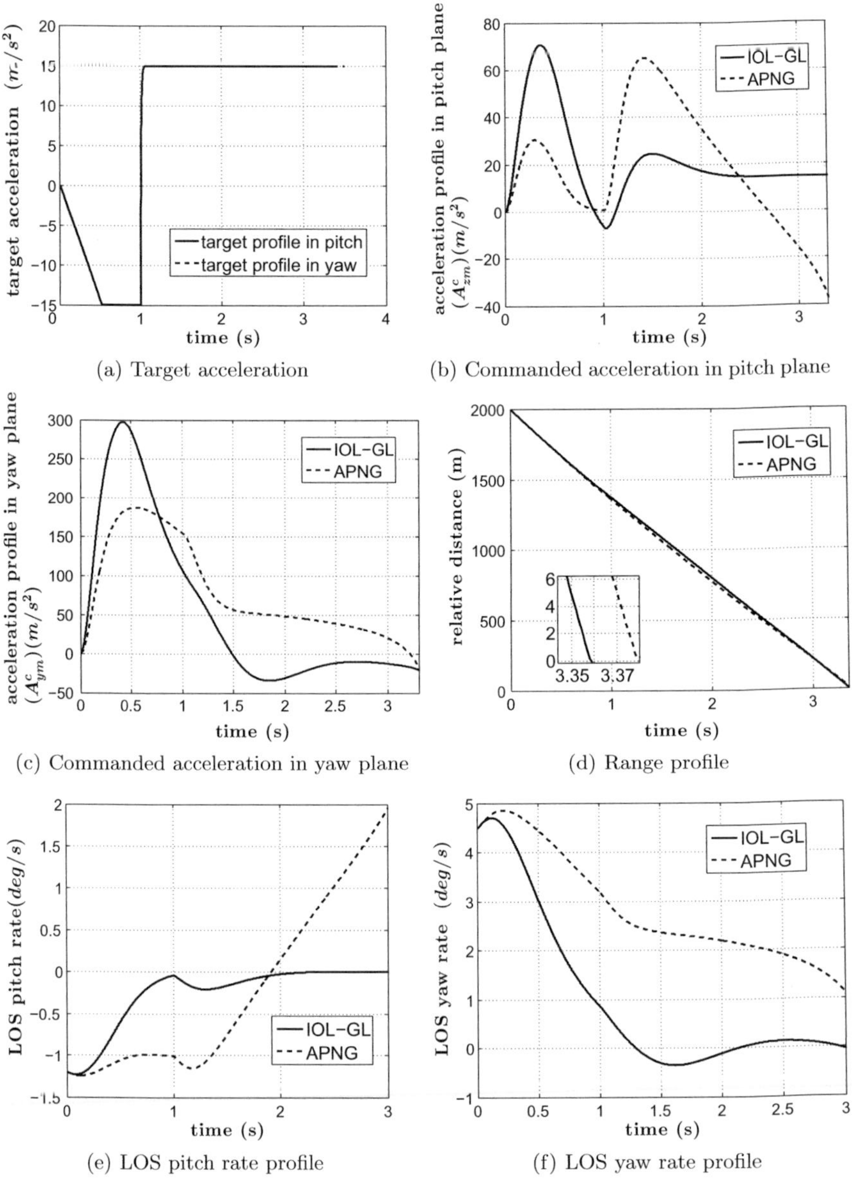

(a) Target acceleration

(b) Commanded acceleration in pitch plane

(c) Commanded acceleration in yaw plane

(d) Range profile

(e) LOS pitch rate profile

(f) LOS yaw rate profile

Fig. 2 Performance comparison of IOL-GL & APNG in case 1

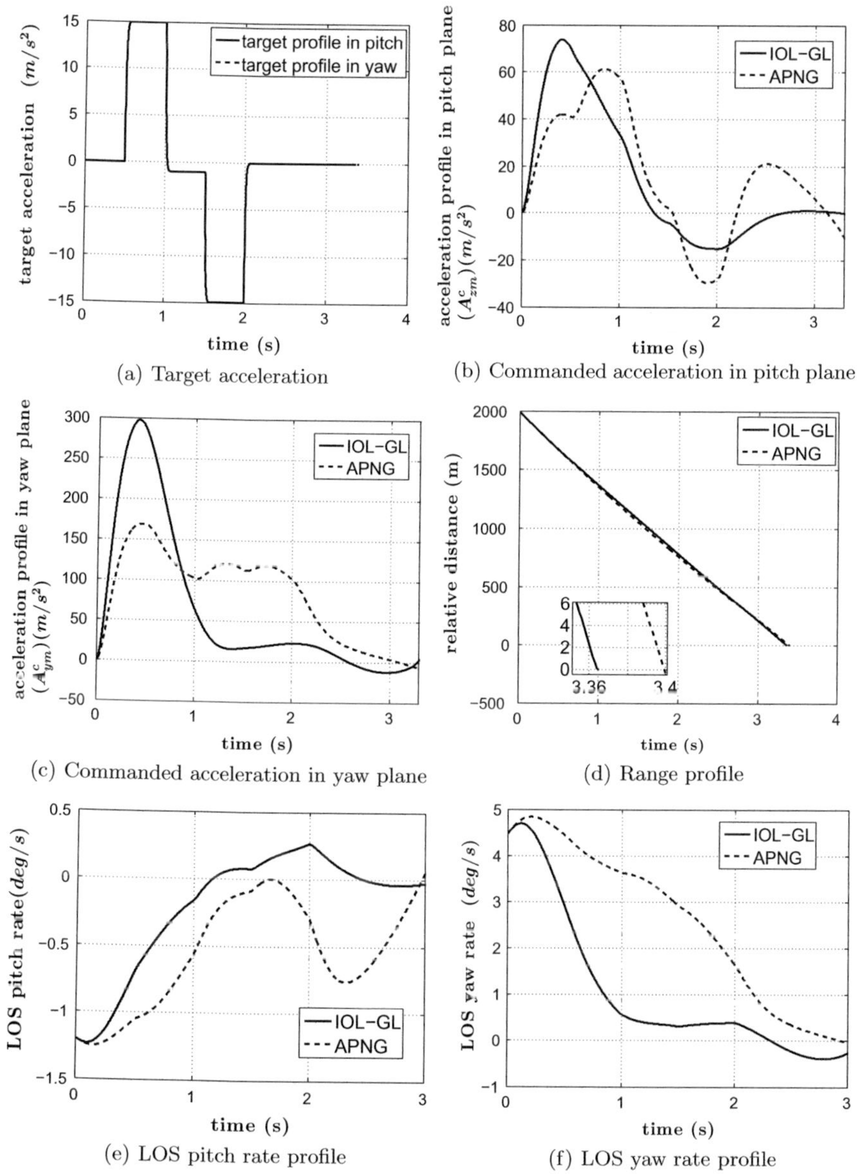

Fig. 3 Performance comparison of IOL-GL & APNG in case 2

damping ratio 0.8 and natural frequency 10 rad/sec is used. For APNG, the navigation constant is chosen 4.5. For IOL-GL, the controller gains are chosen $k_1 = k_2 = 1.6$. Simulations are carried out, and the results are presented in Figs. 2 and 3. From Figs. 2e–f and 3e–f, it can be seen that in IOL-GL the LOS rate converges to zero quite fast in comparison with the APNG wherein the LOS rate is diverging in early stage which can lead to higher miss distance. From Figs. 2b–c and 3b–c, it can also be seen that the control energy in IOL-GL is lesser than APNG. From Figs. 2d and 3d, it can be observed that the interception time is lesser in IOL-GL than APNG.

5 Conclusions

In this work, a 3D guidance law based on the input-output linearization technique is proposed. The effectiveness of the guidance law is shown through simulation results. To show the effectiveness of the proposed design, comparison of performance of the proposed law is carried out with APN guidance laws and it is shown that the proposed law offers superior performance.

References

1. Zhang Z, Man C, Li S, Jin S (2016) Finite-time guidance laws for three-dimensional missile-target interception. Proc Inst Mech Eng Part G: J Aerosp Eng 230:392–403
2. Choi HL, Park Y, Lee HG, Tahk MJ (2001) A three-dimensional differential game missile guidance law using neural networks. In: AIAA GN&C conference, Montreal, Canada, A01–37172
3. Liao F, Ji H, Xie Y (2015) A novel three-dimensional guidance law implementation using only line-of-sight azimuths. Int J Robust Nonlinear Control 25:3679–3697
4. Du R, Meng K, Zhou D, Liu J (2015) Design of three-dimensional nonlinear guidance law with bounded acceleration command. Aerosp Sci Technol 46:168–175
5. Kumar SR, Ghose D (2014) Three dimensional impact angle constrained guidance law using sliding mode control. In: American control conference (ACC), IEEE 2474–2479
6. Marquez HJ (2003) Nonlinear control system. Wiley-Interscience
7. Isidori A (2013) Nonlinear control systems. Springer Science & Business Media
8. Zarchan P (2002) Tactical and strategic missile guidance. Progress in Astronautics and Aeronautics
9. Shneydor NA (1998) Missile guidance and pursuit: kinematics, dynamics and control. Elsevier
10. Siouris GM (2004) Missile guidance and control systems. Springer Science & Business Media
11. Song SH, Ha IJ (1994) A lyapunov-like approach to performance analysis of 3-dimensional pure png laws. Aerosp Electron Syst IEEE Trans 30:238–248

MATLAB-Based Graphical User Interface Development of RADAR with the Implementation of Noise Under Various Bands

Gouresh Sood, Ayush Shah, Aashish Choudhary, Hitesh Arora, Shagun Bishnoi, Sudhir Kumar Chaturvedi and Anirudh Katyal

1 Introduction

We know that RADARs are being widely used for detecting and tracking airplanes, space crafts and transports at sea and also birds and bugs in the atmosphere. It not only measures the speed of vehicles but also maps the surfaces of Earth from outer space, and RADAR is capable enough to measure extra planetary assets of the atmosphere. Codes of RADAR have permitted us only detection of objects at/in the seas and in optically pure air using devices such sonar, lidar. Due to these reasons, we think RADAR has an at most importance in avionics and electrical applications and so we have proposed this project to gain knowledge about RADAR technology and in what ways we can improve it.

Effects of noise have been studied, and how they can be minimized. Thorough study towards detection criteria and working of RADAR has been done [1].

The GUI is divided into two phases: the first phase is the design phase, and second phase is the implementation phase with noise. The design phase involves determining the functionality of the GUI and using GUIDE in MATLAB to develop a simple form of it. The implementation phase involves the editing of main file in the MATLAB script, and the functions from the previous phase are to be defined. When the simulations are completed, the results can be studied to construct the

G. Sood
UTC Aerospace Systems, Bangalore, India
e-mail: gouresh.sood@utas.utc.com

A. Shah · A. Choudhary · H. Arora · S. Bishnoi · S.K. Chaturvedi (✉)
Department of Aerospace Engineering, University of Petroleum & Energy Studies, Dehradun, India
e-mail: sudhir.chaturvedi@ddn.upes.ac.in

A. Katyal
Rockwell-Collins, Hyderabad, India
e-mail: anirudh.katyal@rockwellcollins.com

S. Singh et al. (eds.), *Proceedings of the International Conference on Modern Research in Aerospace Engineering*, Lecture Notes in Mechanical Engineering,
https://doi.org/10.1007/978-981-10-5849-3_8

RADAR for optimum performance in the GUI. According to mathworks, a GUI is a practical interface to a programme. A GUI can make programmes easier to implement with a consistent appearance and with intuitive controls such as push-buttons, pop-up buttons, sliders [2].

2 Methodology

2.1 Design and Processing

The Graphical User Interface Design Environment or GUIDE offers tools for crafting graphical user interfaces so as to make or customize applications. The GUIDE Layout Editor is used, and there we can graphically design our interface and its looks. GUIDE can then autonomously generate the MATLAB code as an .m file for constructing the user interface, which we are able to alter so as to package the behaviour of our application [3]. Methodology for developing the GUI of RADAR with the implementation of noise under various bands is described in the flow chart given in Fig. 1. Basic stages for a GUI construction includes designing the GUI, laying out the GUI and programming the GUI (i.e. writing callbacks in Main-file Editor). Developing a GUI begins by creating a figure with file name.fig file, by typing 'GUIDE' at command window. Components can be added into the GUI by dragging the intuitive controls from the component toolbar. GUIDE stores the graphical user interface in two different files, which are generated for the initial time. The code controls the GUI and includes the callbacks for its components that are in the m-file with extension .m. A callback is a sequence of commands that are executable when a graphics object is activated [4].

From the flow chart, one can infer the design procedure for the development of GUI. The main reason behind this process is that one can easily determine the main components of the systems so that they can be easily implemented (Figs. 2 and 3).

2.2 RADAR Range Equation

The RADAR range is linked to the characteristics of the transmitter (Tx), receiver (Rx), antenna, objectified target and the surroundings by the RADAR range equation. It is not just useful for measuring the maximum range from the RADAR to the object, but it also serves as a tool for studying RADAR operations so as to design a basic RADAR model. In this section, we have derived a simple form of the RADAR equation [5].

Fig. 1 Methodology for GUI (from GUIDE) for modelling RADAR

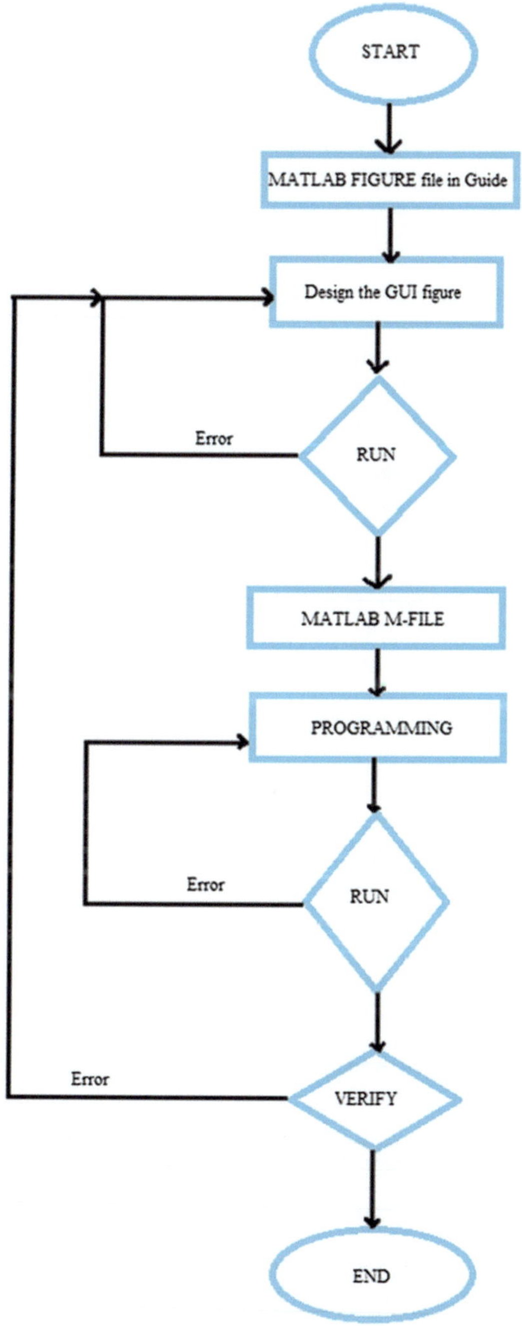

The simple RADAR range equation is as follows:

$$P_s = \frac{Pt * G * Ae * \sigma}{(4\pi)^2 * \mathrm{R}^4} \tag{1}$$

Noise power generated by a resistor is directly proportional to operating bandwidth above which the noise is to measured and also proportional to the equivalent temperature.

$$P_n = k * T_e * B \tag{2}$$

Alongside signal power, there is a reception of noise power signal as well at the antenna. The signal-to-noise ratio (SNR) may be defined as the ratio of signal power and noise power at the same point.

$$SNR = \frac{P_s}{P_n} \tag{3}$$

The ratio of signal-to-noise ratio at input $(\mathrm{SNR})_\mathrm{i}$ to signal-to-noise ratio at the output $(\mathrm{SNR})_\mathrm{o}$ is known as noise factor. The noise factor $(\mathrm{F_n})$ of a device specifies how much additional noise the device will contribute to the noise already from the source. Noise figure is the conversion of noise factor into decibel (dB).

$$F = \frac{SNR}{(SNR)_o} \tag{4}$$

It is the measure of degradation of signal-to-noise ratio. Noise temperature is a means of specifying noise in terms of equivalent temperature [6]. Noise temperature is related to noise factor and absolute temperature by [7].

$$T_e = (F_n - 1) * T \tag{5}$$

$$P_s = P_n * F * (SNR)_o \tag{6}$$

After the implementation of noise, the simple RADAR range equation becomes

$$R = \left(\frac{P_t * Ae^2 * \sigma}{(4\pi) * \lambda^2 * \mathrm{k} * T_e * B * F * (SNR)_o} \right)^{1/4} \tag{7}$$

3 Results

Bandwidth is selected from the pop-up menu. The bandwidth options are from HF (high frequency) to mm band (i.e. 3.0 MHz to 300 GHz). Other than that, here the RADAR operating data inputs are assumed as per assumptions. The transmitted power is assumed to be equal to 1.5e + 6 W. The aperture area is 7.20 m square. The gain of the RADAR is taken to be 45.0 dB. The RADAR cross section or RCS is equivalent to be 0.1 m square. We have considered the noise temperature as 290 K and the noise figure as 3.0 dB. The signal-to-noise ratio at the output section of the receiver, i.e., SNR output, is taken as 6.0 dB. Starting from HF band, the signal-to-noise ratio is decreasing from 220 dB to −250 dB in the mm-band. This transition is linearly detrimental in nature with respect to increase in the frequency.

The results for the HF band are shown in Fig. 4. The HF band is ranging from 3,000,000 Hz to 30,000,000 Hz. The range is differing as per RADAR range equation, and for each frequency value, it is changing. This change is visible through the results shown graphically on the GUI in Fig. 4. The results are shown in the GUI for every 1,000,000 Hz and have 28 increments till the HF band ends at 30,000,000 Hz.

The results for the VHF band are shown in Fig. 5. The VHF band is ranging from 30,000,000 Hz to 300,000,000 Hz. The range is differing as per RADAR range equation, and for each frequency value, it is changing. This change is visible through the results shown graphically on the GUI in Fig. 5. The results are shown on the GUI for every 10,000,000 Hz and have 28 increments till the VHF band ends at 300,000,000 Hz.

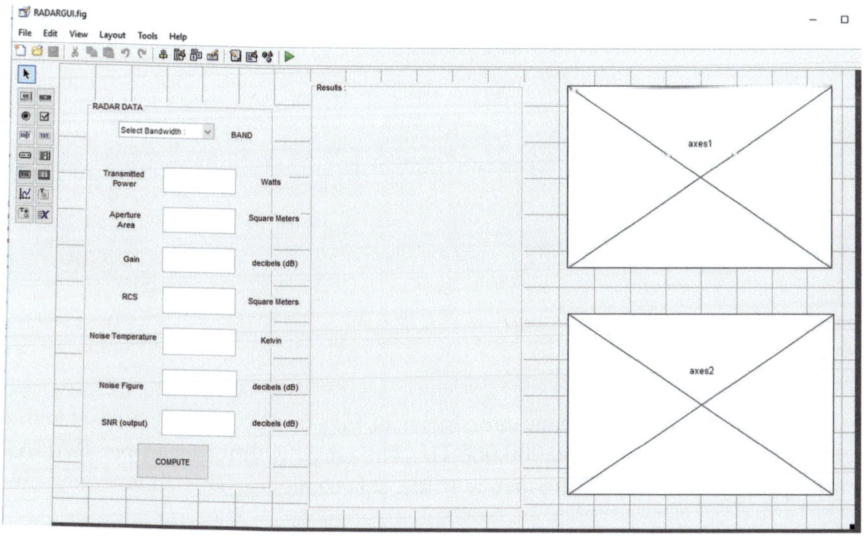

Fig. 2 Finalized GUI created in MATLAB GUIDE

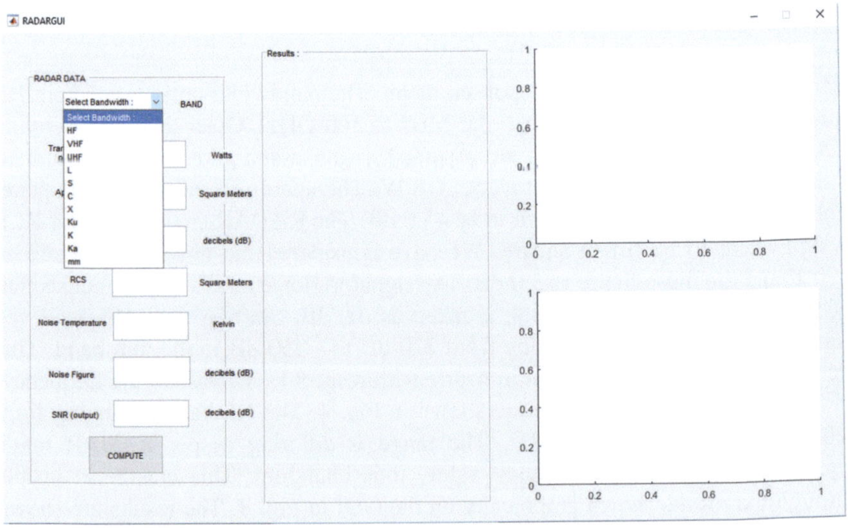

Fig. 3 Selecting bandwidth from the pop-up menu

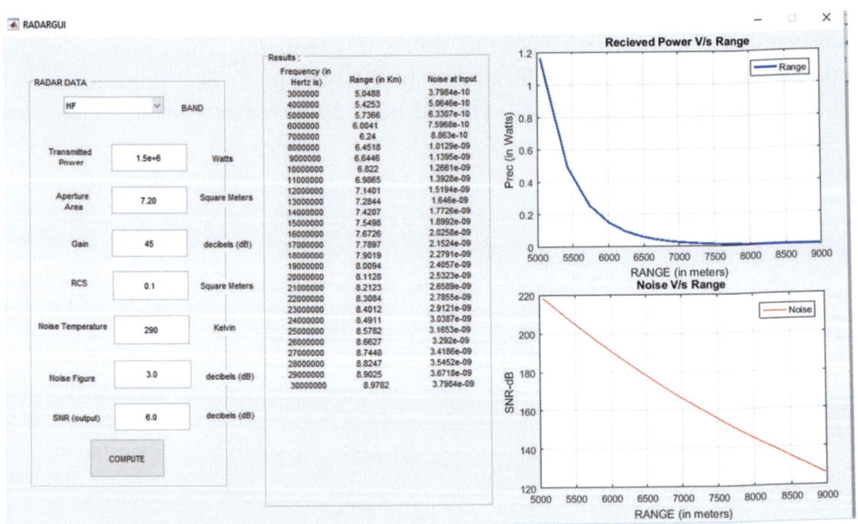

Fig. 4 Finalized GUI working for HF band

The results for the UHF band are shown in Fig. 6. The VHF band is ranging from 300,000,000 Hz to 1,000,000,000 Hz. The range is differing as per RADAR range equation, and for each frequency value, it is changing. This change is visible

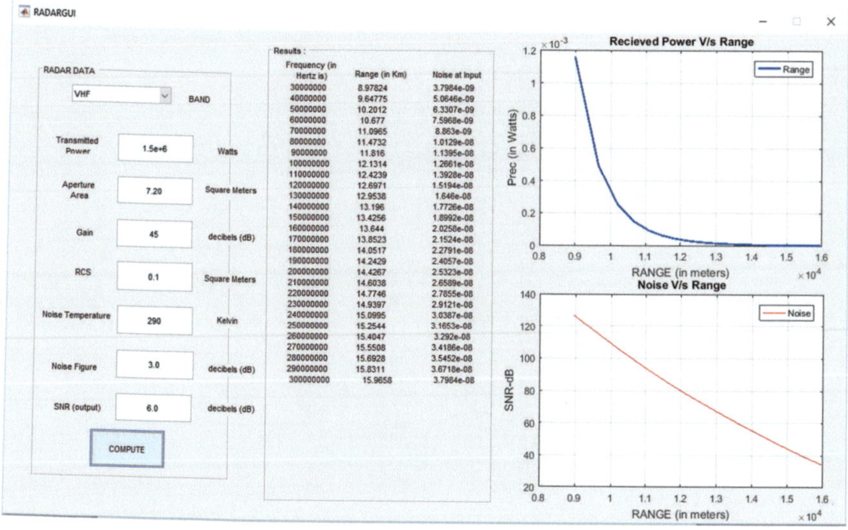

Fig. 5 Finalized GUI working for VHF band

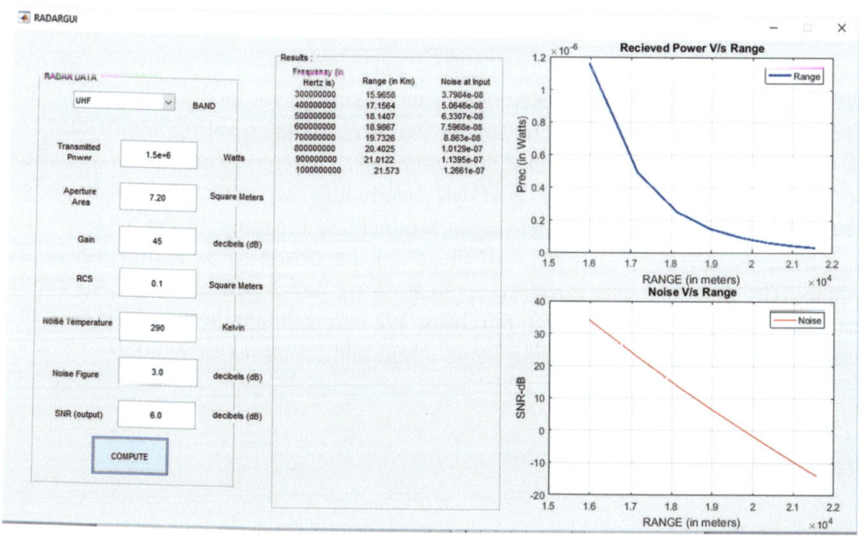

Fig. 6 Finalized GUI working for UHF band

through the results shown graphically on the GUI in Fig. 6. The results are shown on the GUI for every 100,000,000 Hz and have 7 increments till the UHF band ends at 1,000,000,000 Hz (Fig. 7).

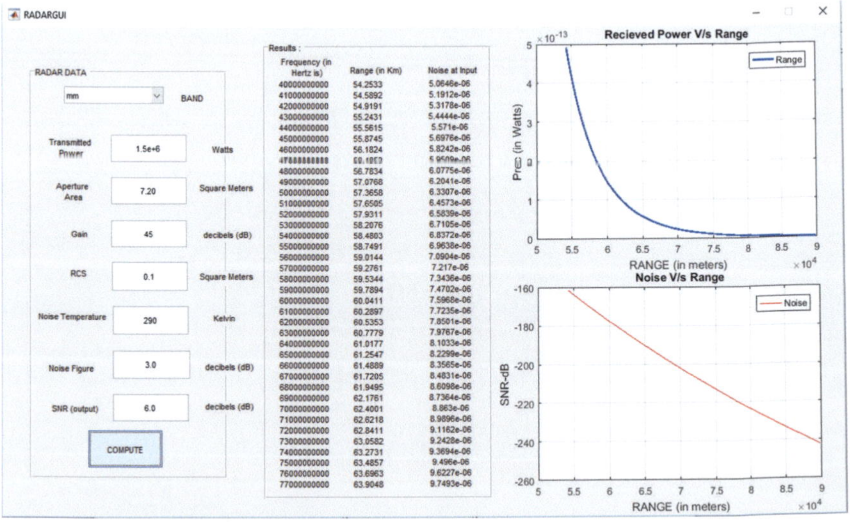

Fig. 7 Finalized GUI working for mm band

4 Conclusions

We infer that the range of the RADAR is increasing as we increase its bandwidth. The noise in the RADAR is increasing as we increase its operating bandwidth; i.e., as we increase the range, the noise also increases. The signal received is somehow decreasing as we increase the RADAR bandwidth as it decreases with range. The SNR in dBs is linearly decreasing as we increase the range of the RADAR (i.e. increasing RADAR bandwidth). All these can be observed as a function in the graphs. The graphical user interface application we developed in MATLAB can be used for educational and research purposes. We can promote the capabilities of this graphical user interface by adding some additional functions or features.

References

1. Merrill IS (1981) Introduction to RADAR systems. McGraw-Hill
2. http://sourceforge.net/projects/fers
3. http://in.mathworks.com/help/matlab/creating_guis/about-the-simple-guide-gui-example.html
4. Nagajyothi A, Raja Rajeswari K (2013) Graphical user interface based signal generator for radar/sonar pulse compression codes. Int J Comput Appl 82.16
5. Mahafza, BR (2022) Radar systems analysis and design using MATLAB. CRC Press
6. Mahafza, BR (2009) Radar signal analysis and processing using MATLAB. CRC Press
7. Scheiner, B (1999) A MATLAB radar range equation and probability of detection evaluation tool. No. ARL-TR-1806. Army research lab adelphi md

Control of UAV Using GSM Technology

Shubhrasmita Mandal, Krishna Maheta, Vivek Kumar and M.S. Prasad

1 Introduction

The life of today's world is highly dependent on the new inventions in the field of science and modern-day technology. It is not possible to have all the benefits of technology without implementing it in the sectors such as medicines, education, infrastructure, electricity, aviation, information technology and other field. Nowadays, the aviation field has grown up so vast that today's world is nothing without aviation, and it is the most innovative sector today. It has developed so much that there are many unmanned aerial vehicles developed which are used for many military and civil purposes both [1]. Also, Android is been flooding today's market with its multitasking options. As android and drones both are the most used inventions and since these would be ruling the world in the future for many new applications, we thought of coming up with an idea of controlling the drones with the help of an Android application. The Android application will communicate with the UAV at the receiving end through GSM technology. To aid this idea, we have to install a GSM modem at the receiving end, which will be connected to the flight controller on the UAV. The block diagram representation for the technique has been shown in the Fig. 1.

The main idea of the paper is to control a drone from any remote part of the world no matter where your drone is seated. This can be most useful feature for military purposes for target killing, surveying, search and rescue, going to terrorist's secret place without the person being present there, and it will also be useful for

S. Mandal (✉) · K. Maheta · V. Kumar
Amity Institute of Aerospace Engineering, Amity University,
Noida, Uttar Pradesh, India
e-mail: shubi_09@ymail.com

M.S. Prasad
Amity Institute of Space Science, Amity University,
Noida, Uttar Pradesh, India

S. Singh et al. (eds.), *Proceedings of the International Conference on Modern Research in Aerospace Engineering*, Lecture Notes in Mechanical Engineering,
https://doi.org/10.1007/978-981-10-5849-3_9

Fig. 1 Block diagram representation

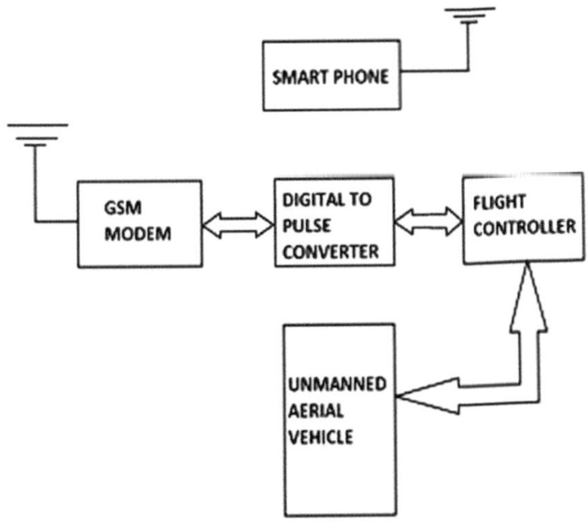

civil purposes such as agriculture and survey if only we can eliminate the problem of lack of continuous telephone network in any region. We know the telephone networks are sometimes not present continuously, so at that moment, we may lose contact with the UAV, but if the problem is solved, one of the advantages is as the person need not be present nearby the drone it would save time and make all the difficult tasks easy. Basically, by using this technique, we are replacing the radio-controlled transmitter and receiver of the UAVs with an Android mobile phone and hence adding up many advantages. The flow chart representation for the whole idea has been given in the Fig. 2.

2 Development of the Android Application

Android Studio is the official integrated development environment (IDE) for Android application development. Using Android Studio, the required application to control the UAV has been developed. The application comprises of various activities [2]. After starting the application, the user needs to sign-up or register first. The user can then provide the required information and login to the application. At the receiving end, there will be a SIM card installed in the GSM modem, which has a 10-digit number. After the user has logged in, he has to enter the 10-digit number of the UAV and press the "dial" button. The mobile phone will be automatically connected to the GSM modem installed on the UAV. Afterwards, the user can give commands such as, ARM, DISARM, ROLL, PITCH, YAW. Using Android Studio and JAVA programming language, the application has been developed.

The user interface has been designed by XML language in the Android Studio, and the application can be made alive by the background JAVA programming done

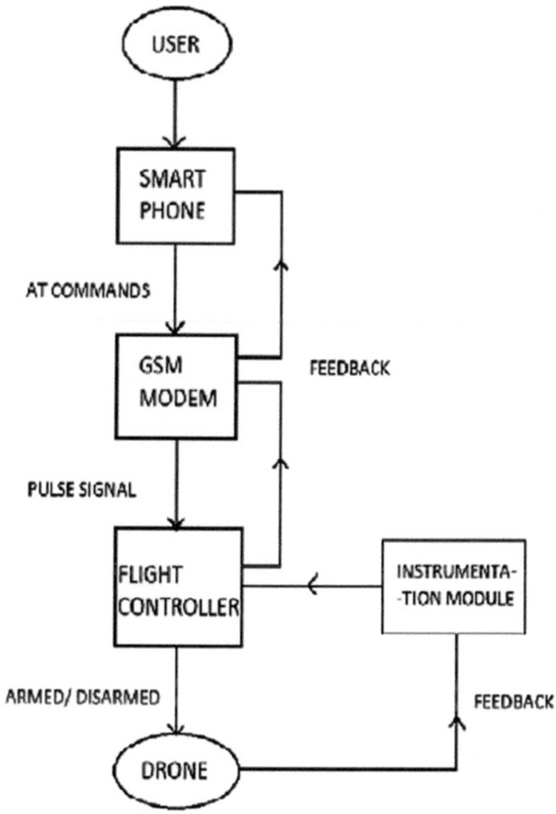

Fig. 2 Flow chart representation

for the individual activities. The Android Studio provides the users with various predefined tools to make the development of the application easier [3]. To make a customized user interface, one has to add his own library files within the project. The activities can be interrelated by adding intents in the JAVA class of the corresponding activity. Furthermore, the application to be developed for this purpose has to be a root application, i.e. the application can be launched only after the device is rooted. The application has been shown by dividing it into various activities in Fig. 3.

3 Android Device Architecture

The modern smartphone consists of two mutually cooperating systems: one is the user interface comprising of the touch screen, audio input and output etc. and the other one is the cellular modem, which controls all the cellular communications.

Fig. 3 UAV_CONTROL application

The interface between the two systems is highly dependent on the device manufacturer. Commonly used interfaces are serial peripheral interface (SPI), universal serial bus (USB) and universal asynchronous receiver/transmitter (UART) [4–8].

3.1 Radio Interface Layer

The radio interface layer is like the middleman of the Android that handles the cellular communications such as provides interface to the cellular modem, responsible for voice messages, phone calls, mobile Internet. RIL can handle all radio-type communications such as GSM, CDMA, 3G and 4G. The RIL can be directly reachable from outside world. All data sent from the cellular network to the mobile device pass through the RIL. The RIL daemon is the core part of the RIL, and it runs on a native Linux process [9, 10].

In smartphones, the ability to communicate with the mobile network is made possible through the help of the cellular modem. This modem is also called as the baseband processor. The software that runs on this processor is termed as baseband firmware.

The whole telephony process in Android is divided into four components such as the phone applications, the application framework, the RIL daemon, the kernel-level device drivers. In this system, the applications are written in JAVA, and the RIL daemon and drivers' code are written in native code [11, 12].

The Linux kernel constitutes the lowest layer of the telephony stack. It contains the drivers, which provide an interface for the user-land applications to communicate with the baseband hardware.

The stepwise telephony process can be well understood by the aid of the Fig. 4.

3.2 RIL to Baseband Communication

The basebands manage the physical radio link between the phone and the communication infrastructure. The interaction between the RIL and baseband depends upon the vendor and type of baseband. The communication may be a proprietary protocol or a standardized text-based GSM AT command set. If the way of communication is GSM AT command set, then the accompanying Linux kernel driver provides a serial device in the /dev file system. In this case, the RILD just opens the device and speaks the GSM AT protocol.

3.3 Rooting of the Android Device

Rooting is the process of acquiring the superuser privilege over an Android device. Rooting means basically gaining the root access to the phone. Once the device is rooted, the user will be able to control the operating system and can uninstall the pre-installed applications.

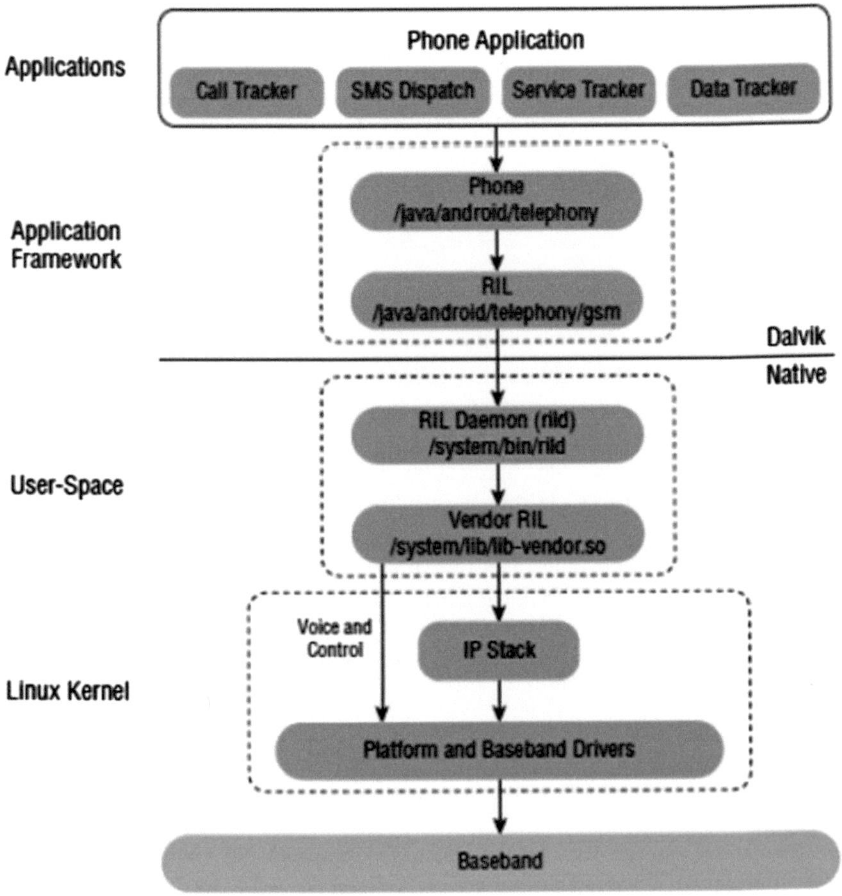

Fig. 4 Android telephony stack

The rooting also has a disadvantage, as the security of the device remains compromised. The applications having root permissions have access to all the user data on the device, and also it can be an open door to steal all the user data.

3.4 Android Debugging Bridge (ADB)

The ADB comprises of the ADB daemon on the android device, ADB server on the host computer and the corresponding ADB command-line client. The ADB server handles the connection between the host computer and the target android device for various functions such as file transfer, installing/uninstalling applications, executing a shell.

ADB shell commands: The ADB shell commands give the shell access to the android device connected to the host computer. After gaining the shell access, user can get the Linux command-line access to the underlying file system of the targeted android device.

The following ADB shell commands can be used to gain the root access on an android device connected to a host computer through USB debugging enabled:

```
adb root adb remount
adb push su/system/xbin/su
adb shell chmod 06755/system
adb shell chmod 06755/system/xbin/su
adb shell chmod 06755/system/xbin/busybox
adb install superuser.apk
```

To check whether the device is rooted, type "adb shell" command on the command window of the computer, and it should return "#". This shows that the device is rooted. After the device is rooted, we can launch the application, as it is using AT commands to communicate with the modem of the device directly. Therefore, to launch the application, we need a rooted android device.

4 GSM Technology

Global System for Mobile Communications (GSM) network consists of the mobile phone working as a transceiver, base station subsystem acting as an interface between the mobile device, and the network subsystem and the network subsystem providing the access to different networks.

4.1 GSM Modem

A GSM modem is a special type of modem, which acts just like a mobile phone. It accepts a SIM card and operated over a subscription to a network operator. It operates at the 850.900/1800/1900 MHz frequency bands. A GSM SIM900A modem is shown in Fig. 5.

4.2 AT Commands

AT commands are the instructions used to communicate or control a GSM modem. AT is the abbreviation of attention. Every command line starts with the "AT". It is the prefix that informs the modem about the starting of a command line.

Fig. 5 GSM SIM900A modem

There are various types of AT commands:

Test Commands: It is used to check whether a command is supported or not by the modem.

Syntax: AT<command line>=?

Read Command: It is used to get modem settings for an operation.

Syntax: AT<command name>?

Set Commands: It is used to modify modem setting for an operation.

Syntax: AT<command name>=value1, value2…value

Execution Commands: It is used to carry out an operation.

Syntax: AT<command name>=parameter1, parameter2… parameter

For example, to dial a phone number using AT commands:

ATD9718046994

5 GSM Modem and the Flight Controller Communication

The GSM modem is connected to the flight controller through the RS232 DB9 connector or TTL connectors. If the flight controller takes digital signal, then the modem can be connected to the controller directly, and if the flight controller takes pulse signal, then a digital to pulse converter has to be installed between the modem and the controller. As the controller gets the signal, it works accordingly and controls the UAV.

6 Conclusion

The control of UAV using Android application will be of great use in the future. As controlling of UAV using RC transceiver works only for a limited range, but by the use GSM modem, we can connect from anywhere to any part of the world, provided the GSM network is present. So controlling an UAV sitting at any part of the world is one of the advantages. Making use of camera and GPS can give the location and video for easy controlling. This technology can be used for many crucial purposes such as, military purposes, agriculture purposes, fieldwork, survey, research and many civilian purposes, if only the network-related difficulties can be overcome. Another disadvantage of this GSM technology is that it does not work above a certain height. One of the solutions can be that we can use this technology for indoor drones.

Acknowledgements The authors wish to give special thanks to Prof. M.S. Prasad and Prof. Vivek Kumar for their noteworthy inputs and suggestions. The authors also wish to thank the Amity Institute of Aerospace Engineering for the opportunity.

References

1. Home Appliances Control System Based On Android Smartphone; Sachin Kishor Khadke Student, Department of Electronics and Telecommunication NBN Sinhgad School of Engineering, Pune, India. (2014) IOSR J Electron Commun Eng (IOSR-JECE), Ver.III 9(3), 67–72, e-ISSN: 2278-2834, p-ISSN: 2278-8735
2. Drake JJ, Fora PO, Lanier Z, Mulliner C, Idley S, Wicherski G. Android Hacker's Handbook
3. Meier R. Professional Android 4 Application Development
4. Schildt H (ed.). The complete reference JAVA, 7th edn. New York
5. Conder S, Darcey L. Android wireless application development, 3rd edn. Features, Android 4.0, vol 1
6. Hoffmann GM, Rajnarayan DG, Waslander SL, Dostal D, Jang JS, Tomlin CJ (2004) The Stanford testbed of autonomous rotorcraft for multi agent control (STARMAC)
7. Arducopter—Arduino-based autopilot for multirotor craft, from quadcopters to traditional helis—Google Project Hosting. Code.google.com
8. Chakravorty R, Cartwright J, Pratt I (2002) Practical experience with TCP over GPRS
9. Cho YC, Jeon JW (2008) Remote robot control system based on DTMF of mobile phone. IEEE international conference INDIN, July 2008
10. Sunday AO, Ibrahim VM, Joshua, A (2012) Remote control of electrical appliances using GSM networks. Intern J Eng Res Develop 1(9):38–45, ISSN: 2278-067X
11. Banerji S (2013) Design and implementation of an unmanned vehicle using a GSM network with microcontrollers. Intern J Sci Eng Tech Res 2(2):367–374
12. Kelkar A (2014) Implementation of unmanned vehicle using GSM network with arduino. Intern J Adv Res Comp Sci Softw Eng 4(4):509–516

Orbital Magnetic Interplanetary Launch System (OMILS)

Samuel Luis and Mukit Azad Khan

1 Introduction

The current requirement by space programs of today is to launch an interplanetary probe for research and to conduct science on different planets in our solar system. So, for this the launch vehicle will have to have a Δv of more than 12 km/s. So this magnetic module will provide a boost to the spacecraft in the form of an impulse maneuver by electromagnetic repulsion force. The paper discusses the way the probe will boost a spacecraft launched by a medium lift launch vehicle by anchoring technique.

2 Oberth Effect

In Astrodynamics, the Oberth effect [1, 2] is experienced when a spacecraft enters into a gravitational well and accelerates when its fall reaches maximum speed. The resulting maneuver helps gain kinetic energy than applying the same impulse outside of a gravitational well. The Oberth effect [1, 2] simply explains that when the spacecraft is at the periapsis of the orbit and hence when it is the fastest in the orbit, when an impulsive maneuver or thrust is given the energy required to change the orbit to the desired orbit or escape orbit is the least as compared to the other points in the orbit.

S. Luis (✉) · M.A. Khan
Department of Aerospace Engineering, Karunya University,
Coimbatore 641114, India
e-mail: luissamuel@karunya.edu.in

M.A. Khan
e-mail: mukitazadkhan@karunya.edu.in

S. Singh et al. (eds.), *Proceedings of the International Conference on Modern Research in Aerospace Engineering*, Lecture Notes in Mechanical Engineering,
https://doi.org/10.1007/978-981-10-5849-3_10

The module mentioned here will use Oberth effect [1, 2] when it is at the periapsis or perigee of the orbit by boosting the spacecraft or giving it the required thrust.

3 Equations of Motion of Oberth Effect

If an impulsive maneuver of Δv is performed at periapsis or perigee of the orbit, then the velocity at periapsis before the burn is equal to the escape velocity, and the specific kinetic energy after the maneuver is:

$$\begin{aligned} \mathrm{K.E} &= 0.5\,\mathrm{mv}^2 \\ &= 0.5(\mathrm{V}_{\mathrm{esc}} + \Delta \mathrm{v})^2, \end{aligned}$$

where $\mathrm{V} = \mathrm{V}_{\mathrm{esc}} + \Delta\mathrm{v}$.

When the vehicle leaves the gravitational field, the loss of specific kinetic energy is:

$$0.5\,\mathrm{V}_{\mathrm{esc}}^2.$$

So it retains the energy:

$$\Delta \mathrm{v} \mathrm{V}_{\mathrm{esc}} + 0.5\,\mathrm{V}_{\mathrm{esc}}^2,$$

which is thus larger than the energy from a thrust maneuver outside the gravitational field by $\Delta\mathrm{v}\mathrm{V}_{\mathrm{esc.}}$

4 Rendezvous Mode

In the *Rendezvous* mode [2], once a medium lift launch vehicle will place a probe or satellite in orbit of around 1000 km by 4000 km altitude it will rendezvous [3] with this electromagnetic rail gun module and pass right through it and gain velocity by the electromagnetic repulsion force. The module will take advantage of the ***Oberth effect***, momentum gain, and ***electromagnetic repulsion force*** to make an impulse maneuver to boost its speed.

This will boost its speed up to twice and it will have enough speed to get an interplanetary trajectory to conduct an interplanetary mission. The spacecraft/probe will first get into catchup orbit of around 250 km as normal probes do when they have to rendezvous and dock with the ISS. Then at the proper time phase when they need to get the velocity boost, the probe will get into a rendezvous orbit with the module and pass through it and get a boost of velocity at a proper phasing angle in order to intercept the planned planet's Sphere of Influence [1, 2].

5 The Anchoring Concept

The Anchoring concept that is presented here is similar to that of NASA's asteroid hitch-hacking concept which proposed harpooning a passing by large celestial body and used its momentum to accelerate without burning fuel. In this case, instead of an asteroid the module will act as the faster body.

The probe will be launched into a trajectory similar to that of a missile (i.e., it would not need to achieve orbit) and at its apoapsis (likely **900–950 km**) it will make contact with the magnetic module's orbit. As soon as the module passes near, the probe will be grabbed by a connecting cable anchor (Figs. 1, 2 and 3) which will be fired from the magnetic module. The spacecraft will be anchored. Momentum transfer will ensure that its relative velocity will be zeroed out and later it will be slowly pulled toward the magnetic module by coiling the cable. By this, the module will gain velocity without using its own fuel. We want to gain a **velocity of 7–9 km/s** with the anchoring using mass moment transfer. It will go through the module and the probe's speed will be boosted based on the principle of the *Electromagnetic Rail gun* [4–6]. The orbiting module will have its own thrusters attitude control/station keeping which is necessary because of the sudden reserve pull made by the slower probe in time of first contact with the cable. The mass of the orbiting module has to be at least 100 times greater than the probe or spacecraft that will be anchored to maintain its own velocity (Figs. 1, 2 and 3).

Fig. 1 Hitch Hiking (Anchoring) an astroid using hook and grapple

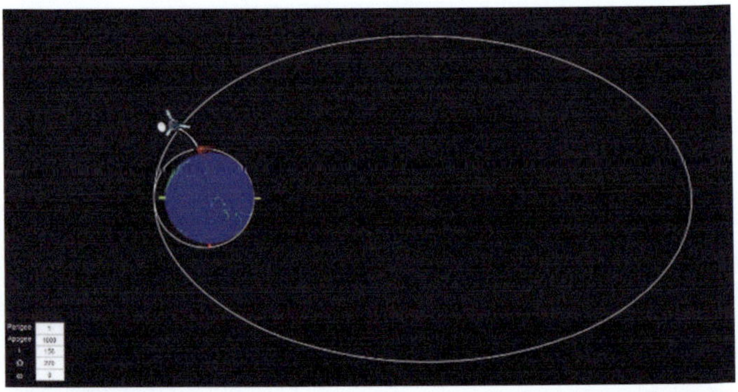

Fig. 2 Probe anchoring with the orbiting module

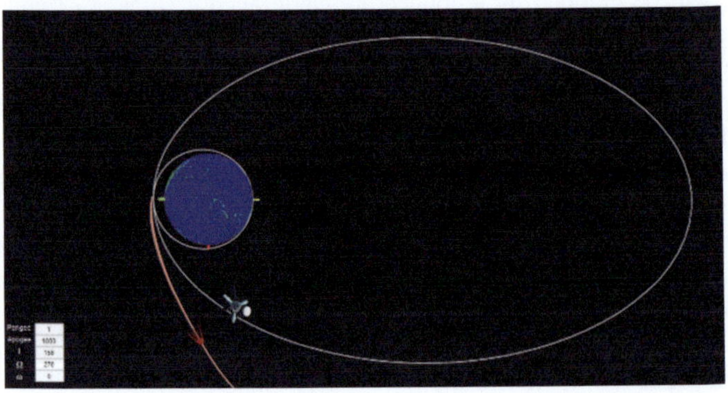

Fig. 3 Probe has accelerated from the module and on its away to preplanned and the module is orbiting on its designated orbit path

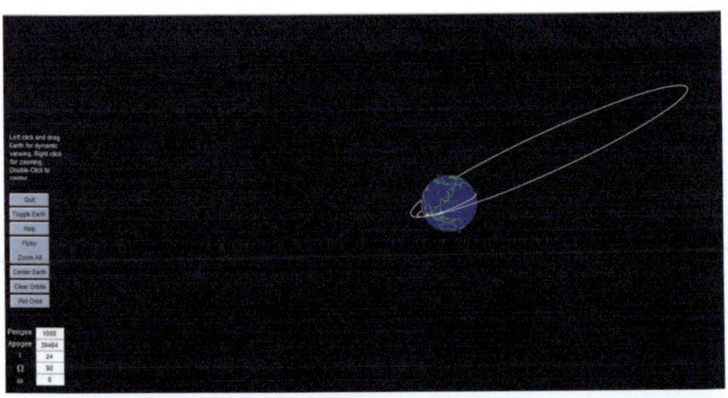

Fig. 4 The *side view* of the probe's orbit

6 The Electromagnetic Module

The electromagnetic module shall be placed in an elliptical orbit of perigee altitude 1000 km and apogee altitude 39464.4 km (Figs. 4 and 5) since its primary goal is to boost the probes launched by medium lift launch vehicles (i.e., Falcon 9, GSLV, PSLV, Ariane 5, etc.). Its *orbital period* will be 12 h and will make 2 orbits in a day. The module mass will be **100 times greater** than the probe or spacecraft that will be anchored to maintain stability. The structure will be approximately 500 m long with a launching platform for the payload (5 times bigger than the ISS).

It will operate based on the principle of the electromagnetic rail gun [4–6] (Fig. 6) which will produce a repulsive force that will boost the speed of the spacecraft/probe (Figs. 7 and 8). It will operate similar to the way that the electromagnetic rail gun system (Fig. 6) works to boost spacecraft. We are expecting at least a **4 km/s** gain through the module. The module will have thrusters for attitude control/station keeping and to adjust against the perturbations while in orbit. The module will have solar panels that will power the electromagnet [7]. They are expected to give an output of about 300 kW (3 times more than ISS and can be improved in future with nuclear energy to produce 1 MW).

The mass of the orbiting module has to be at least 100 times the greater than the probe or spacecraft that will be propelled to maintain its own velocity because it will experience deceleration or reverse thrust due to the magnetic repulsion similar to a gun when it fires a bullet. More than one module can be put into orbit to maximize the launch window.

The *Orbital parameters* of *OMILS* will be:

Gravitational parameter (km^3/s^2) = 398,600

State vector:

r (km) = [0 7378 0]

v (km/s) = [−7.2598 0 3.2323]

Angular momentum (km^2/s) = 58632.3

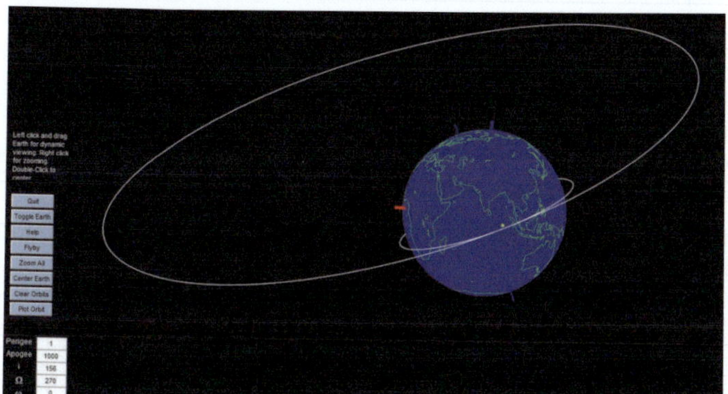

Fig. 5 The *front view* of the probe's orbit

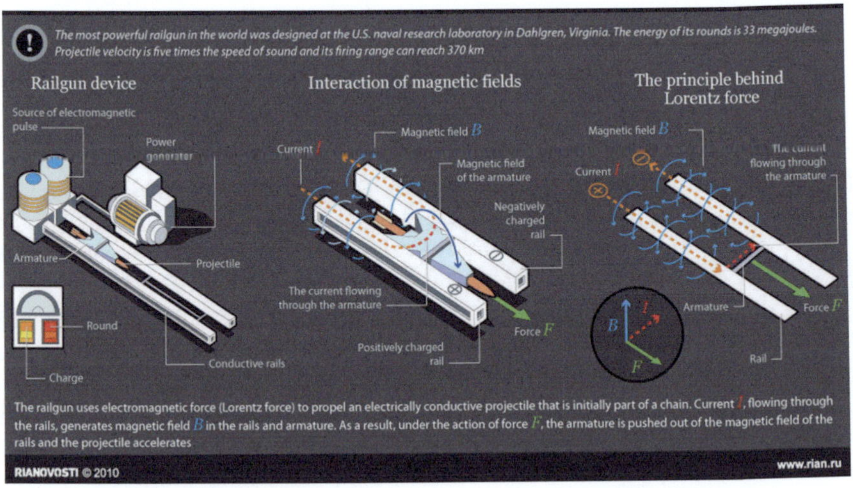

Fig. 6 Working of magnetic rail gun

Fig. 7 Probe is reaching the magnetic module (Artist's concept)

Eccentricity = 0.1689
Right ascension (°) = 90
Inclination (degrees) = 24
Argument of perigee (degrees) = 0
True anomaly (degrees) = 0
Semi-major axis (km): = 8878
Period:
Seconds = 8324.99
Minutes = 138.75
Hours = 2.3125

7 Equation for the Magnetic Acceleration and Calculation

$$F = BI\mathcal{L}$$

Here,
F = Force accelerating the payload,
B = Strength of Magnetic Field,

$\mathcal{L}$ = Length of the armature or the platform,
I = Current required.

For payload of 1 ton to **accelerate** up to **4 km/s²**, we are assuming the length of the platform is 500 m,

$$\mathcal{L} = 500\ \text{m},$$
$$F = ma = 1000\ \text{kg} \times 4000\ \text{m/s}^2 = 4000\ \text{KN},$$
$$I = \frac{P}{V}(\text{Here, } P = \text{Power, } V = \text{voltage}).$$

The current will be depended on the power supply. Based on the power and voltage, the current and magnetic field strength will vary.

For example, if the supplied power is 300 kW and voltage is 5 kV.

$$I = \frac{300}{5} = 60\ \text{amp}$$
$$B = \left\{\frac{4000 \times 103}{500 \times 60}\right\} = 133.33\ \text{Wb}.$$

Fig. 8 Probe is accelerating (Artist's concept)

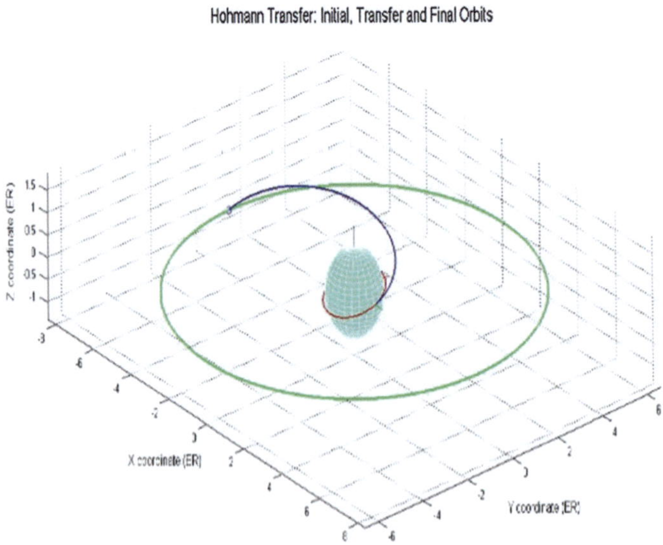

Fig. 9 OMILS providing first phase Δv impulse to GTO

Magnetic shielding is necessary for the payload because of the possibility of damage from strong magnetic field.

8 Conclusions

The paper suggests a noble way to save fuel used in normal heavy lift rockets and an efficient way to boost spacecraft to interplanetary trajectories and geostationary orbits (Fig. 9).

The **Rendezvous** mode [2] seems to be easier and safer than the *Anchoring* mode and thus is more likely to be implemented and used.

The *Anchoring* technique is very much fuel efficient but it has to be precisely launched so that its apoapse meets the module at the particular time interval as to when it has to be anchored so as to avoid flying past it if it misses. But this method presented on the paper will save time as well fuel since the probe doesn't have to achieve orbit as it will be already done by the module itself. The magnetic launch module will accelerate the payload even further thus increasing the fuel efficiency for deep space missions.

References

1. Curtis H (2005) Orbital mechanics for engineering students
2. Wiesel WE (2011) Spaceflight dynamics
3. Breger L, How JP. Safe trajectories for autonomous rendezvous of spacecraft
4. Baum CE (2007) Some electromagnetic considerations for design of railgun electromagnetic projectile launchers
5. McNab IR (2003) Launch to space with an electromagnetic railgun
6. Ma P, Yang M, Hu Y (2012) A whole trajectory simulation for the electromagnetic rail gun
7. Bernardes JS, Stumborg MF, Jean TE (2003) Analysis of a capacitor-based pulsed-power system for driving long-Range electromagnetic guns. IEEE Trans Magn 39:486–490

Design and Analysis of Two-Axis Seeker Stabilization System

Dommeti Rajesh, A.V. Praveen and Mallikarjuna Rao Pasumarthi

1 Introduction

Systems that work in airborne environments that rely on the resolution contributed by optical sensors require a stabilization system to keep apart the line-of-sight (LOS) from the operating conditions. For systems employing seekers, optical, infrared sensor stabilization accuracy is of prime factor in maintaining sufficient resolution to allow target identification and recognition at maximal ranges. The process of system modeling that meticulously predicts stabilization performance, which are important in both design tradeoffs and testing of the system [1, 2].

The basic concepts are convenient for achieving LOS stabilization are (i) momentum stabilization, which employs a twirl mass and (ii) rate stabilization which utilizes inertial rate sensors. Previously, rate stabilized platforms have been employed for speedy laser designator systems mounted in an aircraft, while ISP has been used in tactical missiles. Currently, various sensors are being proposed for tactical missiles with long standoff range fulfillment, hence greater LOS stabilization performance. The relative performance potency is inherent in the two types of systems and has not yet been rigorously treated. This subject is increasing importance due to principle application of LOS stabilization in missile seekers. At the same time the cost/performance trade studies that must be performed to select stabilization perceptions for a defined application [3, 4].

D. Rajesh (✉) · A.V. Praveen · M.R. Pasumarthi
Department of Electrical Engineering, College of Engineering (A), Andhra University, Visakhapatnam 530000, AP, India
e-mail: rajesh.dommeti@gmail.com

A.V. Praveen
e-mail: venkatapraveenrs@gmail.com

M.R. Pasumarthi
e-mail: electricalprofessor@gmail.com

S. Singh et al. (eds.), *Proceedings of the International Conference on Modern Research in Aerospace Engineering*, Lecture Notes in Mechanical Engineering,
https://doi.org/10.1007/978-981-10-5849-3_11

This paper presents the development of the system models prerequisite to evaluate the LOS stabilization performance. The models are presented in detail for each concept to provide an understanding of the basic physical phenomena which determines the LOS stabilization performance achievable with each concept. Since the basic LOS stabilization models are developed in a linear fashion, the extension to the nonlinear case is also included as a technique for determining system performance. Also the conventional servo motor actuation is replaced by BLDC actuator including the variation of geared and direct drive. To demonstrate the validity of the models developed, a case study is presented covering both momentum and rate stabilized systems which are verified by multi-body dynamics (MBD) simulation results [5]. The primary contribution of this paper lies in the model derivation and the detailed presentation of these models which provides insight into the relationship of the system parameters with the techniques necessary to predict LOS stabilization performance for both rate and momentum stabilized platform.

2 Stabilization System Description

Generally, momentum and rate stabilization systems can be implemented with a two degree of freedom gimbal platform which provides LOS stabilization about the two axes orthogonal to the LOS. Momentum stabilization is accomplished through the design of symmetrical structure and suitable drive. But in case of missile seeker, the structure is mostly unbalanced type due to seeker mass and angle constraints. A rate stabilized platform utilizes two rate gyros mounted on the inner gimbal with their input axes orthogonal to stabilize the LOS in inertial space. The primary control problem is not that of commanding the platform LOS, but rather that of minimizing LOS motion induced by undesired torques. This paper is concerned with deriving the models necessary to demonstrate the effect of the system parameters on the degree of LOS isolation achieved [6].

The seeker actuation system is to meet the dynamic characteristics of an active Radar seeker. It acts as an allay between the missile and the seeker dynamics. The mechanical assembly incorporates to house the seeker, control electronics, and feedback sensors. The PMBLDC servo system modeled to operate in two modes viz. position mode and track mode. The seeker rotates individually in both azimuth and pitch planes in any mode. The body disturbance rejection can be achieved by the stabilization Loop, in-loop to track Loop. The innermost actuator loop which physically imparts the rotation to the seeker actuation. It was same in position as well as track mode. The block diagram of the actuation system describing all the control loops is shown in Fig. 1. The system has a switching logic to change from position mode to track Mode [7–9].

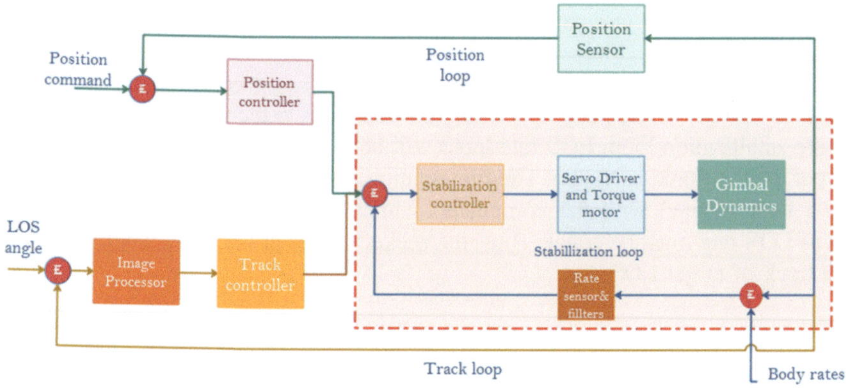

Fig. 1 Block diagram of seeker control system

2.1 *Position Mode*

In this Mode, the initial orientation of the seeker is toward the target LOS. The designation command is developed based on the ground radar information to search the specified target. The seeker servo system responds to the high rate, orient designation commands. Based on angle sensor feedback, the compensators are tuned as per design requirements of position loop.

2.2 *Track Mode*

This mode is to track the target with respect to angle command. The LOS error is measured by the radar applying the mono-pulse techniques. The servo system acts to the angular error measured by the radar and activates the seeker in the direction to negate the LOS error. The corrections in the angular error are being made the LOS rate varies in account with the angular error correction rate. Servo system makes accurate measurement of LOS rates in the track loop and provides position feedback to homing and guidance system. The stabilization loop preserves the spatial rigidity to seeker LOS. Stabilization loop consists of a rate gyro as the feedback sensor, and the effective body rates compelling on seeker LOS will be countered by the stabilization loop controllers in the presence of the track loop constraints.

As the track loop consists of invariable delays the signal input to stabilization loop and it will be proportional to the true LOS rate. The relationship between the LOS rate (λ) and LOS angular error (ξ) as follows

$$\frac{\xi}{\lambda} = \frac{\tau}{\tau S + 1} \tag{1}$$

At low frequencies, ξ is proportional to λ and leads to the required rate command to the stabilization loop. LOS rate input will be given to the missile guidance loop as per the measured LOS rate. The true rotation rate feedback by the Rate Gyro will have to correlated with the computed λ. Ideally, both computed LOS rate and Gyro LOS rate must be equal. Thus, the seeker rotates at the same rate to maintain the track of target LOS.

2.3 Decoupling and Isolation Factors

The stabilization system should isolate the sensor LOS from base body rates caused by the vibration and the maneuvering of the dynamic platform. Therefore, the sensor will keep steering in the direction of the target and generate LOS tracking path. In a working system, the essentiality of the decoupling ratio between the sensor-mounted ISP and the body rates of ISP is determined by considering the dynamics of ISP carrier such as an unmanned air vehicle or a missile

$$Decoupling\,factor = \frac{LOS\,rate}{Body\,rate} \tag{2a}$$

$$Isolation\,factor = \frac{monopulse\,error}{Body\,rate} \tag{2b}$$

In a system, the decoupling performance indicates to the stabilization characteristics of the platform. For the stabilization loop of a two-axis gimbal sensor system against the body rates of the ISP should be less than −40 dB at1 Hz . This implies that the designed stabilization loop can reduce below 100th part of the amplitude of the sinusoidal body rates exerted on the sensor system by the ISP at1 Hz [9].

2.4 BLDC Servo System

The overall control system is constructed utilizing two identical stabilization loops as shown in Fig. 2 for the inner frame (elevation) and outer frame (azimuth) gimbals. In Fig. 2, the equivalent gimbal inertia is taken into consideration for two independent channels. This control system aims to isolate the stabilized seeker LOS from the base body rates. So, the elevation gimbal's angular velocities should be zero,. i.e., the sensor optical axis must be preserved as non-rotating in the inertial space despite torque pulsation from the external environment.

The control loops in the elevation and azimuth channels are mutual which depends on each other by the cross-coupling unit by replacing the inertia as shown

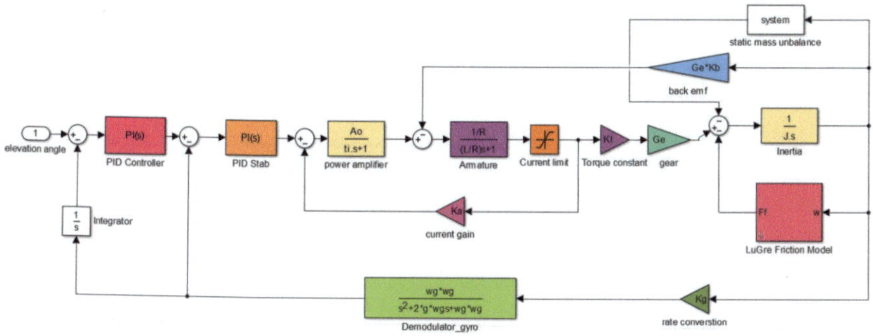

Fig. 2 Block diagram of BLDC servo system with stabilization and position loop

in Fig. 2, and it is built based on the relationships of the torques affected on the two gimbal frames. The cross-coupling expresses the inertial properties of the gimbal joint dynamics. It shows the fact that the azimuth gimbal frame can influence the elevation gimbal frame in the absence of base body rates. In addition, there is a similar impact on the part of the elevation gimbal on the azimuth gimbal. So, the cross-coupled block is known as the mutual effect of one frame on another frame. The BLDC actuation system provides smooth torque speed characteristics as well as quick response to the position input. Also, these actuation systems occupy less volume envelope and weight due to its high torque to weight ratio. For the both channels, the actuation modeling is same, but equivalent parameters vary according to their torque requirements.

3 Dynamic Mass Unbalance and Kinematic Couplings

The dynamic mass unbalance is the consequence of unsymmetrical mass dissemination in physical design also called 'product of inertia.' The dynamic unbalance concept indicated by the inertia matrix elements about each fame axes,. i.e., if the design gimbal frame is symmetrical with reference to its frame axes, then the gimbal has no dynamic unbalance and its inertial matrix is a diagonal matrix. If the gimbal is asymmetrical with reference to its frame axes, then the gimbal has dynamic unbalance and it has an inertial matrix with non-diagonal elements. Also, the gimbal frames congregation provides the connection between the base carrier and sensor LOS.

To estimate the ISP dynamics, the mechanisms that couple the base body rates and the LOS angular motion must be interpreted. Also the cross-coupling block is the ultimate result of the kinematic relationships between the gimbal inertial parameters and the base rates. With the help of Lagrange's and Newton's laws [2], the cross-coupled equations are derived and implemented in MATLAB/Simulink

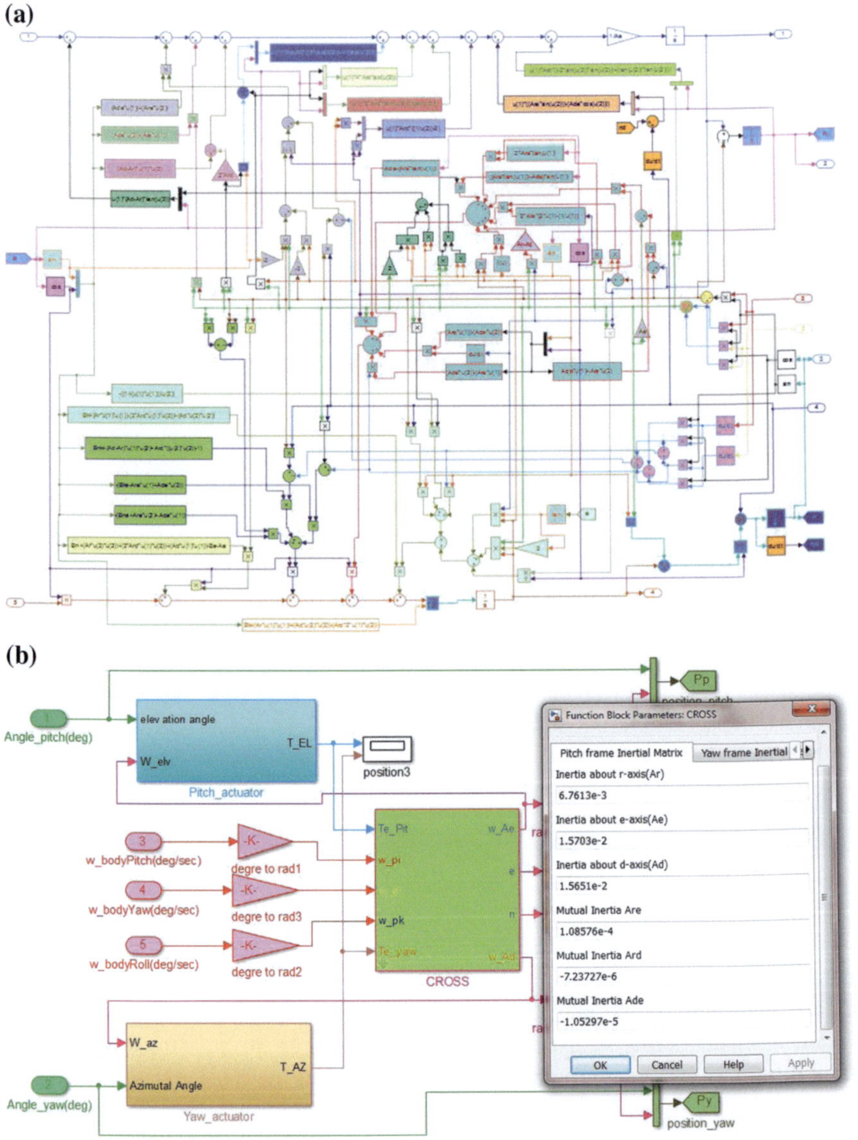

Fig. 3 **a** Internal clock diagram of cross-couple block. **b** MATLAB simulation diagram of the system with body rates

environment as shown in Fig. 3. Hilkert and Hullender [10] model was used and found to be adequate after several experiments. The spring restraint torque induced by the cables and flexible tube is also considered.

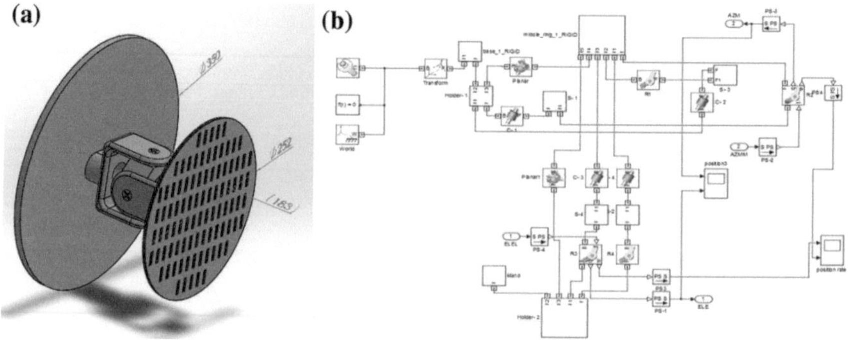

Fig. 4 **a** CAD model of the seeker. **b** Simscape MBD model corresponding to the CAD model

4 Multi-body Dynamic and Modeling

In the design methodology, CAD-multi-body dynamic (MBD) simulation in the initial design phase of Prototype modeling. Commercial engineering simulation software has adapted to changing user requirements by being more user-friendly and interactive. Simulation using combination of CAD-dynamic simulation packages also enables the engineer to access the specific capabilities of different software packages. Using the appropriate combination of design and analysis packages reduces design cycle times and permits the analysis of complex mechanical systems. In addition, the design engineer and control engineer can work concurrently, thus making the overall design process more efficient. The designed CAD model with true dimensions is imported to Simscape environment as shown in Fig. 4 and then model is subjected to rated conditions [5].

Actuation torque from BLDC motors is connected through the revolute joints and the blocks. There is some saturation which is assigned with boundary conditions. The connected model is simulated, and the results are observed in mechanics explorer and scopes. Also, this simulation has provided with frequency domain analysis same as earlier cases in the literature. The results show concurrency with linear simulation results in cross-coupled design for the same set of values.

5 Simulation and Results

The simulation incorporates with existing system values with 650 gm seeker plate and true dimensions. The simulation is done with pitch channel—42 W, 360 mNmts direct drive and yaw channel—50 W, 1300 mNmts direct drive considering all nonlinearities. Also for the same gimbal dynamics, gear motors with 55 W, 45 mNmts, 7:1 gear pitch channel actuation and 99 W, 115 mNmts 9:1 gear yaw

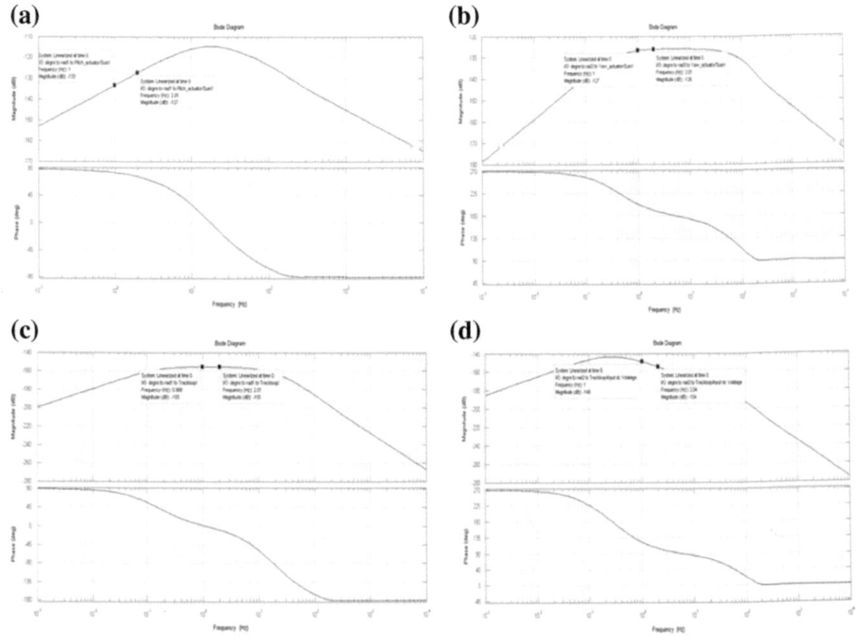

Fig. 5 **a** Isolation factor response for pitch channel. **b** Isolation factor response for Yaw channel. **c** Decoupling factor—frequency plot for pitch channel. **d** Decoupling factor frequency plot of yaw channel

channel actuator motor are considered. The simulation results correlate with practical designed specifications.

5.1 Frequency Response

Figure 5 shows the frequency response of position loops of both channels with respect body rates in order to indicate the Decoupling and Isolation factors at 1 Hz. In Fig. 5a, b subplots shows the isolation factors −133 dB, −127 dB for pitch and yaw channels respectively. In Fig. 5c, d subplots shows the decoupling factors −155 dB, −149 dB for Pitch and yaw channels respectively, which are less than −40 dB. The optimum PID parameters for rated conditions are tabulated in Tables 1 and 2.

Figure 6 shows bode plot response of position, stabilization loop for both channels, which has −3 dB frequency at 7.3, 35 Hz for pitch channel and 6, 27.7 Hz for yaw channel at rated conditions against 4, 25 Hz as rated practical values. From the results, the elevation channel response is more active and the bandwidth can be increased up to 35 Hz. The track loop frequency is designed to be

Table 1 PID parameters and corresponding position bandwidths

Channel	Position loop			Position loop bandwidth (Hz)
	Proportional gain (Kp)	Integral gain (Ki)	Differential gain (Kd)	
Pitch channel	0.19818	46.468	−0.0044	8.5
	0.1550	32.4966	0.00015	8.3
Yaw channel	0.2918	47.4550	−0.00947	8.5
	0.29046	46.1092	−0.00966	11.3

Table 2 PI parameters and corresponding stabilization bandwidths

Channel	Stabilization loop		Stabilization loop bandwidth (Hz)
	Proportional gain (Kp)	Integral gain (Ki)	
Pitch channel	51.6805	2003.885	18.3
	75.5944	2532.656	27.6
Yaw channel	74.49	144.08	21.1
	106.0969	279.0651	30

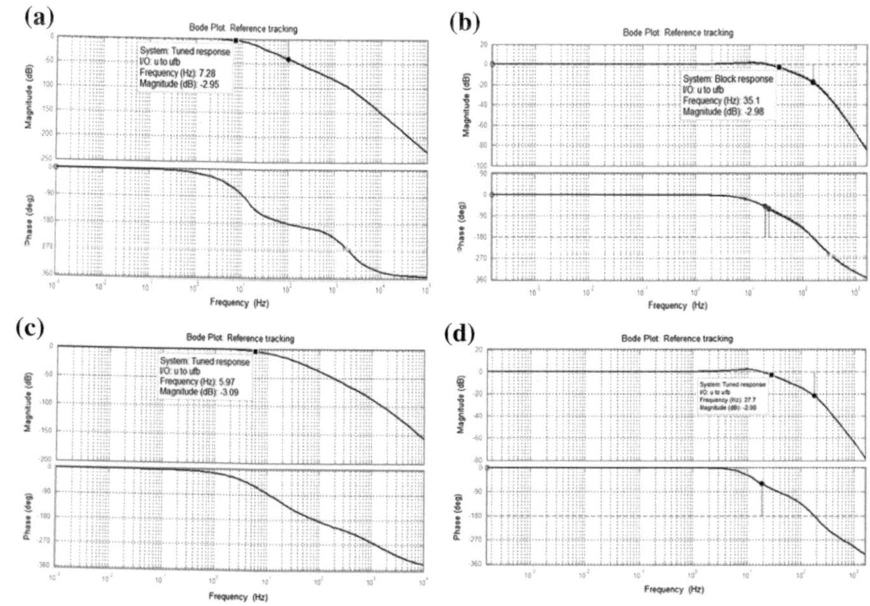

Fig. 6 **a** Bode response for pitch channel position loop. **b** Bode response for pitch channel stabilization loop. **c** Bode plot for yaw channel position loop. **d** Bode plot of the yaw channel stabilization loop

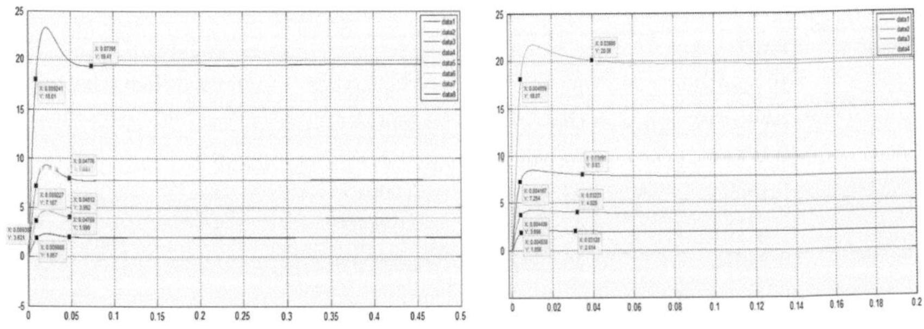

Fig. 7 Shows the step response of pitch, yaw channels for step inputs of 2, 4, 8, and 20

1 Hz from the design specifications. For gear drives, the bandwidth is lower, but the transient response characteristics are superior.

5.2 *Time Response (Body Rate Variation on Step Response)*

The unbalanced cross-coupled model is tested with different amplitude step at t = 0 at both elevation and yaw channels, considering the base angular velocity of 2 Hz sine wave with amplitudes of 2 each at i, j, k—*axes,* respectively, (30, 30, 60 in the second case). The PI parameters are

$K_{PI\_position}(s) = 53.622 + 16.398/s;$
$K_{PI\_stable}(s) = 4.683 + 239.601/s$ for elevation channel and
$K_{PI\_position}(s) = 36.712 + 11.76/s;$
$K_{PI\_stable}(s) = 3.171 + 203.712/s$ for yaw channel actuator.

In Fig. 7, the approximate rise time and settling times are shown in data points. It can be observed that the rise time is almost same for all responses, but the small variation in settling time is observed, which is proportionate to magnitude of step signal. Also, there is considerable error in the output in case of step input with a magnitude of 20, i.e., the error is proportion to the amplitude of the input signal. In the second case, the body rates are increased to 30, 30, 60, and the corresponding outputs observed that the effect of base rate change is 0.10 at lower amplitude which are in permissible limits.

6 Conclusion

The net torque required for gimbal system is governed by the unbalanced mass to compensate the asymmetric nature which caters the actuator power requirements. Also the torque generated by the actuation system is used to track the target with required slew rates. The main objective of the actuation system for seeker system stabilization should be quick in response with high precision and resolution control. So application comes with low torque and highly accurate speed and position resolution. Geared drive is the most feasible solution to achieve fast response with improved transient and steady state response. Also, the geared drives are more compact with less weight and reduce the structural requirements. Hence, the supporting structure will be of less weight, which leads to low mass budget requiring less torque. It is observed that multi-body dynamic simulation gives the optimum feasible solution considering the all possible nonlinear dynamics of the actuation motor, gear system, and gimbal.

References

1. Rue A (1974) Precision stabilization systems. IEEE Trans Aerosp Electron Syst. AES-10: 34–42
2. Ekstrand B (2001) Equations of motion for a two-axes gimbal system. IEEE Trans Aerosp Electron Syst 37:1083–1091
3. Smith B, Schrenk W, Gass W, Shtessel Y (1999) Sliding mode control in a two-axis gimbal system. In: 1999 IEEE aerospace conference. Proceedings (Cat. No.99TH8403)
4. Bigley WJ, Tsao SP (1989) Optimal motion stabilization control of an electro-optical sight system. Tracking, and Pointing III, Acquisition
5. Liu J, Chen G, Gong Y, Chen H (2011) Modeling and simulation of loader working device based on SimMechanics. In: Proceedings 2011 international conference on transportation, mechanical, and electrical engineering (TMEE)
6. Kim S, Kim S, Kwak Y (2010) Robust control for a two-axis gimbaled sensor system with multivariable feedback systems. IET Control Theory Appl 4:539–551
7. Rue A (1969) Stabilization of precision electrooptical pointing and tracking systems. IEEE Trans Aerosp Electron Syst. AES-5:805–819
8. Khodadadi H, Motlagh MRJ, Gorji M (2011) Robust control and modeling a 2-DOF inertial stabilized platform. Int Conf Electr Control Comput Eng 2011 (InECCE)
9. Otlowski DR, Wiener K, Rathbun BA (2008) Mass properties factors in achieving stable imagery from a gimbal mounted camera. In: Airborne intelligence, surveillance, reconnaissance (ISR) systems and applications V
10. Hilkert JM, Hullender DA (1990) Adaptive control system techniques applied to inertial stabilization systems. Tracking, and Pointing IV, Acquisition

UAV Advancement: From Increasing Endurance, Route Re-Plan and Collision Avoidance, to Safe Landing in Critical Conditions

Kalpesh Garvit Pandya and Kevin Nesamani

1 Introduction

It has been very well known that UAVs have got many functionalities now, in comparison with the very first model, and these functionalities were or are being updated just because of the inability to achieve something. We just had a simple model of a fixed wing UAV, which then got a new size, shape, materials, etc., to overcome the drawbacks occurring in the previous version. After its success, it was analyzed that something can be added to it in terms of the payload, which in turn gave an idea of adding certain payloads like weapons, camera for surveillance, goods package. But, the next time, surveillance raised a question that, how to surveil from a single location in air? The answer to the question came as multi-copter vehicle, with hovering capacity which was/is an inability of fixed wing vehicle. So, in the same way, even today, some or the other drawback overcomes with the wise implementation and usage of advanced technologies. Here, we too are using some new techniques and technologies to get the solution of some challenging inabilities of drone and the system used in it.

The performance of any vehicle, while testing, is one of the most important factors to watch. This performance is based on range, endurance, altitude, attitude control, vertical take-off and landing (VTOL), battery life, and other autonomous functionalities. The major problem we are facing in the field of UAVs is battery life. A UAV requires higher mAh battery, but that increases the weight, which in turn can destabilize the UAV. So, the solution needed to be found, in this matter,

K. Garvit Pandya (✉)
Department of Aeronautical Engineering, Veltech University, Chennai, India
e-mail: garvitpandya28@gmail.com

K. Nesamani
Department of Electronics and Communication Engineering,
Veltech University, Chennai, India
e-mail: kennith.n@gmail.com

S. Singh et al. (eds.), *Proceedings of the International Conference on Modern Research in Aerospace Engineering*, Lecture Notes in Mechanical Engineering,
https://doi.org/10.1007/978-981-10-5849-3_12

can be something like providing a battery of another type or something that can charge up the UAV on the way. This change in battery is not going to solve our problem as the only better option that remains with us is hydrogen fuel cell. So, we have only the latter option, charging facility, because we require higher endurance to fly up to the duration what an operator or an owner wants their vehicle to fly. The application of an UAV cannot be in the time limitation of an hour or two that too including the completion of the task assigned and coming back. In this regard, what we can do is, add a charging facility for the UAV battery, in the path itself.

In addition to the endurance, another major issue we found was, attitude control and an autonomous function, at the time of any obstacle on the path. So, we are adding a new sort of technique—on path obstacle avoidance system. The real problem occurs when the obstacle is avoided but what if something else happens? For instance, the drone comes across another obstacle immediately after avoiding the previous one. This problem can be solved with the new technique of collision avoidance. Moreover, route re-plan is the step which is an important one to accomplish the mission assigned to any UAV.

In spite of all these above implementation to aggrade the UAV performance level, it is necessary to take care of the UAV physically. The important fact is, an UAV can lose connectivity at any point of time, under any circumstances, and this will result in UAV falling from a certain altitude to the ground, thereby ending up breaking the vehicle. Hence, we are also adding a feature for its safe landing under critical conditions.

2 Lithium—Polymer Battery

It is very common that larger the battery, longer the flight time will be. But, there is a limitation too. With the increase in the size of the battery, the increase in flight time is not proportional. In other words, after some level of battery capacity, the flight time gets constant and does not make any difference with higher capacity battery. There are certain numbers which corresponds to the battery's specification like 2200 mAh is the capacity of the battery, 20–30 c is its discharge rate, and 5v will be its voltage (Fig. 1).

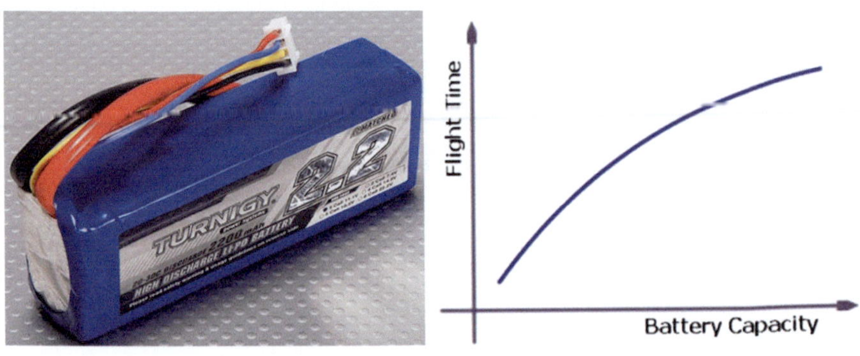

Fig. 1 Sample Lithium-Polymer battery and its efficiency graph. (*Source* www.oscarliang.com)

In the selection of the battery, we have got a pool of higher mAh batteries like 5000 mAh, 10,000 mAh, or even 12,000 mAh. But, it becomes ineffective for higher flight time due to its heavy weight and the bigger physical size. The table below gives an idea about how to select the battery from the pool of choices, based on the capacity. The below given table is an example of how a mathematical model can be built to estimate the flight time.

2.1 Calculation

In a battery, if we know the mAh, i.e., is the capacity of the battery (C) and the current drawn in ampere (I), we can find out the time duration of the battery or flight time (T).

$$T = C/I$$

Also, the weight of the battery will affect its efficiency. So, the battery efficiency needs to be calculated with respect to its weight. The calculation done by a source with other parameters is shown in Tables 1 and 2.

Table 1 Sample table of batteries available with several mAh capacities (*Source* www.oscarliang.com)

Turnigy Nano 4S Battery (25C–50C)				
Capacity (mAh)	Weight (g)	Cap vs weight	Price ($)	Cap vs price
1600	181	8.84	15.39	103.96
1800	207	8.70	17.25	104.35
2200	239	9.21	21.13	104.12
2650	275	9.64	30.25	87.60
3000	299	10.03	37.36	80.30
3300	337	9.79	40.97	80.55
4000	433	9.24	38.31	104.41
4500	467	9.64	46.27	97.26
5000	523	9.56	56.33	88.76
6000	623	9.63	68.78	87.23

Table 2 Sample table of the data collected after increasing the load step by step (*Source* www.oscarliang.com)

Tricopter hover time test against weight						
Standard setup with all equipments and same 2200 mAh battery						
Total initial weight: 1193 + 239						
Round	Load (g)	Time	Time (s)	Actual capacity used (mAh)	Hover throttle	mAh/second
1	0	08:14	494	1922	63	3.89
2	110	07:18	438	1900	68	4.34
3	220	06:13	373	1878	70	5.03
4	340	05:55	355	1908	73	5.37
5	405	05:20	320	1900	79	5.94
6	515	N/A	N/A	N/A	85+	N/A

3 On Path Charging Facility

This is one of the most important phases of this paper. Currently, what is going on in the field of UAV is flying, changing the battery, flying, etc. The new technology advancement can be done to the problem of changing the battery again and again. And, what if the drone is half the way of accomplishing the mission? It can neither go further and finish the work nor come back. So, what we have come up with this is a charging dock [1]. This charging dock will be placed on several distances on buildings. The passer by drones will be able to identify that in the nearby region charging dock is available. After the signal of detecting the charging dock, it will check its battery level. If it is 40% or below, it will go to the dock, get it charged, and will take-off again to continue its mission. If not, it will continue flying. Here, we can set the cutoff percentage level according to our requirement and the distance between two consecutive charging docks can be preprogrammed into the vehicles, so that it can check that after how much distance the next dock is, when it is checking its battery percentage level near a dock (Figs. 2 and 3).

- **Stages: -**
- **Stage 1** Drone lands on the charging platform
- **Stage 2** Drone wireless charging and monitoring (2.4G Wi-Fi communication) [1]
- **Stage 3** Drone take-off.

Many companies are coming with their new techniques in charging pads as an option for increasing endurance which can be used until we get the hydrogen fuel cells regular for use.

Fig. 2 Skysense company charging pad. (*Source* Skysense datasheet)

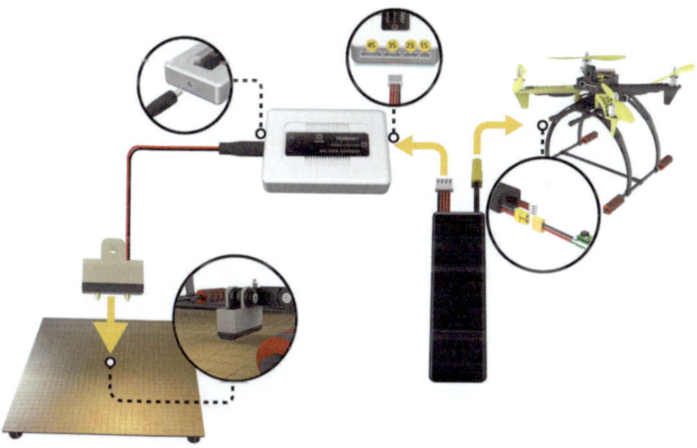

Fig. 3 Skysense charging pad connections. (*Source* www.erlerobotics.com)

4 Collision Avoidance System

For implementing a highly intelligent UAV, there should be a system which provides safety to humans as well as to the UAV itself. For this, there is a provision of collision avoidance in our proposed system. The UAV is assigned a particular path which has to be followed by it to reach the destination. However, in this pre-defined path, set by the user, there may be some obstacles which may be temporary or permanent such as birds and trees. The UAV which is fully autonomous should be able to detect these obstacles and avoid them by using sensors and an algorithm. The most widely used sensor for this application is the ultrasonic sensor which provides high reliability, accuracy, range and is also cost-effective. For a quad-copter, it is necessary to monitor the whole 360° area around the UAV; hence, four of these sensors are to be used for this application which would suffice our need. There are several works focused on navigation and obstacle avoidance procedures for helicopters. Some of the pioneers in autonomous navigation for helicopters worked at NASA Ames Research Center. In 1980 and 1990 decades, they have published a series of papers highlighting some techniques developed for automatic nap-of-the-earth flights such as computer vision, integration of active and passive sensors, design of control strategies tested in 3D computer simulations. In the beginning, the authors developed 2D models of the environment, and later on they extended the path search techniques to 3D in order to obtain a low-altitude guidance system for military helicopters [2].

Out of some important options like ultrasonic detectors, LIDAR, Laser detector, etc., we have here chosen ultrasonic sensors because of its higher predictability. The ultrasonic sensors are used to avoid the obstacles with real-time information. This kind of sensor, by default, is very sensible to objects whose position is out of the line of vision [3]. Our goal is to prevent collision, where we don't know the exact position and the speed of the quadrotor, the obstacle itself is the reference point for evasive closed-loop maneuvers. In such a case, it is more advantageous that the ultrasound sensors are able to detect obstacles from a long range of distance and with accuracy than that it has a large field of vision, so that the collision can be precisely avoided. An important aspect considering the use of the ultrasonic sensors refers to its position in the structure of the quadrotor. The propellers of the quadrotor should not interfere with the sensor, this may cause error readings, and the system will malfunction, for this proper measure has to be taken to limit the sensor range so as to focus only in the straight path. For detection of moving objects, we are using a cross-check method which will verify whether the encountered obstacle is a stationary object or a moving object. This is done by monitoring the obstacle for two times in a fixed interval of time. If the system detects an obstacle which is in the path of the UAV, it won't change its path right away, but wait for the sensor to detect the obstacle for the second time after a certain time. If the same obstacle is observed for the second time, then the path of the UAV will be changed depending on the algorithm, whereas if the obstacle is not detected on the second instance, it will consider the obstacle as a moving object and continue on its same path. This will avoid false detection of objects which will prove to be an advantage as the UAV would not have to reroute unnecessarily (Figs. 4, 5 and 6) [4, 5].

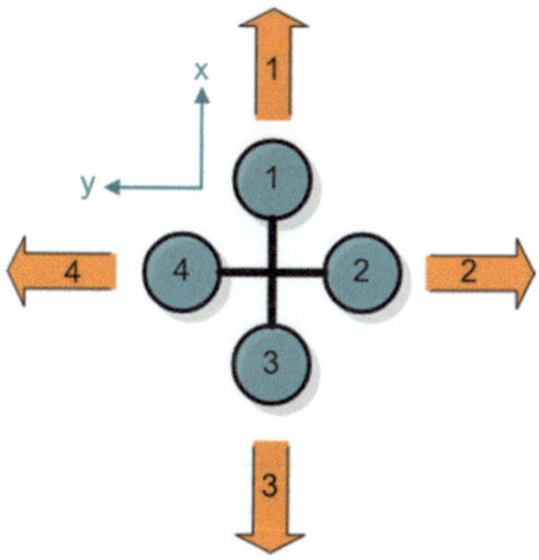

Fig. 4 Position of sensors and direction of propagation [6]

Fig. 5 Ultrasonic sensor [7]

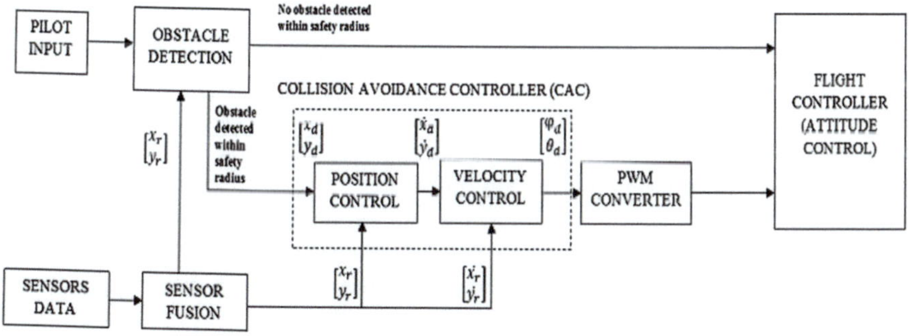

Fig. 6 Block Diagram of Collision Avoidance [8]

5 UAV with Autonomous Level 4

Autonomous system is one of the systems which describes the difference between a drone and a UAV. A drone can be semi-autonomous or mostly a remotely piloted vehicle (RPV) be called an UAV, until and unless it can fly without RC or with an automatic controls. The autonomous levels of UAVs are shown in Table 3.

Table 3 Sheridan's scale of degrees of automation [9]

Level	Description
High	10. The computer decides everything, acts autonomously, ignores the human
	9. The computer informs the human, if it wishes to
	8. If asked, computer informs the human
	7. The computer executes automatically, then informs the human necessarily
	6. The computer allows the human a limited time period to veto before it executes automatically
	5. The computer executes the suggestion if approved by the human
	4. The computer suggests only one alternative
	3. The computer narrows the selection down to a few
	2. The computer offers a complete set of decision or action alternatives
Low	1. The computer offers no assistance; all the decisions and actions must have to be taken by the human

Table 3 describes the work sharing between the human and the computer. Also, it shows the capability of the computer to control a flight automatically without any input from the human. That is the beauty of this technology. The levels are described here, below, from 1 to 10. [9]

- **Level 1** Remotely piloted vehicle. Here, the 100% input is required from the human.
- **Levels 2, 3** High level of human input is required in this level. It can be operated in simple environmental condition. Example—Predator, Global Hawk, etc., are being operated using ground control stations.
- **Level 4** Onboard route re-plan. It has the capability to change its path when needed and come back to its original flying path, without the interference or input from the human.
- **Levels 5, 6, 7** Group coordination, tactical re-plan, and goal. These all requires medium level of human input. All of these can work in the moderate environment and can operate multifunction missions.
- **Levels 8, 9** Distributed controls and group strategic goals. Low-level human input is enough for this level. Moreover, a big mission can be achieved with the help of group planning on air, without the input from the human.
- **Level 10** Fully autonomous. This is an extremely high-level operation, where the computer will not take input from the human. It has no environmental limitation and can be used in any kind of missions [9].

It has been seen that almost up to level 3 we have reached easily. Yes, even up to level 9 some research is being done, but considering the regular flying purpose, we are near to the level 4. So, we are including the level 4 of autonomous system in which the UAV will change the route accordingly, when needed. As, we are using on path charging facility and the collision avoidance, route re-planning becomes a necessary part to be carried out. Because it will change its path to avoid collision or

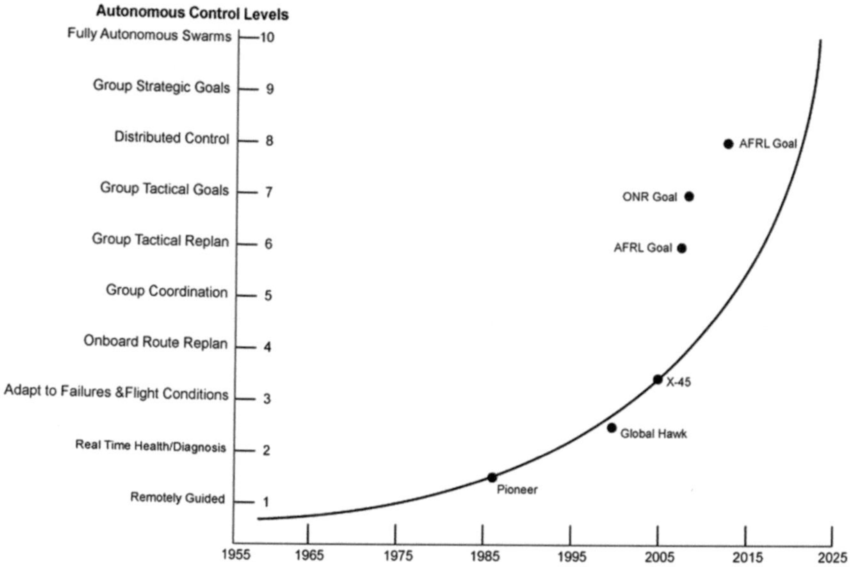

Fig. 7 Autonomous control levels [10]

for charging itself, it has to come back to its original mission path to accomplish the mission. The same follows in the charging operation (Fig. 7).

The characteristics of the vehicle will be:

1. The autopilot onboard the drone is to be designed in such a way that it can autonomously fly from one location to another.
2. The drone have take-off, landing, and collision avoidance methods implemented on board [11].

This function is not so easy to perform, but the on board should be that efficient that it can perform the route re-plan functionality [12].

6 Failsafe Parachute Landing

We have vehicles in the market which has got a very high-quality materials that does not seem to break if fell on the ground due to some failure. But, we cannot be dependent on that. So, in case of a malfunction in the system such as motor failure, sensor failure, or the most common battery failure, the UAV has no control over itself and can cause disastrous results, when it crashes. The scenario that we are describing here is when the craft has a problem, and it is difficult to land or there isn't much time to land safely on your own. To avoid crash landings, quadcopters can be equipped with parachutes. The flight controller has an option of adding a

Fig. 8 Sample of parachute on a multicopter vehicle. (*Source* www.books.google.co.in)

parachute. The hardware mechanism of a parachute must be connected via servo wire to the flight controller, and the trigger must be configured through the software. The software is made in such a way that when such a malfunction is detected by the controller, it will trigger a parachute which will deploy immediately and ensure safe landing, and later the location where the UAV has landed can be found out and recovered back. This safety measure can help in reducing the cost incurred in case of a damage to the system.

If we look at the principle for attaching the parachute, it is found to be simple. The folded parachute is inserted in a tube. On the top of the tube, a lid is there which is kept closed by the servo. From the bottom of the tube, a big spring pushes the parachute upside. So, the moment servo opens up the lid and the parachute will open up [13] (Fig. 8).

6.1 Calculation

The surface area required for a parachute depends on the weight of the vehicle and the required descending speed.

$$\text{Diameter} = \left(70^* \sqrt{\text{m}}\right)/\text{V}$$

where m is mass of the vehicle in grams and V is the velocity in km/h.

According to the current availability in the market, a parachute with the capacity of handling 4.5 kg weight weighs 310 grams (approximately). Weight is quite a serious factor to be concerned about. For that purpose, the lightest material can be used which is graphene aerogel. It has got a very high strength while being the lightest material. This can be the game changer for our safe landing purpose during emergencies [13].

By selecting the right parachute, it will be easier to avoid the crashes that take place with the failure of the vehicle due to any reason.

7 Conclusion

The presented techniques and technologies in this paper are those with which we can overcome the drawbacks that are occurring in some UAV functionalities. The concepts for high endurance and low failure of UAVs focused in this research will play a significant role in real-time application. The reason behind adding all the important features is to increase its capability in long duration flight without facing any problem on path and with which an assigned mission can easily be accomplished.

References

1. Skysense Company—Charging Pad Datasheet 201601
2. Victor HL (1990) Cheng: concept development of automatic guidance for rotorcraft obstacle avoidance. IEEE Trans Rob Autom 6(2):252–257
3. Becker M, Coronel R, Sampaio B, Bouabdallah S, Vincent de perrot, roland Seigwart: In: Flight Collision Avoidance for a Mini UAV Robot Based on On-Board Sensors
4. Gageik N, Müller T (2012) Obstacle detection and collision avoidance using ultrasonic distance sensors for an autonomous quadrocopter
5. Sabatini R, Gardi A, Richardson MA (2014) LIDAR obstacle warning and avoidance system for unmanned aircraft, world academy of science, engineering and technology. Int J Mech Aerosp Ind Mechatron Manuf Eng 8(4)
6. Bouabdallah S, Becker M, Perrot V, Siegwart R (2007) Toward obstacle avoidance on quadrotors. In: XII DINAME—International Symposium on Dynamic Problems of Mechanics, 2007, Ilhabela—SP. Proceedings of DINAME 2007, 2007. vol 1, pp 1–10
7. Hossain A (2013) Undefined obstacle avoidance and path planning. In: 120th ASEE Annual Conference and Exposition (2013)
8. Rambabu R, Bahiki MR, Md Ali SA (2015) Relative position-based collision avoidance system for swarming uavs using multi-sensor fusion. ARPN J Eng Appl Sci (2015)
9. Proud RW, Hart JJ, Mrozinski RB (2013) Methods for determining the level of autonomy to design into a human spaceflight vehicle: a function specific approach. In: NASA Johnson Space Center
10. Vachtsevanos G, Ludington B, Reimann J, Antsaklis P, Valavanis K (2005) Modeling and control of unmanned aerial vehicles—current status and future directions. In: Workshop on Modeling and Control of Complex Systems (MCCS), Ayia Napa, Cyprus

11. Goyal P, Smeur E, de Croon G (2016) Mission planning for sensor network deployment using a fleet of drones. In: Delft University of Technology, Delft, Zuid-Holland, 2629 HS, The Netherlands
12. Boggero L (2015) Autonomous planning and replanning of a single Unmanned Aerial Vehicle: strategies and simulations. In: CEAS
13. Favini E, Niezrecki C, Chen J, Willis D, Niemi E, Manohar S, Desabrais K, Charette C (2012) Sensing performance of electrically conductive parachute fabrics and suspension lines. In: AIAA conference and seminar, Dublin, Ireland. B

High Angle of Attack Analysis of Cascade Fin in Subsonic Flow

Manish Tripathi, M.S. Mahesh and Ajay Misra

1 Introduction

While designing control devices for missiles, the designer needs to optimize parameters such as aerodynamic efficiency, weight, actuator requirement, and stowability. At high angles of attack, fin stall poses an additional challenge towards creating an optimized design. Introduced in 1980s by the Russians [2, 18], grid or lattice fins are a relatively new aero-mechanical technology used for tail controlled missiles. These have been used as control and stability devices in many Soviet and American-borne missiles and bombs for steering and stability. They have been utilized as drag brakes on the SOYUZ TM-22 spacecraft. Recently, these fins were used by the Falcon 9 Reusable Launch Vehicle developed by SpaceX (USA) [16] for manoeuvring in the final stage of re-entry.

A grid fin (also known as lattice control surface) is an unconventional lifting [13] and control surface that consists of an outer frame supporting an inner grid of intersecting planar surfaces of small chord. Unlike conventional fins, they are aligned normal to the airflow allowing air to pass through. Their honeycomb structure provides high strength-to-weight ratio, thereby allowing them to be made thinner, reducing weight and cost of materials. Their small chord results in low hinge moments, thereby mitigating the requirements for control actuators, leading to further reduction in weight. They also provide enhanced aerodynamic control at high angles of

M. Tripathi (✉) · M.S. Mahesh · A. Misra
Department of Aerospace Engineering, Defense Institute of Advanced Technology (DU), Girinagar, Pune, Maharashtra, India
e-mail: manish.tripathi1189@gmail.com

M.S. Mahesh
e-mail: maheshms7@gmail.com

A. Misra
e-mail: ajaymisra@diat.ac.in

S. Singh et al. (eds.), *Proceedings of the International Conference on Modern Research in Aerospace Engineering*, Lecture Notes in Mechanical Engineering,
https://doi.org/10.1007/978-981-10-5849-3_13

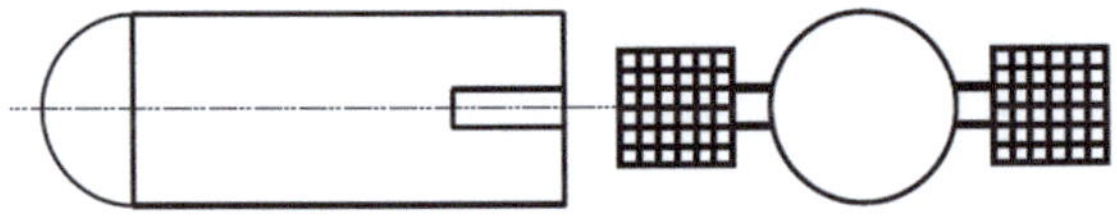

Fig. 1 Front and side view of a missile having grid fins with cross and horizontal members [9]

attack and wide Mach number ranges [4]. In fact, due to their high stall angles compared to conventional planar fins, grid fins do not show a sharp break in lift even up to angles of attack as high as 50°. However, the biggest disadvantage related to these fins is the enhanced drag force leading to numerous world wide studies to reduce it in order to increase efficiency [10].

One of the measures to increase the aerodynamic efficiency was given by Misra et al. [7–9] who proposed a new category of grid fin termed as the cascade fins. A cascade fin has planar members placed parallel to each other at an optimized gap-to-chord (g/c) ratio [8]. Unlike grid fins, these lack cross members. This results in reduced drag. Theoretical study [2, 5] and experimental observations [10, 11] point to an enhanced lift at higher angles and deviation from the conventional stall of a flat plate making them a viable candidate for highly manoeuvrable vehicles. Some of the parameters effecting the aerodynamic efficiency are g/c, number of planar members, the presence of end plates, and the cross-sectional shape of the planar members [9] (Fig. 1).

Present research aims at obtaining optimized geometry specifications for maximum aerodynamic efficiency, reducing drag and retaining the lifting characteristics. This was done by carrying out a sequence of CFD simulations for cascade fins at low subsonic speed for varying angles of attack using different gap-to-chord (g/c) ratios between fin planar members and the results were compared against a single flat plate. Further, results were validated against experimental results obtained from the National Wind Tunnel Facility [8] available in the Department of Aerospace Engineering at the Indian Institute of Technology, Kanpur. Consequently, to get insight into the physics of the flow leading to stall delay, flow visualization of the pressure coefficient (C_p) and velocity contours will be used.

2 Computational Approach

2.1 Geometry and Flow Parameters

Our cascade fin consists of three flat plates (planar members) placed parallel to each other bounded by two vertical end plates as shown in Fig. 2b. The plates have a span (b) of 0.2 m and chord length (c) of 0.1 m, thus an aspect ratio of 2. Thickness (t) of the plate is 0.0025 m, and gap-to-chord ratio (g/c) of plate is 0.4, 0.5, 0.6 and 0.7, thus giving it a gap (g) of 0.04, 0.05, 0.06 and 0.07 m between the plates. The flat plate geometry specifications are the same as the plates of the cascade configuration as shown in Fig. 2a.

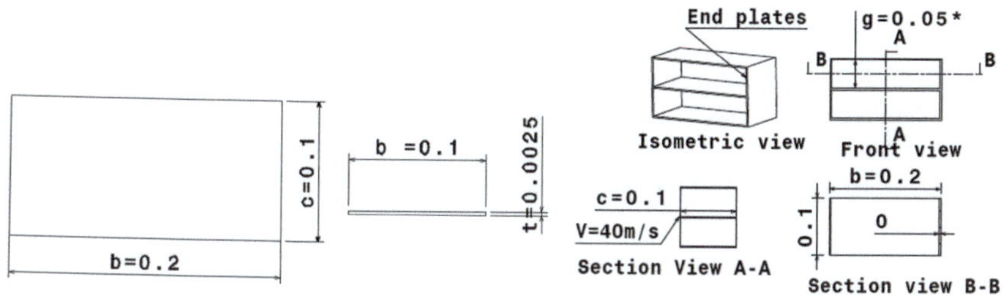

(a) Flat plate (b) Cascade fin geometry (*variable)

Fig. 2 Geometry for simulation (dimensions in m)

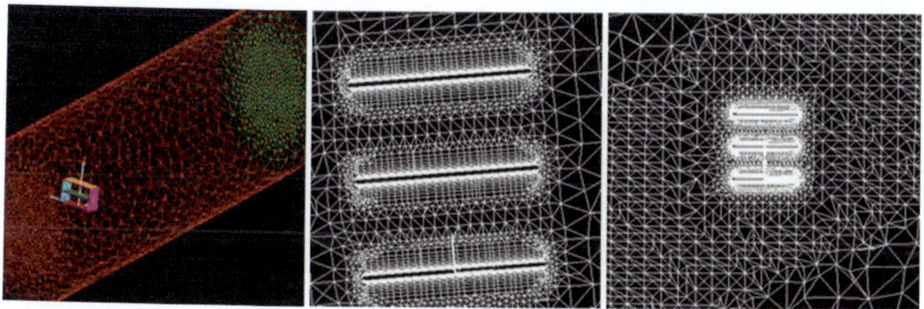

(a) Mesh (Inner + Outer Domain) (b) Mesh Refinement with prism layers close to walls (c) Mesh aligned with the freestream

Fig. 3 Computational hybrid mesh

Computational domain was a cylinder of variable radius. The radius was adjusted so that the frontal area of the cascade to domain cross-sectional area remains less than 5%. The domain was aligned with the free stream at different angles of attack to assist in getting accurate results (Fig. 3c). Domain length was taken to be 0.5 m upstream and 2.0 m downstream in order to capture the flow variations behind the fin accurately.

The flow conditions are taken to be standard ambient conditions, i.e. pressure 101325 Pa and temperature 288 K. The flow is maintained at 40 m/s, i.e. at Mach 0.1176, with Reynolds Number of 2.855×10^5 and varying angles of attack from 0° to 50° in steps of 5°, and an additional 53°. Free-stream turbulence was taken as 0.08% as per the specification given for the wind tunnel in the literature [14]. The turbulence length scale was taken as 0.1778 based on the hydraulic diameter of the wind tunnel. The reference length was taken as chord length (0.1 m) and reference area as the area of a single plate (0.02 m^2).

2.2 *Solution Methodology*

Numerical analysis (as shown in subsequent sections) was done using 3D Navier–Stokes equation coupled with a turbulence model to cater for the turbulent flow field.

Solution was obtained using commercial CFD++ solver [3]. The governing equations were solved in this software using the finite volume method. Implicit, pressure-based solver was adopted for the incompressible low-speed flow numerical investigation. The flow variables were discretized using the second-order spatial discretization scheme. In order to capture the flow separation, rotational flow and adverse pressure gradient conditions due to high α, we used the RANS based one equation Spalart–Allmaras (SA) turbulence model [17]. This model was selected as it is relatively simple (involving one transport equation for eddy viscosity) and has a specific design for aerospace applications involving wall-bounded flows. These have been found to produce good results for low Reynolds number [1] using mesh configuration with the wall $y^+ \leq 5$, resolving the viscous sublayer. This model produces numerically stable solutions and has been found to be more robust than the Menter SST model [6]. In order to reduce the calculation time, the domain is decomposed into multiple parts (8). Calculations were performed on each individual part in separate core, and MPI (Message Passing Index) method was used to exchange data between the cores.

2.3 *Computational Mesh and Boundary Conditions*

Mesh was created using MIME mesher [12] software. The mesh was divided into two regions one encompassing the other (Fig. 3a). The inner region was closer to the fin geometry with a relatively denser mesh. The mesh was a hybrid mesh consisting of tetrahedral cells away from the body and prismatic layers in close proximity to the fin walls. The mesh was further refined closer to the walls of the fins to efficiently capture the variation in the flow parameters in the boundary layer of the fins. The mesh took care of the sharp turns of the fin corners and edges by using the sharp-turn blending feature of MIME which creates degenerate prismatic elements that are projected from edges or vertices to more effectively mesh the areas of high geometric curvature (Fig. 3b). Mesh for the cascade fin consists of 1.7 million cells consisting of 1.1 million tetrahedrals, 0.6 million triangular prisms and 8000 pyramids. For the single flat plate, mesh consisted of 0.8 million cells divided into 0.56 million tetrahedrals, 0.18 million triangular prisms and 27 thousand hexahedrals. Mesh close to the walls was refined using prismatic layers such that the value of y^+ remained less than 5 to capture the flow gradients in the turbulent viscous sublayer.

Characteristic-based inflow/outflow boundary conditions at the farfield, inlet and outlet boundaries utilize the supplied data in Riemann solvers to determine the local Mach number and flow direction. This is very helpful when the boundaries of the free stream are close to the geometry. The walls have been assigned the adiabatic no-slip boundary condition. The solver uses solve to wall methodology to capture flow variations inside the turbulent boundary layer of the fins.

3 Results and Discussions

3.1 Validation of CFD Results

In order to achieve optimum mesh density for accurate results, calculations were performed for two different mesh configurations using wall y^+ as a guidance to select the appropriate mesh configuration close to the wall by comparing C_L results for fine and coarse mesh with experimental data. y^+ is a non-dimensional distance [1, 15], used in turbulent flow simulations to delineate how fine or coarse the mesh is close to the walls. y^+ for the two mesh case configurations were taken as $y^+ \approx 20$ (coarse mesh) and $y^+ \approx 5$ (fine mesh). The coarse mesh was resolved up to the buffer layer, and the fine mesh resolved the mesh up to viscous sublayer of the turbulent boundary layer. As shown in Fig. 4, the CFD results show an appreciable increase in accuracy for fine mesh. The difference between the simulated and experimental results became less than 10% compared to the coarse mesh where it was found to be 18%. This improvement in results can be attributed to the fact that the Spalart–Allmaras model has been designed to be valid throughout the boundary layer, thus requiring fine mesh resolution close to the walls. Thus, our analysis produces accurate results for the fine mesh configuration.

To validate the present solution methodology, the calculated lift and drag coefficients (C_L, C_D) for flat plate (FP) and 3 planar member cascade with g/c 0.5 (3MC g/c 0.5) were compared with experimental data as given by Misra in [7, 8]. Relative error for C_L was found using:

$$Error = \frac{C_L(si) - C_L(ex)}{max(C_L(ex))} \times 100. \tag{1}$$

where $C_L(si)$ = simulated C_L, $C_L(ex)$ = experimental C_L and $max(C_L(ex))$ = maximum value of experimental C_L. Similar formula was used for calculating C_D simulation error by replacing all the C_L terms with the corresponding C_D values. The numerical results were found to be in close agreement with the experimental data as shown in Fig. 5. The error is found to be less than 11% for C_L, less than 7% for C_D for

Fig. 4 C_L for two different mesh configurations

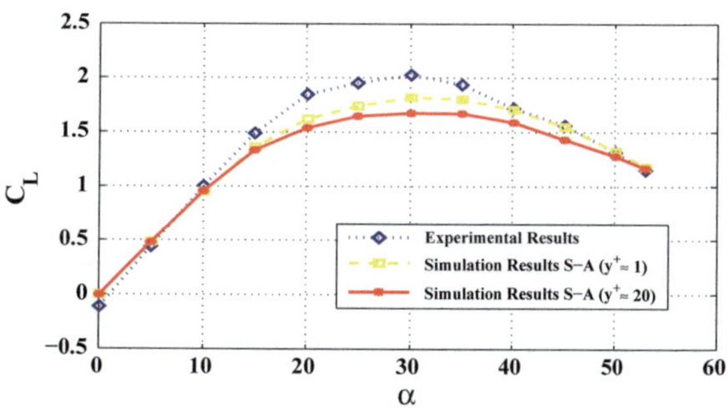

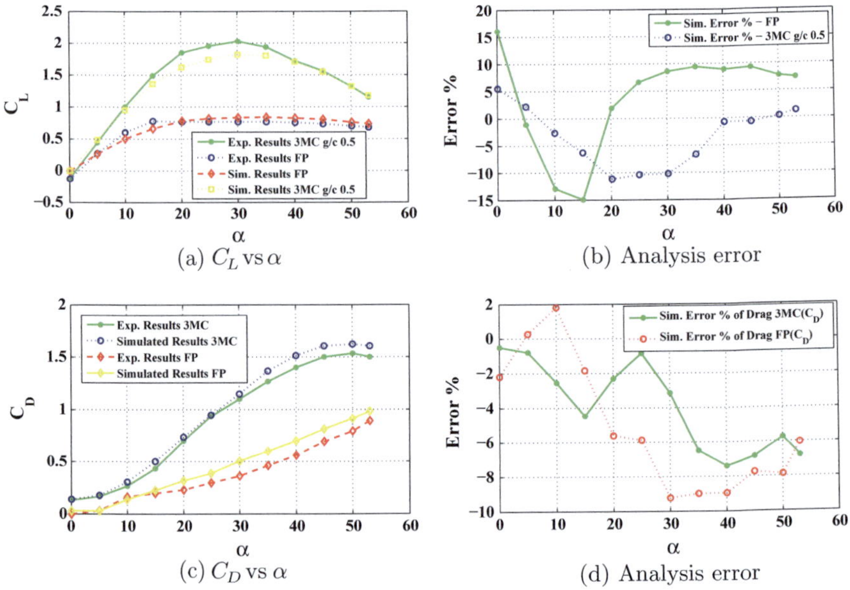

Fig. 5 Validation of simulation results

cascade fins case and less than 15% (C_L and C_D) for FP, indicating that the numerical methodology adopted is accurate enough to capture flow physics related to cascade fins at low subsonic speeds within the particular angle of attack (α) range. Absolute error for FP is less comparatively, yet due to the low values of experimental C_L and C_D, its percentage error is high.

As shown in Fig. 5a, as α increases, C_L increases linearly up to 20°. After this, the C_L increases nonlinearly due to the production of vortices and eddies. Further, beyond 30°, lift starts reducing slowly. However, this reduction does not exhibit the conventional abrupt reduction in C_L. Thus, fins are able to provide reasonable lifting force up to $\alpha = 53°$, making them suitable candidates as control surfaces on highly manoeuvrable missiles. Albeit, Fig. 5c clearly emphasizes that the increased number of planar members leads to an increase in drag. This further increases as α increases. This can be reduced by virtue of geometric alterations such as cross section, thickness which will be dealt in future research.

3.2 *Comparison of Aerodynamic Coefficients of 3 Planar Member Cascade of Different g/c Versus Flat Plate*

The three most important aerodynamic parameters while investigating the performance of any control device are C_L, C_D and C_L/C_D values. Figure 6 shows variation of these aerodynamic parameters for a cascade fin of three parallel plates placed at

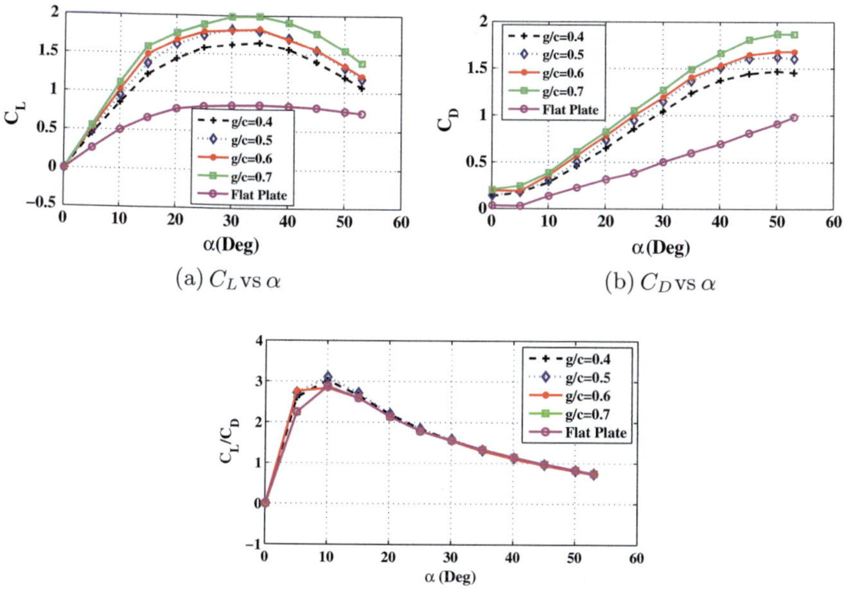

(a) C_L vs α

(b) C_D vs α

(c) C_L/C_D vs α

Fig. 6 Aerodynamic performance comparison of cascade fins of different g/c with flat plate

a variable g/c(gap-to-chord) ratio of 0.4, 0.5, 0.6 and 0.7. The values for flat plate inclined at different angles of attack are also provided for reference. As pressure on lower and upper surfaces for all the three plates is same at $\alpha = 0°$, no lift is generated and hence lift coefficient is zero. It can be noted that lift (Fig. 6a) is always higher than planar fin for all angles. As α increases, lift coefficient increases linearly, but unlike conventional fins, it lacks a sudden drop in lift even until 30°. Interestingly even beyond this high angle of attack, the lift coefficient does not reduce abruptly, again displaying deviation from the conventional stall characteristic of planar fins and thus enhanced lift at high angles of attack is produced, making them suitable for highly manoeuvrable vehicles. It is noteworthy that as g/c ratio between the plates increases, C_L also increases, apparently due to the reduction in interference between lower and upper plate flow parameters (explained later). The stall angle increases as the g/c ratio reduces owing to the fact that flow between the members get energized due to the cascade effect, leading to delay in flow separation. Moreover, the presence of end plates leads to flow being restrained, and this further contributes to increased lift.

Although due to the presence of end plates and sharp edges, drag of cascade fin is found to be higher than the flat plate for all cases as shown in Fig. 6b. Moreover, as g/c increases, drag increases (maximum for g/c 0.7). This can be due to the increase in pressure drag attributed to increased flow separation as plates move apart. Skin friction drag also increases due to the increase in end plate size.

The ratio of C_L and C_D is an indication of the aerodynamic efficiency of the control surface. Greater the C_L/C_D ratio, better is the aerodynamic performance. As shown in Fig. 6c, the aerodynamic efficiency effectively remains the same with no significant changes for all configurations. Though, at lower angles due to lower drag, flat plate gives better efficiency. Yet, increase in lift due to the cascade configuration was earlier observed in Fig. 6a, thereby making them a befitting candidate for a control surface on ammunition and missiles.

Flow-field visualization across the cascade fins in comparison with the flat plate was carried out using the C_p(coefficient of pressure) and velocity contour (U) distribution plots.

3.3 *Pressure Contours at Different α (10°, 30°)—Cascade Fin Versus Flat Plate*

C_P contours over the upper and lower surfaces of three plates of cascade fins for different g/c and flat plate are given in Table 1 followed by an enlarged view for 3

Table 1 C_P Contour

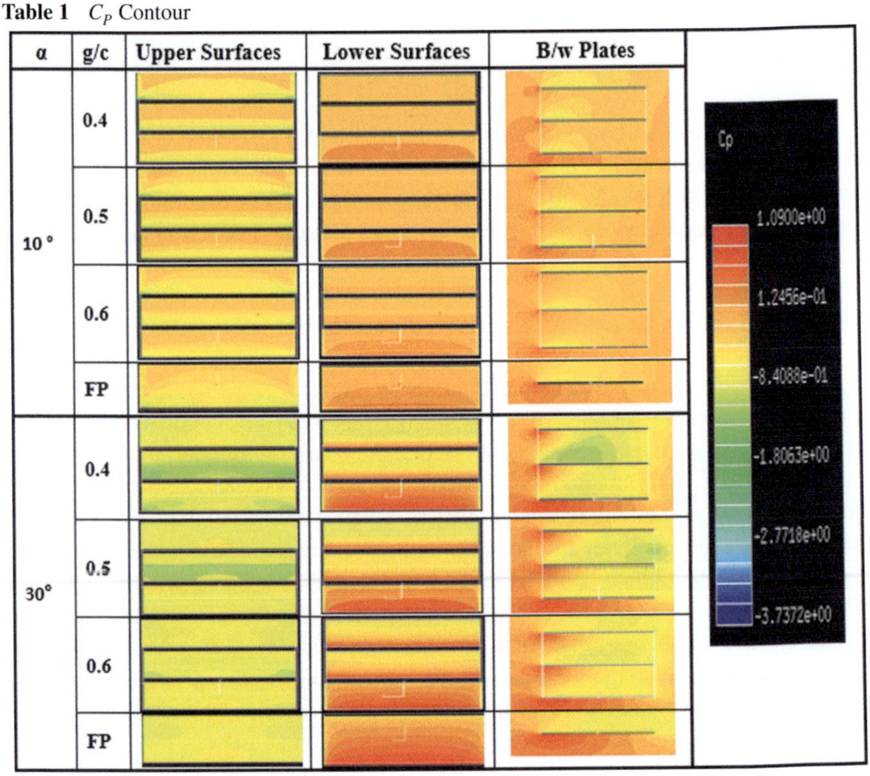

Table 2 Enlarged view of C_P Contour for $\alpha = 30°$, $g/c = 0.5$

Upper Surfaces	Lower Surfaces	B/w Plates
	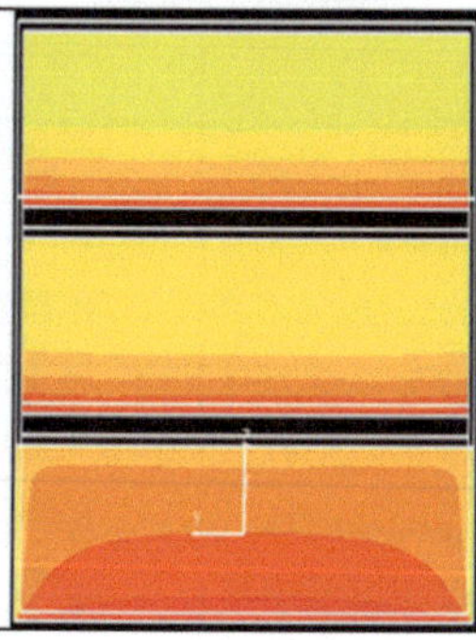	

member cascade of g/c 0.5 at $\alpha = 30°$. These results tend to indicate a stark difference in pressure difference between upper and lower surfaces in middle and bottom plates as compared to the upper surface of top plate and lower surface of the lowest plate, respectively. High pressure of the lower surface partially neutralizes the low pressure created on the upper surface of the plate below as seen by the increased C_P value on the upper surfaces (Fig. 2) of 2nd and 3rd plates. Thus, as plates are brought closer, this effect increases and ultimately leads to reduced lift as shown in Fig. 6a. Although, due to constrained geometry of cell, flow gets energized and adverse pressure gradient reduces as shown in Table 1 for g/c 0.4 at 30°, as is evident by the reduced value of C_P for the upper surface of second plate (Table 2). This leads to attached flow and delayed stall as evident by the increase in α for which C_L starts reducing for g/c 0.4 in Fig. 6a. This effect becomes more prominent as α increases.

3.4 *Velocity Contours at Different Angles of Attack (10°, 15°, 40°)—Cascade Fin Versus Flat Plate*

Table 3 shows cross-sectional view of the cascade fin and flat plate of velocity distribution across the geometry at various angles, followed up by an enlarged view of 3 member cascade at g/c 0.5 and $\alpha = 30°$. As seen for 0°, flow remains the same for all plates. But as α increases, flow between the plates becomes more energized, which further leads to a reduction in flow separation region (Fig. 7). Also as g/c reduces, this effect increases, leading to further reduction in flow separation. This leads to increased stall angle as discussed earlier. These results suggest that decision on selection of a particular g/c ratio must be done based on optimizing the amount of lift required (for control effectiveness) and flow separation control at high angles.

Table 3 Velocity contours of FP versus cascade fin at different α and g/c

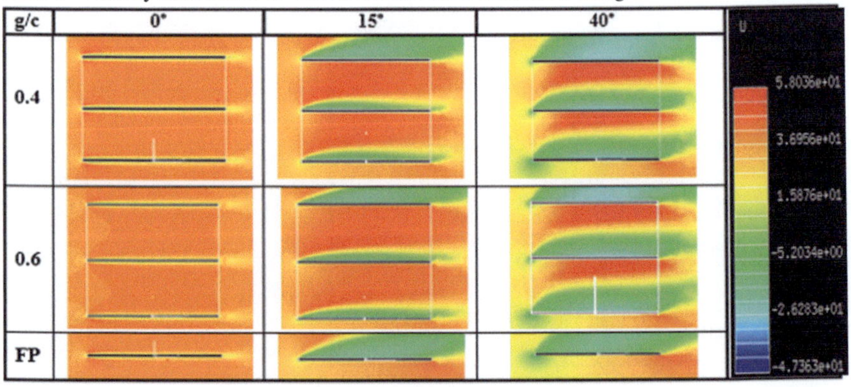

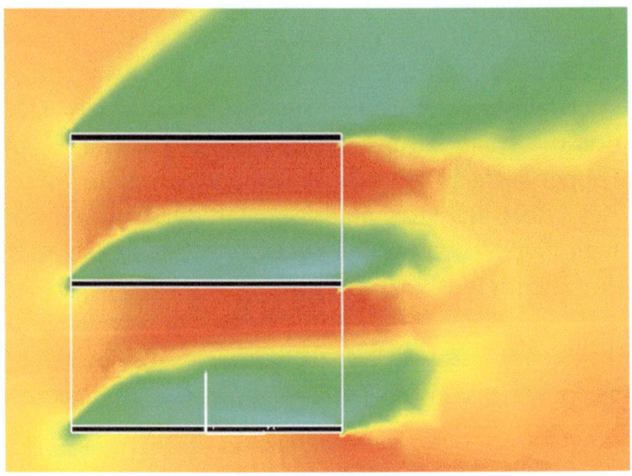

Fig. 7 Velocity vector $\alpha = 30°, g/c = 0.5$

4 Conclusions

Numerical results using CFD analysis have been calculated for cascade of three parallel plates placed at different g/c ratios and compared with a single flat plate result. The results were further validated with experimental data. Some of the major conclusions are as follows:

1. Lift coefficient of cascade fins was found to be more than that of the flat plate with same aerodynamic efficiency, even at high angles of attack, making them suitable candidates for manoeuvring missiles at high angle of attack.
2. Cascade effect leads to interaction between the pressure variation of the lower surface of upper plate and upper surface of corresponding lower plate, thus reducing

the effective lift. Thus, bringing the fin planar members too close is detrimental to its aerodynamic performance.

3. As plates are brought closer, the flow gains more and more kinetic energy causing reduced flow separation region. Thus, this increases the aerodynamic control at high angles of attack. Thus, geometry needs to be optimized keeping in mind the desired overall lift and stall delay.
4. Drag is an important parameter for these fins which requires further investigation.

References

1. Ariff M, Salim SM, Cheah SC (2009) Wall y^+ approach for dealing with turbulent flow over a surface mounted cube: part 1-low reynolds number. In: Proceedings of seventh international conference on CFD in the minerals and process industries, Melbourne, Australia
2. Belotzerkovsky SM (1987) Wings with internal framework. In: Machine translation, FTD-ID (RS) T-1289-86, Foreign Technology Div
3. CFD++ 15.1.1 (2015) CFD++ user manual
4. Eugene LF (2001) Tactical missile design. In: AIAA education series, pp 1–18
5. Ledlow TI, Burkhalter JE, Hartfield RJ (2015) Integration of grid fins for the optimal design of missile systems. In: AIAA atmospheric flight mechanics conference
6. Menter FR (1994) Two-equation eddy-viscosity turbulence models for engineering applications. AIAA J 32(8):1598–1605
7. Misra A (2009) Investigation of grid and cascade fins for missile flight stabilization and control. Ph.D. Thesis. In: Indian institute of technology, Kanpur, India
8. Misra A, Ghosh AK, Ghosh K (2008) Cascade fins—an alternate tail stabilization unit. In: AIAA atmospheric flight mechanics conference and exhibit, vol 6884. Honolulu, Hawaii
9. Misra A, Ghosh AK, Ghosh K (2008) Stability and control of aircraft bombs at high angle of attack using cascade fins. In: NCAAT proceedings, pp S1.3/1–S1.3/7
10. Miller MS, Washington WD (1998) Experimental investigations of grid fin aerodynamics: a synopsis of nine wind tunnel and three fight tests. In: Proceedings of the NATO research proposal, Page 21 RTO-MP-5. Missile aerodynamics, NATO research and technology organization
11. Miller MS, Washington WD (1993) Curvature and leading edge sweep back effects on grid fin aerodynamic characteristics. AIAA
12. MIME User Manual (2015) Metacomp technologies
13. Munawar S (2010) Analysis of grid fins as efficient control surface in comparison to conventional planar fins. In: 27th International congress of the aeronautical sciences
14. NWTF Faculty & Management. http://www.iitk.ac.in/infocell/flier/nwtf1.pdf
15. Salim SM, Cheah S (2009) Wall y^+ strategy for dealing with wall-bounded turbulent flows. In: Proceedings of the international multiconference of engineers and computer scientists, Hong Kong
16. SpaceX Falcon 9. http://www.spacex.com/falcon9
17. Spalart PR, Allmaras SR (1992) A one-equation turbulence model for aerodynamic flows. In: AIAA 30th aerospace sciences meeting and exhibit, Reno, NV, USA
18. Zaloga S (2000) The Scud and other Russian Ballistic Missiles, New Territories. Concord Publications Co., Hong Kong

CFD Investigation of Supersonic Free Jet from a CD Nozzle

S. Sathish Kumar and C. Muthuraj

1 Introduction

Knowledge of jet mixing aerodynamics is vital to several areas of commercial and military aircraft design, such as jet propulsion efficiency, propulsion integration, aero-acoustics and jet interference with aircraft structure. Once jet flow leaves the nozzle, it becomes a free shear layer, and action of turbulence dominates flow developments farther downstream. Therefore, jet flow contains large interactions with surrounding medium. These combinations include turbulent mixing and compressibility effects such as isentropic expansion and shock, and as such jet flow properties are difficult to measure or predict analytically.

Further, shocks formed near the nozzle exit may reflect repeatedly at the sonic line in the shear layer. Although the convicted turbulence interacts with shocks in the jet, the position of the reflected shock depends on the location of the sonic line in the turbulent shear layer. Such interdependence of flow interactions can become very complex.

In this paper, work has been carried to investigate supersonic jet flow field mixing with surrounding medium using commercially available software FLUENT 12.1 v. Further study has been carried on a selection of turbulence model for accurate prediction of supersonic jet flow field as it mixing with surrounding medium.

S. Sathish Kumar (✉) · C. Muthuraj (✉)
Aeronautical Development Agency, Bangalore, India
e-mail: sathishkumar@jetmail.ada.gov.in

C. Muthuraj
e-mail: muthuraj@jetmail.ada.gov.in

S. Singh et al. (eds.), *Proceedings of the International Conference on Modern Research in Aerospace Engineering*, Lecture Notes in Mechanical Engineering,
https://doi.org/10.1007/978-981-10-5849-3_14

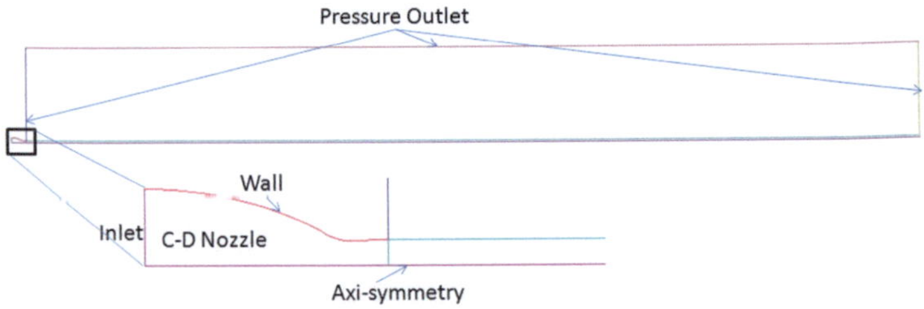

Fig. 1 Computational domain

2 Numerical Method

For CFD analysis, we consider 2D nozzle geometry details from NASA/TM—2004–212391. Coordinates of the nozzle geometry details are available in the test report [1]. Exit diameter of the nozzle is 1.0 inch at which the flow leaves from the nozzle. Figure 1 shows 2D flow domain used for the CFD analysis.

2D flow geometry was modelled assuming the axisymmetric condition. The computation domain was shown in Fig. 1 and was approximately 300D downstream of the nozzle exit, and the top plane was about 30D. The boundary condition for the nozzle inlet was set to pressure inlet with a prescribed total pressure, static pressure and total temperature. The turbulence intensity and length scale at the nozzle inlet kept default values. The far-field upstream and outlet boundary were set as pressure outlet with a prescribed ambient boundary condition provided in the report NASA/TM—2004–212391 [1].

3 Mesh

Meshing is accomplished in ICEM CFD, Version 12.0–1 Ref. [12]. ICEM CFD is a ANSYS tool used for modelling and meshing geometries. The reason for this tool selection is because of its easiness and powerful, unstructured grid generation technique for non-turbo machinery components. It has the unique capability to generate unstructured mesh with good quality. Figures 2 and 3 show computational

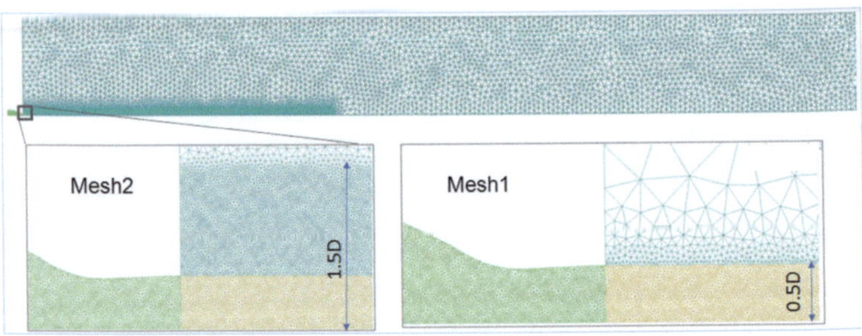

Fig. 2 Computational mesh

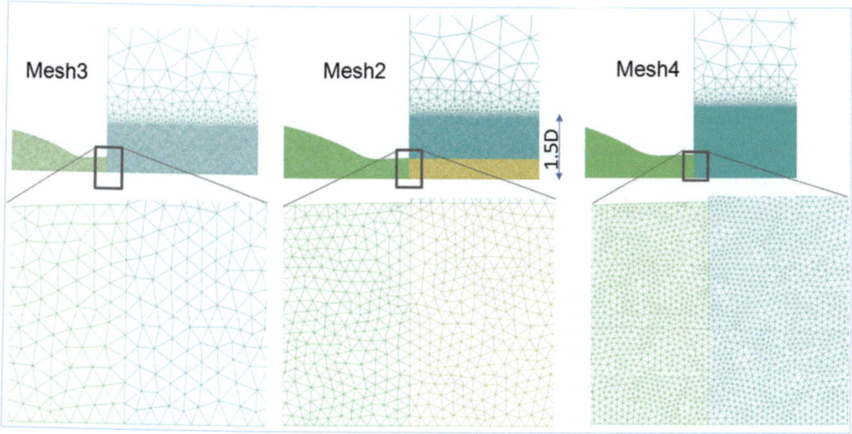

Fig. 3 Mesh sensitivity

mesh and sensitivity mesh, respectively. Details of these meshes will be discussed in Results and Discussion in detail.

4 Boundary Conditions

Mach number at the inlet of the nozzle is approximately 1.4, and the inlet flow is assumed to be axial. Total pressure, total temperature, and static pressure were 312.7 KPa, 300K, and 98.24 KPa respectively at the nozzle inlets. Slip conditions applied to the wall. Static pressure of 98.93 kPa was set at outlet boundary condition.

5 Results and Discussion

5.1 Importance of Mesh Height

For an accurate prediction of the flow field, sufficient grid density must be provided in the mixing region. The grid is divided into two domains: a high-grid density inner domain and near the jet flow, and an outer domain with reduced grid density to cover the free-stream domain away from the jet flow. Unstructured grid has been used in both inner domain and outer domain to accumulate a wide range of jet flow. Significantly less grid density is defined in the outer domain than the inner domain. Furthermore, the adaptive grid procedure provides adequate grid densities to support accurate computations in the jet shear layer and in regions near a shock front.

In the first mesh (Mesh1), height of the fine grid is considered as 0.5D with 1st cell size of 0.025 as shown in Fig. 2. Total no of cells for Mesh1 came to approx 0.6 million cells. In the second mesh (Mesh2), height of the fine grid is considered as 1.5D with the 1st cell size of 0.025 as shown in Fig. 2. Total no of cells for Mesh2 came to approx 1 million cells. In the third mesh (Mesh3), height of the fine grid is considered as 2.5D with the 1st cell size of 0.025, but this mesh has failed to get the converge solution. Further investigation carried on Mesh1 and Mesh2 to see the effect of height of fine grid on supersonic jet flow field.

All CFD computations have been performed in FLUENT 12.1 v Ref. [13] using Spalart–Allmaras turbulence model. Results of velocity, temperature and density plotted along the centre line of the axis are shown in Fig. 4 between Mesh1 and Mesh2. Further velocity profile showed in radial direction from the axis is shown in Fig. 5. Each radial profile is separated with distance of 250D between two test radial profiles. Results obtained with Mesh2 are relatively close to test data as compared to Mesh1. These indicate once the flow leaves from the nozzle exit which has a tendency to mix with surrounding atmosphere, so it requires sufficient mesh elements to capture the mixing layers. Mesh2 captures most of the test points and relatively close to test data as compared to Mesh1. In axial velocity plot, Mesh2 clearly captures the decay of velocity and potential core as it moves away from the nozzle exit as compared to Mesh1. In temperature plot, since the flow is unheated,

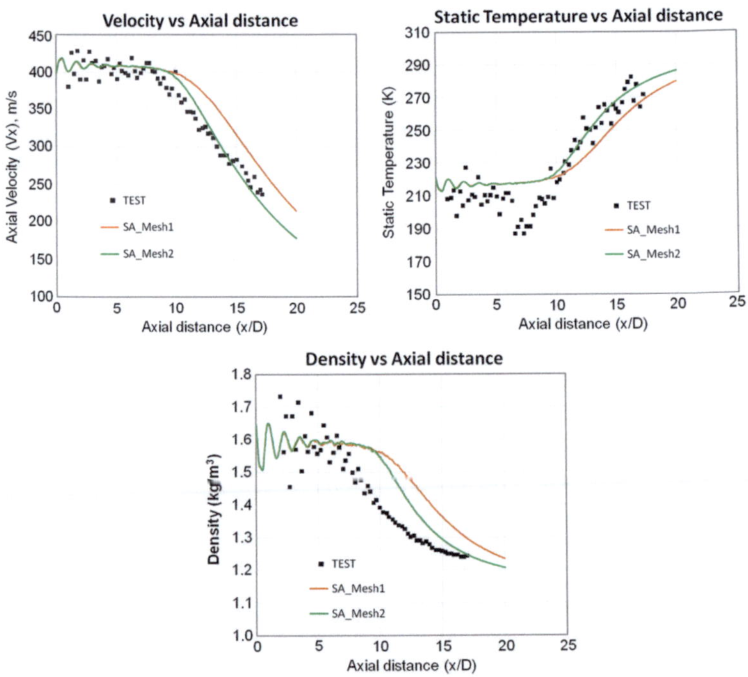

Fig. 4 Velocity, temperature and density along the centre line of the axis

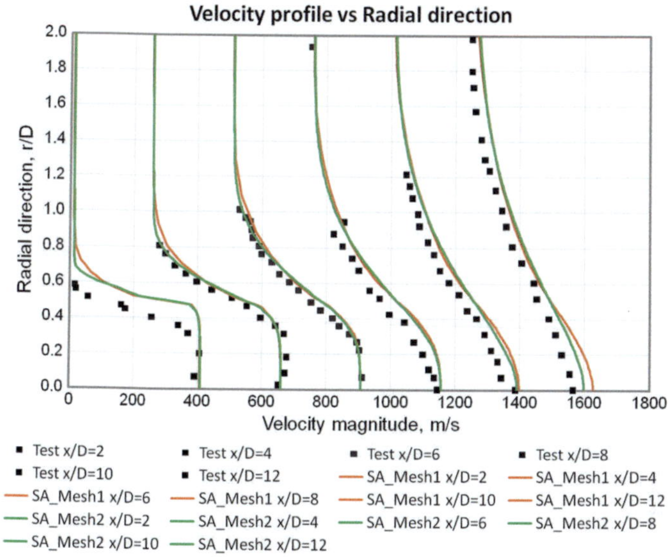

Fig. 5 Velocity distribution along the radial direction

low temperature leaves from the nozzle exit as it mixes with ambient condition and started raising the temperature consequence drop in density. Mesh2 captures the low-temperature profile compared to Mesh1; however, Mesh2 also could not capture the sudden drop in the temperature in the region between x/D of 5–10. CFD cannot capture the local drop in the temperature profile. In density plot, compared to Mesh1, Mesh2 is relatively close to test data. However, Mesh2 also could not capture the region between x/D of 7.5–15, suspecting SA model could not capture local shock effect appeared on flow field, and further details of density deviation will be discussed in the selection of turbulence model. Hence, for further investigation, 1.5D of height of Mesh2 has been used for all CFD simulation.

5.2 *Mesh Sensitivity*

Mesh sensitivity study has been made to arrive a suitable grid at which results will not vary with further refining the grid. For supersonic jet, sufficient grid density is required to capture the high velocity gradient regions in the shear layer and in free shear layer, and also sufficient grid elements required to capture the high pressure gradient regions near shock fronts. The number of grid points in each direction of an unstructured grid is fixed. Local grid density can be varied by redistributing the available grid points in the computational domain to predict the flow characteristics such as velocity gradients, density and temperature.

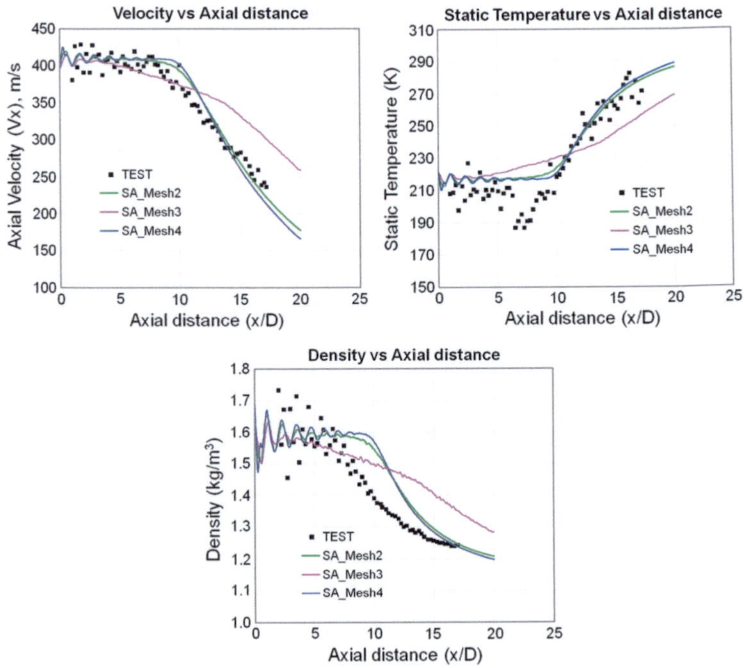

Fig. 6 Effect of mesh on velocity, temperature and density

For arriving suitable grid, three meshes have been used for CFD simulations. Firstly, Mesh2 with 1st cell size of 0.025 is considered as shown in Fig. 3. Hence, total no of cells for Mesh2 came to approx 1 million cells. Secondly, Mesh3 with 1st cell size of 0.05 is considered as shown in Fig. 3. Hence, total no of cells for Mesh3 came to approx 0.5 million cells. Thirdly, Mesh4 with 1st cell size of 0.0125 is considered as shown in Fig. 3. Hence, total no of cells for Mesh4 came to approx 2 million cells. CFD computations for all these three meshes have been performed in FLUENT 12.1 Ref. [13] v using Spalart–Allmaras turbulence model. Results of velocity, temperature and density plotted along the centre line of the axis are shown in Fig. 6. Further velocity profile showed in radial direction from the axis is shown in Fig. 7. From the results, it was shown that no considerable difference was obtained between Mesh2 and Mesh4. Further simulations were continued with Mesh2 to have the advantage of lesser number of cells.

5.3 *Selection of Turbulence Model*

CFD analysis has been carried out using RANS equations. Consider three turbulence models such as Spalart–Allmaras (SA), K-epsilon (KE) and SST by keeping

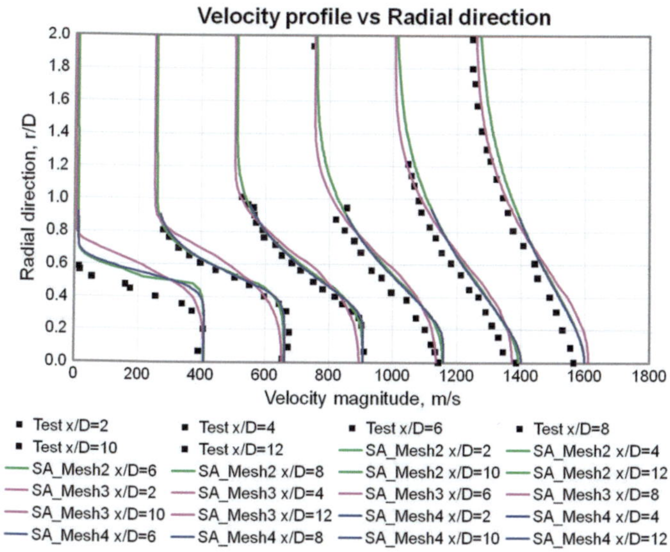

Fig. 7 Effect of mesh on velocity profile in radial direction

default values of turbulent viscosity. These models are validated for the prediction of turbulent mixing with free shear flows. The centreline characteristics and velocity radial profiles in the supersonic plumes of M = 1.4 CD nozzles are shown in Figs. 8 and 9. From density plot, we understood that apparently a weak shock formed at the nozzle exit (i.e. in the initial region of the jet plumes). Such patterns are not clearly distinguished in the velocity and temperature plot due to the random uncertainty.

Spalart–Allmaras (SA) model shows good agreement with test data except density compared to KE and SST models. SA model captures the velocity and temperature profile with test data but deviated in density profile, this is because of pressure gradient which is not captured correctly due to the presence of weak shock using SA model and its effect is shown in density profile. The velocity radial profiles close to the nozzle exit are distorted due to the presence of shock, which was captured through SA model. Further this model deteriorates to capture velocity decay and potential core as the jet moves away from the nozzle exit.

In K-epsilon model, it captured velocity and density profile with test data but deviated in temperature profile. This indicates that temperature rise or drop through mixing with surrounding medium is not captured through KE model. This model captures the flow field from the exit of the nozzle as it exposed to atmosphere causes decay of velocity and density profile. SST model, could not capture the velocity and temperature profile while SA and KE models could properly capture that. However, it captured centreline density flow variation from the nozzle exit. SST model apparently captures density flow field due to presence of shock, near the

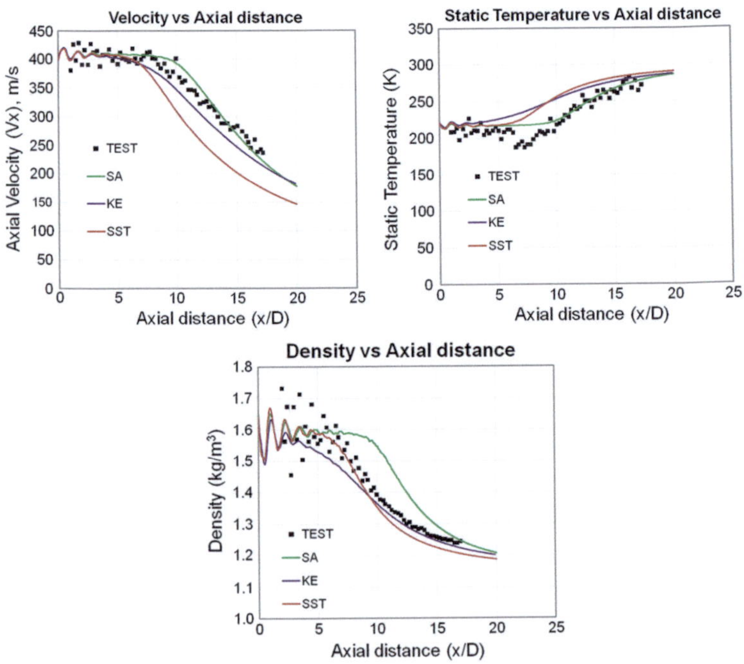

Fig. 8 Effect of turbulence model on velocity, temperature and density

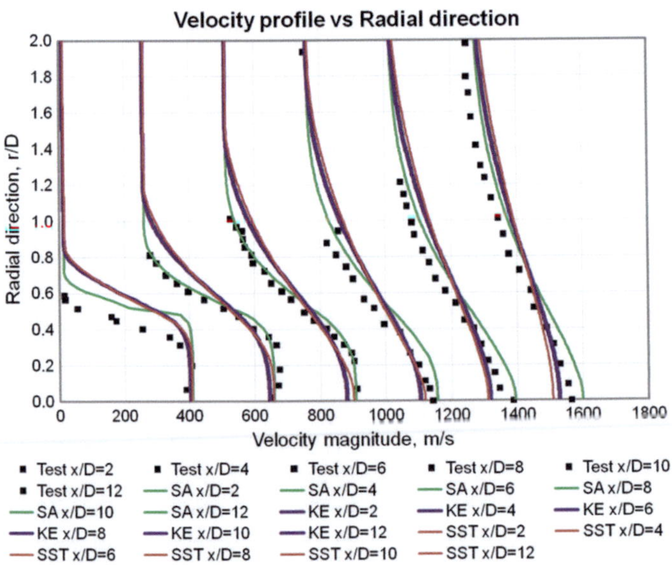

Fig. 9 Effect of turbulence model on velocity profile

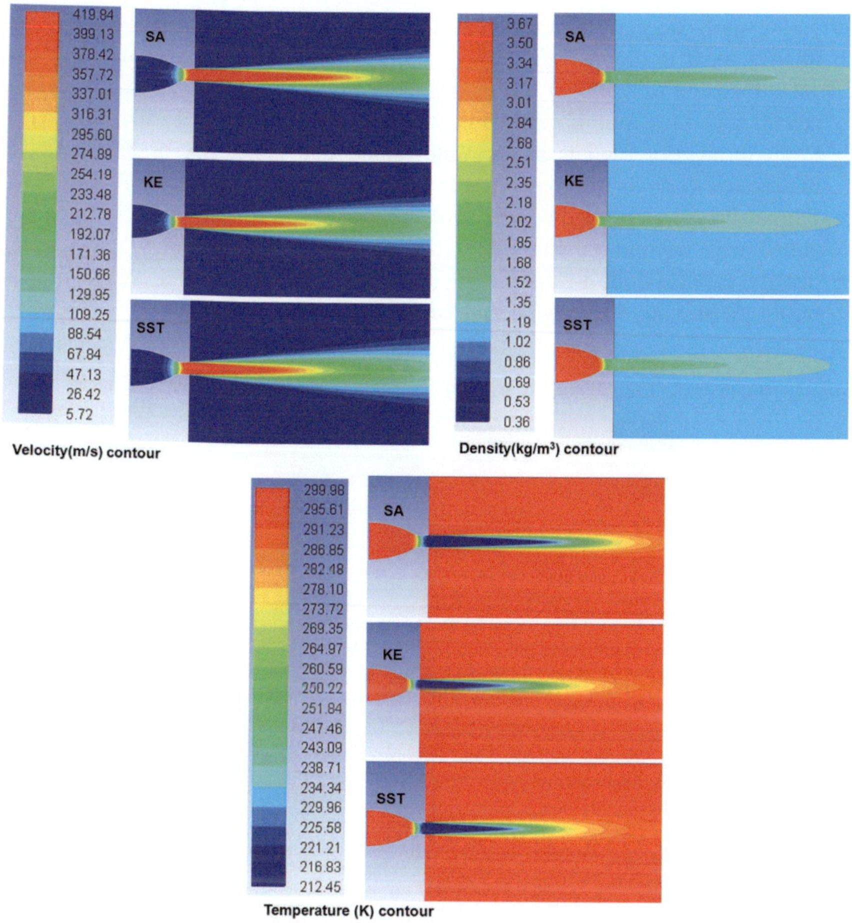

Fig. 10 Contours of velocity, density and temperature

nozzle exit, but this model deteriorate in velocity radial profile near mixing region. Contours are also been shown in Fig. 10 to understand the flow field between turbulence models.

6 Conclusion

Flow jet simulation study has been made to understand the flow field as it mixes with the surrounding medium downstream of the nozzle exit. Quantitatively, the trends of flow field such as velocities, density and temperature downstream of the nozzle were captured using CFD with test data. Based on the present studies, the following conclusions can be derived.

1. 1.5D mesh height of refined grid in the normal direction shows good improvement in the results.
2. 1st cell size of 0.025D has been considered for the CFD simulations to take the advantage of less no. of cells compared with fine grid of 0.0125D cell size.
3. From turbulence models, it was understood that either SA or KE model is preferred model for free jet flows if the shock was not been dominated flow. Further these models show good agreement in capturing centreline velocity and temperature beyond the nozzle exit, whereas SST model captured the centreline density variation due to the presence of shock, but this model decays in capturing the mixing flow region and beside the potential core.

Acknowledgements The authors are thankful to Shri. Rajkanwar Jolly, Group Director (Naval System Engg) and management of Aeronautical Development Agency for supporting this research and permitting the presentation to conference.

References

1. Panda J, Seasholtz RG (2004) Velocity and temperature measurement in supersonic free jets using spectrally resolved rayleigh scattering. NASA TM—2004-212391
2. Reichardt H (1942) Gesetzmässigkeiten der freien Turbulenz. VDI-Forschungsh 414
3. Görtler H (1942) Berechnug von Aufgaben der freien Turbulenz auf Grund eines neuen Näherungsansatzes. Z. Angew. Math Mech 22(5):244–254
4. Warren Jr WR (1957) An analytical and experimental study of compressible free jets. Publ. No.: 23,885, Univ. Microfilms, Inc.
5. Abramovich GN (1963) The theory of turbulent jets. M.I.T. Press
6. Love ES, Grigsby CE, Lee LP, Woodling MJ (1959) Experimental and theoretical studies of axisymmetric free jets. NASA TR R-6. (Supersedes NACA RM L54L31 by Love and Grigsby; RM L55J14 by Love; RM L56G18 by Love, Woodling, and Lee; and TN 4195 by Love and Lee.)
7. Dash SM, Wolf DE (1984) Fully-coupled analysis of jet mixing problems, Part I: Shock-Capturing Model, SCIPVIS. NASA CR-3716
8. Dash SM, Pergament HS, Thorpe RD (1980) Computational models for the viscous/inviscid analysis of jet aircraft exhaust plumes. NASA CR-3289
9. Seiner JM (1984) Advances in high speed jet aeroacoustics. AIAA-84-2275
10. Seiner JM, Norum TD (1980) Aerodynamic aspects of shock containing jet plumes. AIAA-80-0965
11. Abdol-Hamid KS, Wilmoth RG (1989) Multiscale turbulence effects in underexpanded supersonic jets. AIAA J 27:315–322
12. ANSYS, ANSYS ICEM CFD 12.1, ANSYS, inc, USA, 2015
13. ANSYS, ANSYS FLUENT 12.1, ANSYS, inc, USA, 2015

Study and Design of Golf Ball Like Dimpled Aircraft 2-D Wing and Effects on Aerodynamic Efficiency

Awadh Kapoor and R. Jaykrishnan

1 Introduction

"Featheries", a name given to the golf balls of olden days which made of leather and were stuffed with wet goose feathers. Later, the design evolved and balls began to be made of a gum from the sapodilla tree, heated to form a sphere and became to be known as gutta-percha. Smooth objects offer less resistance to the air flowing around. This, being a common notion for a very long time, the balls designed were very smooth. Over the time, many golfers noticed that old, scratched, and dented golf balls covered a longer trajectory range as compared to a new smoother ball. With dimples, a golf ball travels about twice as far as a smooth ball. This accidentally marked the invention of dimples on a golf ball. By the year 1930, dimpled golf balls were common in practice.

The phenomenon of delayed flow separation and consequent reduction in drag is of particular interest to the aerodynamics research community. The question of interest here is if introducing dimples on a golf ball greatly reduced its drag and improved its flight range, then could similar effect be achieved for surfaces like an aircraft wing or a car. A recent test conducted over a car with dimples by a few enthusiasts in the league to bust the myth about the dimple effect proved a dimpled car to be 11% more fuel efficient than a smooth one [1]. But as researchers, the need to find the science behind this effect is of utter importance. Moreover, as in case of automobiles where drag is the sole cause of concern when dealing with the

A. Kapoor (✉)
Institut Supérieur de l'Aéronautique et de l'Espace, Toulouse, France
e-mail: awadh.kapoor@euroavia.eu

R. Jaykrishnan
Manipal Institute of Technology, Manipal University, Manipal, Karnataka, India
e-mail: jaykrishnan.r@manipal.edu

S. Singh et al. (eds.), *Proceedings of the International Conference on Modern Research in Aerospace Engineering*, Lecture Notes in Mechanical Engineering,
https://doi.org/10.1007/978-981-10-5849-3_15

efficiency of vehicles, the scenario for airborne vehicles is slightly different. In case of aircrafts, it is just not the drag which is to be reduced but the effect of dimples on the lift also requires some attention. If an overall increment in the L/D ratios of the vehicle can be achieved, the dimple effect can be concluded as a favorable phenomenon for the aeronautical industry.

In order to proceed with the application of dimples on an aircraft wing, we need to analyze the cause of "dimples effect" on a golf ball. As shown in Fig. 1, the laminar boundary layer around a smooth sphere separates rapidly creating a very large wake over the entire rear face. This wake is a region of low pressure creating a pressure gradient between the rear and the front face of the dimple and hence increasing the pressure drag for the ball. On the other hand, dimples on a sphere force an early transition of boundary layer from laminar to turbulent adding energy to the flow and delaying the separation further aft. This results in a much narrow wake with a reduced pressure gradient leading to a reduction in the pressure drag around the ball. This is also referred as "tripping" the boundary layer.

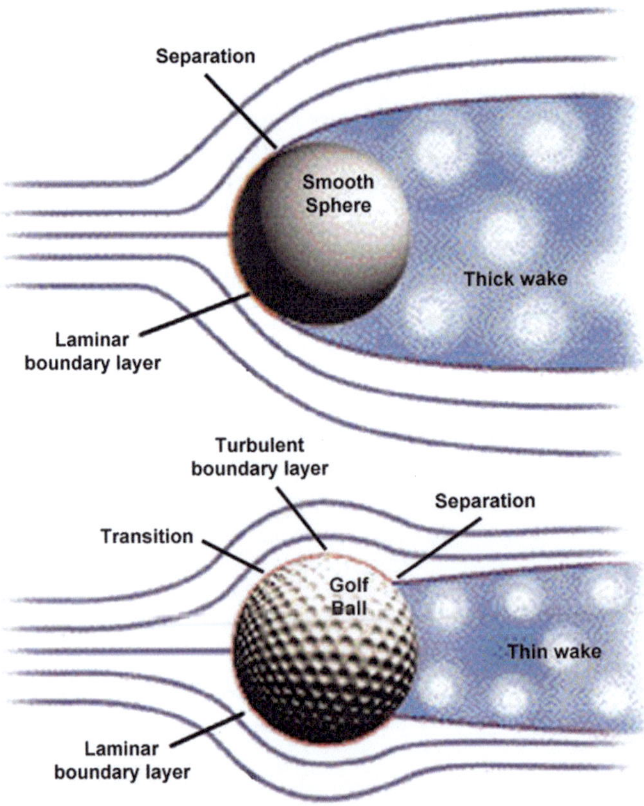

Fig. 1 Flow separation on a golf ball [1]

2 Methodology

2.1 Airfoil Selection and Modeling

The selection of airfoil for this analysis is done keeping in mind the ease of experimental data availability and its application in the industry. Going by these guidelines, NACA 2412 is selected considering its application in Cessna 172, the simplest yet advanced aircraft commonly used to many aeronautical applications for learning and research purposes. Airfoil coordinates are obtained from the UIUC database [2] and imported into ICEM CFD. The chord c of the airfoil is selected to be 1 m, and the trailing edge of the airfoil is considered to be pointed. Airfoil design being not so complicated requires basic modeling options like spline, split, join for which the use of an extensive modeling package is not required. A C-shaped domain is selected with radius of the C-curve being 10c, far field above, below, and behind being 10c, 10c and 20c, respectively. The dimple is modeled based on the information provided by various golf ball manufacturing companies according to which the depth of the dimple, its curvature, and its radius plays a major role in its aerodynamics. The lift coefficient of the golf ball increases if small dimples are added between the original dimples. When launched at small angles, golf balls with deep dimples were found to have greater lift effects than drag effects [3]. Figure 2 shows the simple parameters considered while designing the dimples over an airfoil surface, where R is the radius of the dimple curve and D denotes the span over which the dimple spreads. The location and number of dimples with respect to the airfoil chord is varied as per different models considered. A fillet radius of R/10 around the two dimple edges can also be provided while modeling the dimples so as to prevent sudden sharp variations in the geometry.

2.2 Grid Generation

The analysis deals with the effect of dimples on the separation point. This fosters a need for sufficient number of grid points to fall inside the boundary layer and hence capture the flow physics properly. Also, in the dimpled model, the grid points along the curvature of the dimple as well in downstream of the dimple should also be placed in a manner that it captures the airfoil wake in detail. This can be achieved

Fig. 2 Geometry of dimple

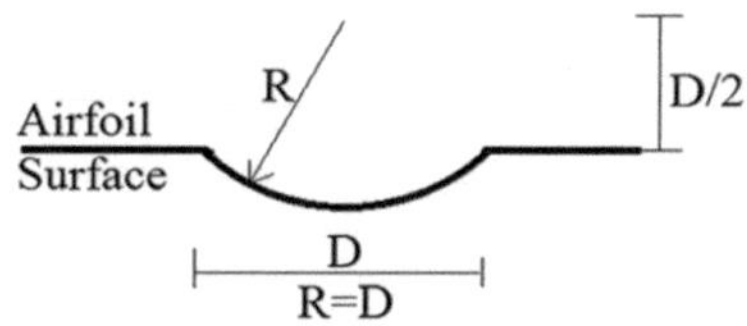

Fig. 3 Generated grid

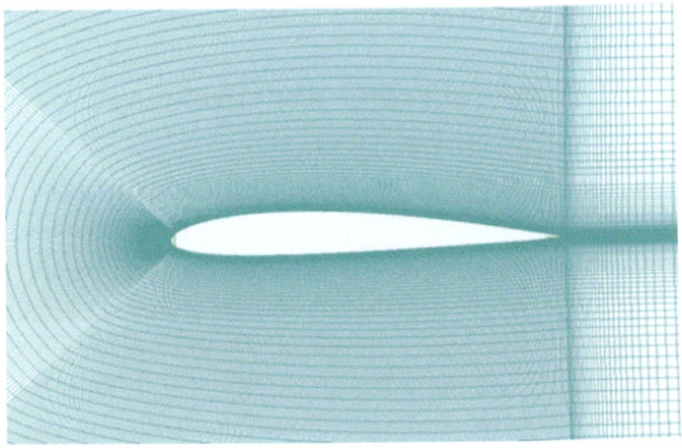

either by inducing wall functions during the problem setup or by generating a mesh which has the nondimensional y+ values close to 1 following "law of the wall" [4]. Figure 3 shows the mesh used for the analysis. The grid generation process also involved the grid dependence test for highest accuracy.

2.3 *Model, Boundary Conditions, and Solver Setup*

For the simulation, it is important to capture the phenomenon of laminar to turbulent transition for accurate drag prediction. For this, either the transition SST model or the k-kl-omega model which are available in ANSYS Fluent V15 can be used. These models facilitate the computation of both laminar and turbulent regions of the flow as well as the intermittent turbulence which is inherent with the physics of transition. By running initial simulations for a simple airfoil, the k-kl-omega model is found to produce slightly more accurate results and hence this model is selected for further analysis. More details about this model can be found in [5, 6]. The three transport equations used in this model are as follows:

$$\frac{Dk_T}{Dt} = P_{k_T} + R_{BP} + R_{NAT} - \omega k_T - D_T + \frac{\partial}{\partial x_j}\left[\left(v + \frac{\alpha_T}{\sigma_k}\right)\frac{\partial k_T}{\partial x_j}\right], \tag{1}$$

$$\frac{Dk_L}{Dt} = P_{k_L} - R_{BP} - R_{NAT} - D_L + \frac{\partial}{\partial x_j}\left[v\frac{\partial k_L}{\partial x_j}\right], \tag{2}$$

$$\frac{D\omega}{Dt} = C_{\omega 1}\frac{\omega}{k_T}P_{k_T} + \left(\frac{C_{\omega R}}{f_W}\right)\frac{\omega}{k_T}(R_{BP} + R_{NAT}) - C_{\omega 2}\omega^2 + C_{\omega 3}f_{\omega}\alpha_T f_W^2\frac{\sqrt{k_T}}{d^3} + \frac{\partial}{\partial x_j}\left[\left(v + \frac{\alpha_T}{\sigma_k}\right)\frac{\partial \omega}{\partial x_j}\right]. \tag{3}$$

Considering the need for a density-based solver, the compatible boundary conditions are chosen as pressure far field for all around the domain with values calculated using the isentropic relations. The far field values are taken for 13,500 ft with a velocity of 0.193 Mach which are the standard service ceiling and cruise Mach number values, respectively, for Cessna 172. Though the compressibility effects are low at this Mach number, the energy equation and viscous heating terms are still given sufficient importance and the case of compressible flow is also included for any minor changes in the values. The turbulence intensity is considered to 0.5% which is the standard value for low turbulence cases.

3 Dimple Application

The location as well as geometry of dimples plays a major role in governing the aerodynamic efficiency of the airfoil. As shown in Fig. 4, experimental results suggest the application of dimples near the leading edge of the airfoil for low Reynolds number operation [7]. It is noteworthy how the turbulent kinetic energy behind the airfoil has reduced after separation. There is even an alteration in the separation point due to the application of dimples. This is also analogous to the fact that for gliders, where the Reynolds number is very low, and the use of inflatable wings with symmetrical cylindrical lobes proves to be aerodynamically more efficient than the same inflatable wings without any lobes.

As in case of a golf ball which is symmetric, the effect of dimples on the lift is overpowered by the effect on drag. Magnus effect being a major cause of lift generation in a golf ball makes it a feasible option to apply dimples all around the ball. In case of airfoils, the absence of Magnus effect and the fact that application of dimples near the leading edge for high-speed flows would obviously lead to an early flow transition and a consequent reduction in the lift gives a valid reason to avoid dimple application near the leading edge for high Reynolds number flows. Left with the option of applying dimples near the trailing edge, several models are tested for CFD analysis with varying position and size of dimples as mentioned in Table 1.

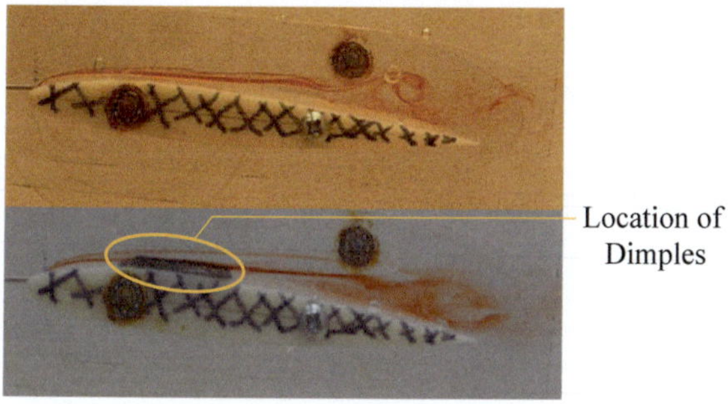

Fig. 4 Smooth airfoil (*top*), dimpled airfoil (*bottom*)

Table 1 Analyzed models

Model name	Dimple size	Location
T1FD70C	Full size	70% chord
T1FD90C	Full size	90% chord
T1HD95C	Scaled half down	95% chord
T1QD95C	Scaled quarter down	95% chord

4 Results

The initial phase of the simulation focuses on the validation of results obtained from the CFD analysis by comparing it to the experimental data available for a smooth airfoil [8]. These results were found to be satisfactory, and the percentage error was well within limits of 4–9% as shown in Fig. 5 which assures the accuracy of mesh generation and correctness of the boundary conditions used. Further, different models as mentioned in Table 1 were analyzed. As the location of the dimple was moved closer to the trailing edge, the reduction in the drag values became more significant. A similar observation could be made with the dimple size. Up to a limit, when the dimple was scaled down, it proved to be more efficient than its bigger counterpart.

It is important to note that the effect of dimples should to be studied for a wide range of angle of attack. Models T1FD70C, TIFD90C, and T1HD95C have not been analyzed for angles of attack between 0° and 9° as these models suggested positive results only for higher angles of attack. In the strive to obtain positive results for a wider range of angles of attack, the model T1QD95C was analyzed which proved to be of more significance than any of the previous models. Figure 6 (left) gives a clear picture of this interpretation. The model gives an increase in the aerodynamic efficiency for angle of attack starting from 4° and ranging up to 15° or even beyond that. Considering that the maximum value of L/D for NACA 2412 is

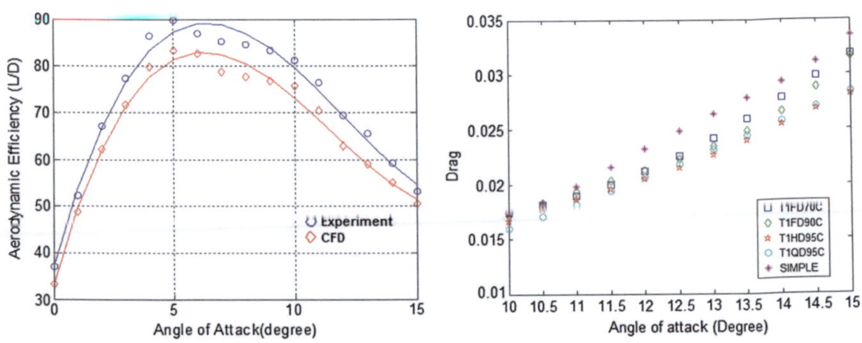

Fig. 5 Experiment versus CFD results (*left*), drag for various dimple configurations (*right*)

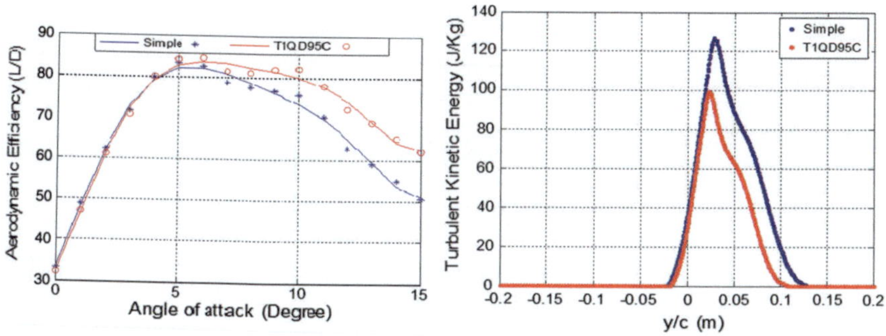

Fig. 6 Aerodynamic efficiency (*left*) and turbulent kinetic energy (*right*) for smooth and dimpled air foil (*left*)

obtained around 4°, it can be selected as the fitting angle for any arbitrary aircraft for maximum aerodynamic efficiency during the cruise phase. Apart from these models, several other models with cases considering the dimple location at the bottom of the airfoil were also analyzed which produced negative results when L/D values were compared.

The results shown in Fig. 5 (right) clearly suggest the prime effect of dimples to be reduction in the drag. The effect is more prominent at higher angles of attack as the lift generating region of the airfoil does not extend to the regions closer to the trailing edge of the airfoil. Hence, in such cases, lift values remain almost unaffected, while the drag drastically reduces due to reduction in the size of the wake. However, at low angle of attacks, where the lift generating region covers even up to 60–70% of the chord, presence of dimples in this region causes a reduction in the value of lift due to the change in pressure inside the dimple which slows the flow down in this region. The drag reduction phenomenon is still present but since the Cl values decrease drastically, the overall change in L/D is reduced. Figure 6 (Right) shows a clear proof of the reduction in wake size due to the presence of dimple near the trailing edge of the airfoil. A quantitative proof of the same is obtained by plotting the turbulent kinetic energy at a distance 0.2c from the trailing edge for a y span of −0.2 to 0.2 m.

Figure 7 gives a qualitative proof about the reduction in wake size. The contours of turbulent kinetic energy also explain the unsymmetrical behavior of its numerical values, top curve being the greater region wetted by turbulence, and the graph in Fig. 6 (Right) is rather shifted toward the positive y/c values. Thus, the results obtained are complying well with the established theory.

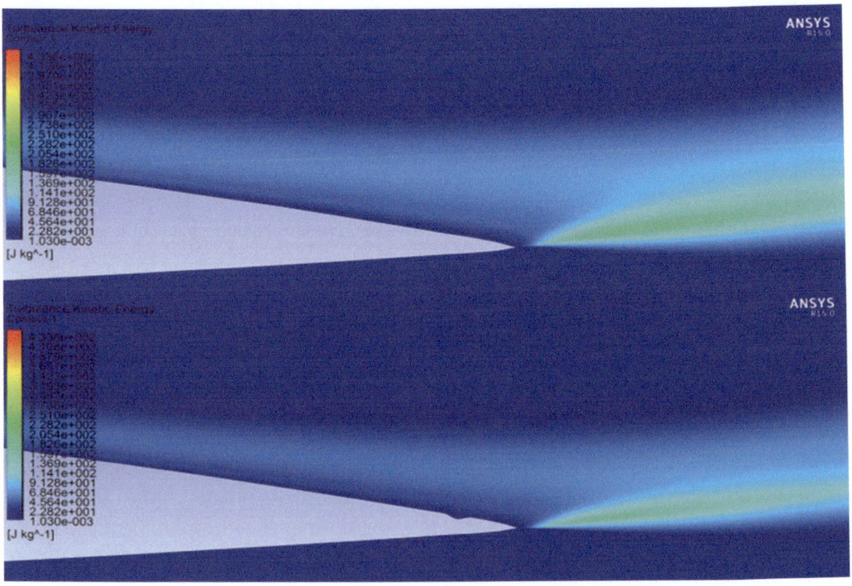

Fig. 7 Turbulent kinetic energy behind the airfoil, smooth (*top*)/dimpled (*bottom*)

5 Conclusions

The presence of dimple on the top curve of the airfoil has varied effect on the flow characteristics for difference angles of attack. The application is found very useful for high angles of attack application where stall is delayed due to increased L/D. At low angles of attack, optimizing the location and size of dimples can lead to an improved L/D which facilitates a shorter takeoff. The need for high lift devices might also be reduced due to this effect. Also the fuel efficiency of the aircraft could be improved by a higher L/D during cruise at low angles of attack. The only challenge to this application is the fabrication of such dimpled panels from composites. Advancements in the composite technology shall play a major role in the feasibility of this concept for the aeronautical industry.

References

1. Chear CK, Dol SS (2015) Vehicle aerodynamics: drag reduction by surface dimples. Int J Mech Aerosp Ind Mechatron Eng 9(1):202–205
2. http://m-selig.ae.illinois.edu/ads/coord_database.html
3. Leong J-C, Lin C-Y (2007) Effects of golf ball dimple configuration on aerodynamics, trajectory and aeroacoustics. J Flow Vis Image Process 14(2):183–200
4. Davidson PA, Nickels TB, Krogstad P-Å (2006) The logarithmic structure function law in wall-layer turbulence. J Fluid Mech 550:51–60

5. Keith WD, Cokljat D (2008) A three-equation eddy-viscosity model for reynolds-averaged Navier–Stokes simulations of transitional flow. J Fluids Eng 130(12)
6. Di Pasquale D, Rona A, Garrett SJ (2009) A selective review of CFD transition models. In: 39th AIAA fluid dynamics conference
7. Vento JL (2011) Analysis of surface augmentation of airfoil sections via flow visualization techniques. California Polytechnic State University, San Luis Obispo, CA, 93405. http://digitalcommons.calpoly.edu/aerosp/38/
8. Anderson JD (2001) Fundamentals of aerodynamics. McGraw-Hill, Boston

Computational Investigation of the Effect of Various Spike Geometries Mounted on Supersonic Vehicles

Aishwarya Kushary, Anjali Chopra, Jayanta Sinha and Vamsikrishna Undavalli

Nomenclature

C_d	Coefficient of drag
l	Length
D	Diameter
l/D	Ratio of spike length and fore-body diameter
s/D	Ratio of axial length and fore-body diameter
B	Blunt body
S1-7.5	Spike of diameter 1″ @ 7.5° semi-cone angle
S2S'1-7.5	Spike 1 of diameter 2″ and Spike 2 of diameter 1″ @ 7.5° semi-cone angle
S3S'3-30	Spike 1 of diameter 3″ and Spike 2 of diameter 3″ @ 30° semi-cone angle
S6S'3-30	Spike 1 of diameter 6″ and Spike 2 of diameter 3″ @ 30° semi-cone angle

A. Kushary · A. Chopra · J. Sinha (✉) · V. Undavalli
Amity Institute of Aerospace Engineering, Amity University,
Noida 201303, Uttar Pradesh, India
e-mail: jsinha1@amity.edu

A. Kushary
e-mail: aishwarya.kushary@gmail.com

A. Chopra
e-mail: anjalichopra1994@gmail.com

V. Undavalli
e-mail: u.vamsikrishna@yahoo.com

S. Singh et al. (eds.), *Proceedings of the International Conference on Modern Research in Aerospace Engineering*, Lecture Notes in Mechanical Engineering,
https://doi.org/10.1007/978-981-10-5849-3_16

1 Introduction

Reduction of drag and aerodynamic heating levels for supersonic and hypersonic vehicles, such as interplanetary mission vehicles, missiles, is vital as they may lead to structural damage as well as lower the fuel efficiency. A strong shock wave is created due to high-speed flows. The flow, during reentry of the vehicle, loses most of its kinetic energy as it passes through the strong shock ahead of the fore body. This causes the temperature and pressure to rise. This temperature rise is known as 'aerodynamic heating,' and the increased pressure downstream of the shock contributes to high-pressure drag. Aerodynamic drag reduction results in greater fuel efficiency thereby reducing propulsion system requirements and maximizing payload capacity. Reducing aerodynamic heating eliminates the requirement of a complex thermal protection system and increases volumetric efficiency. But through several years of research, it is observed that the two go hand in hand and must be optimized as per the mission profile. For reentry vehicles, which are subjected to considerably high temperatures, reduction of aerodynamic heating should be the primary concern as opposed to the increase in drag. Minor changes in the vehicle fore-body structure could prove helpful in achieving required values of the drag coefficient. It may also prevent ablation to an extent by altering the flow field in the vicinity ahead of the front nose hence reducing high drag and heat levels. This may be done in various ways depending upon the free stream Mach number and geometry of the spike, mainly, its *l*/D ratio. At high Mach numbers, there is formation of shock wave/s which results in abrupt increment in pressure, density, and temperature. By varying the geometry of the fore body, the flow field around it can be altered. A blunt fore body induces a bow shock detached from the body at supersonic Mach numbers. As the air passes through this strong shock, its temperature, pressure, and density rise sharply as its velocity falls. Most of the reentry vehicles use a blunt hemispherical fore body, and it has proven helpful in reducing high levels of aerodynamic heating by creating a strong detached bow shock upstream of the flow and distributing the heat over a larger area. Another reason for choosing blunt noses at hypersonic Mach numbers is that they give efficient usage of the available volume—better accommodation of aircraft instruments and crew. A slender or sharp fore body creates a series of weaker oblique shock waves and is characterized by significant drag reduction. Oblique shocks are attached shock waves created as the flow encounters a sharp object or a deviation into the flow at some angle to the incident flow direction. A special case of oblique shock is a 'normal shock' which is normal to the incident flow field. For cases where drag is a greater concern than aerodynamic heating, slender bodies are used to prevent the formation of bow shock.

A shock wave, when created, forms a region of recirculation between the body and the wave where the flow is unsteady. This recirculation zone is enveloped by shear layer which results in drop in surface pressure and temperature. Experimental and computational studies are carried out in order to study the behavior of steady as well as unsteady flows. Typically, the computations involving unsteady flow regimes are a bit complex. This is because unsteady flows consider the time function unlike steady flows. With time steps in the picture, formulations have to be changed to obtain desired results. In transient flows as such, all flow parameters like temperature, pressure, and density are time dependent.

A spike is a thin cylindrical rod with a sharp head which is mounted at the stagnation point of the blunt body where the static pressure is maximum and the flow velocity becomes zero. The spike and spike-head could be of fixed or variable length and diameter. A spike creates weaker oblique shock waves instead of a strong bow shock. The foremost shock is referred to as 'foreshock.' Moreover, it acts as a flow separator and encourages separation of the boundary layer at the surface of the body, creating a shear layer. The shear layer, as the flow advances, propagates downstream of the flow field and attaches itself to the blunt body surface. This is known as the reattachment shock or recompression shock. The reattachment elevates the temperature, pressure, and density locally but results in an overall reduction in drag and aeroheating. To ensure that these parameters do not attain significantly high values, the geometry of the spike has to be optimized. The effectiveness of the spike can be further enhanced by changing the spike tip design. Since 1940, when the idea of mounting a spike was initially conceived, it has been studied extensively by many aeronautics and aerospace researchers. Some of them focused on studying the effects of spike at various Mach numbers and Reynolds's number, whereas some adopted a completely different approach by keeping the Mach number and Reynolds's number constant and changing the geometry of the spike. Various experiments like flow over a sharp pointed spike tip, hemispherical aerodisk mounted on a probe, flat aerodisk, hemispherical probe were conducted. It was observed that changing just the configuration of the spike could alter the flow field and spikes at high angles of attack did not yield desirable results. Hence, optimization has been a vital aspect in this research.

The main objective of this paper, however, remains to analyze the flow over five spike shapes, to assimilate the outcome in order to illustrate the significance of modified fore-body designs for reentry vehicles at Mach 2. Several parameters like drag, coefficients of pressure have been studied for various cases to draw distinctions. CAD models were generated for better visualization of the geometrics. Figure 1 shows the scaled models of all the configurations.

A blunt fore body, B, was created of length 22 mm and radius of the blunt hemispherical nose 7.5 mm; the diameter of the fore body is 15 mm. The geometry used for S1-7.5 is such that the fore body has a length of 14.5 mm and its diameter is 15 mm. The spike runs from the middle of the nose of the fore body, and its length is 15 mm. The diameter of the spike is 1 mm, and the semi-cone angle of the tip is 7.5°. For S2S'1-7.5, the fore body has a length of 22 mm and its diameter is 15 mm. The first spike runs from the middle of the nose of the fore body. The diameter of the first spike is 2 mm. The second spike starts where the first spike tip ends. The diameter of the second spike is 1 mm. The combined length of both the spikes is 15 mm. The semi-cone angle is 7.5° for both spikes. For the S3S'3-30 geometry, the semi-cone angle is 30° for both the spike-heads. Both the spike-heads have a diameter of 3 mm. The shaft of the spike is 1 mm in diameter. The distance

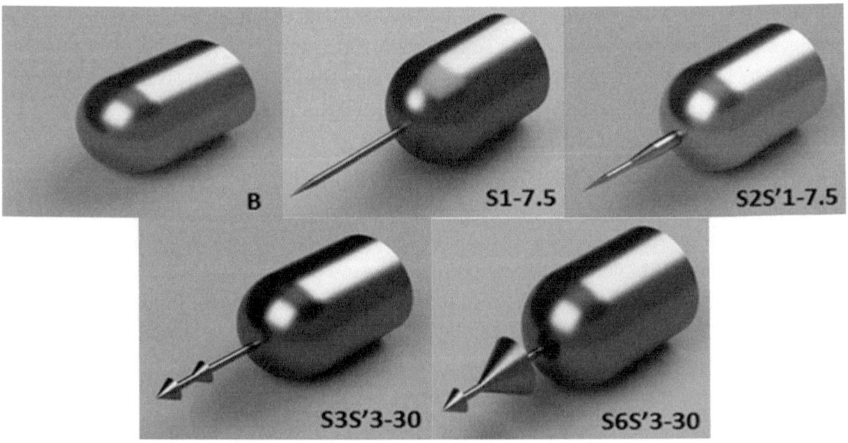

Fig. 1 Model configurations

between both the spikes is 3 mm, and the fore-body length is 14.5 mm. For the S6S'3-30 geometry, the semi-cone angle is 30° for both the spike-heads at a distance of 3 mm from each other. The first spike-head has a diameter of 6 mm, and the second spike-head has diameter of 3 mm. The length of the fore body was maintained at 14.5 mm.

2 Literature Review

In a comprehensive study by Ahmed and Qin [1] in 2011, the authors obtained interesting results by adding a disk fore body to the spike. The assembly of the spike-aerodisk varied from 0.5D to 2.5D, D being the diameter of the main body. The results showed that drag reduction depends on the size and the increase in spike length causes the reattachment point to move downstream but the strength of the shock decreases. Elsamanoudy et al. [2] in 2013 demonstrated drag reduction by the means of a spike-disk-spike regime in the sense that the flow encounters a spike first followed by a blunt aerodisk mounted on another spike followed by the main body. The spike diameter was set at 0.1D, L/D at 1 and simulated in Mach numbers 6, 8, and 10, where L/D is the non-dimensionalized spike length. Spiked aerodisk led to about 56% reduction in the coefficient of drag. The problems related to aerothermodynamics were taken up by Yadav et al. [3] using a series of two spikes with hemispherical caps for nine different configurations at Mach 3. In the cases where the overall length of the spikes was greater than 1.5 times the diameter of the base body, there was a maximum of 40% decrease in heat flux and about 20% reduction in drag. Roveda [4] in 2009 analyzed two models with different spike configurations. His research showed that a difference in surface pressure increased in both the models but the temperature contours remained almost the same. Gilinsky [5] in 2002 conducted a research on spike-nosed bodies in supersonic flows and concluded that aerodynamics of the body can be controlled by integrating one or more needles of the same dimensions on the nose of the blunt body. Decrease in overall drag was observed for the five-needle case, but it also led to an increase in lift as opposed to the single-needle geometry. Sahoo et al. [6] in 2013 carried out computational studies on the effect of a spike on the flow field over blunt body at supersonic speed of Mach 2 to obtain both steady and unsteady flows and observed that the sharp spiked fore body led to reduction in drag by about 31%. Milicev and Pavloviv [7] in 2002 conducted an experiment on a cylindrical model with four different types of spike for Mach number 1.89. They observed that the model without spike gave maximum drag coefficient and least lift coefficient; the model with blunt cylindrical spike was rendered most effective as it gave least drag coefficient and maximum lift coefficient. White [8] in 1993 performed a CFD analysis on a hemispherical IR dome of diameter 5.5 in. concentrically mounted on a missile body of diameter 7 in., in Mach 4 flight at zero angle of attack for altitudes between 10 and 100KFT. He observed shear layer generation and reattachment, recirculation patterns and low dynamic pressure levels within the separated flow region, and bow shock wave geometry. Ma et al. [9] in 2015 analyzed flow over a cylinder equipped with an aerospike of L/D = 1 at Mach 5. They observed three processes in one cycle: collapse, inflation, and withhold. Harish and Rajagopal [10] in 2014 through experimental and computational analysis calculated the values of stagnation temperature for a reentry vehicle with and without spike at Mach 9.1. The CFD model gave 900 K as the bow shock temperature without spike and 625 K with the spiked model. Hence, it was deduced that the geometry with a spike ahead of the reentry vehicle was more effective in reducing aeroheating during reentry.

3 Computation

The problem was modeled using commercial software ANSYS. Geometry and mesh generation was done through Gambit. Flow solution and post-processing was performed using ANSYS Fluent, GetData Graph Digitizer, and Microsoft Excel. Axisymmetric geometries were created to reduce computational time and were meshed using structured meshing scheme. Steady analysis was done using a density-based solver with explicit algorithm to incorporate viscous effects in the governing equations. RANS-based standard k-ω scheme was adopted for turbulence modeling so that the recirculation zone is properly captured. They efficiently solve complex boundary layer flows under adverse pressure gradient.

4 Results and Discussion

The computational setup was validated based on Sahoo et al. [6], and the pressure plots obtained were in good agreement with the reference results. Validation plot is given in Fig. 2.

The pressure contours for various geometries were obtained as given in Fig. 3.

After successfully validating the computational setup for two cases, namely B and S1-7.5, the new geometries were subjected to supersonic flow field analysis and it was observed that with changing geometry shapes, the stagnation point was shifted and a significant pressure drop was obtained as shown in Fig. 4 and Table 1.

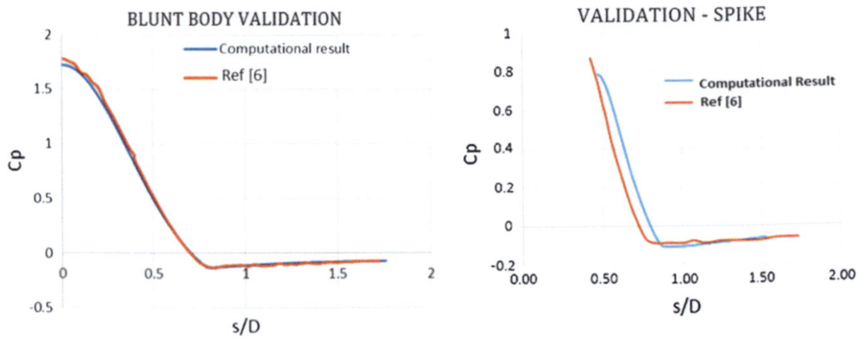

Fig. 2 Validation plots—blunt and spike

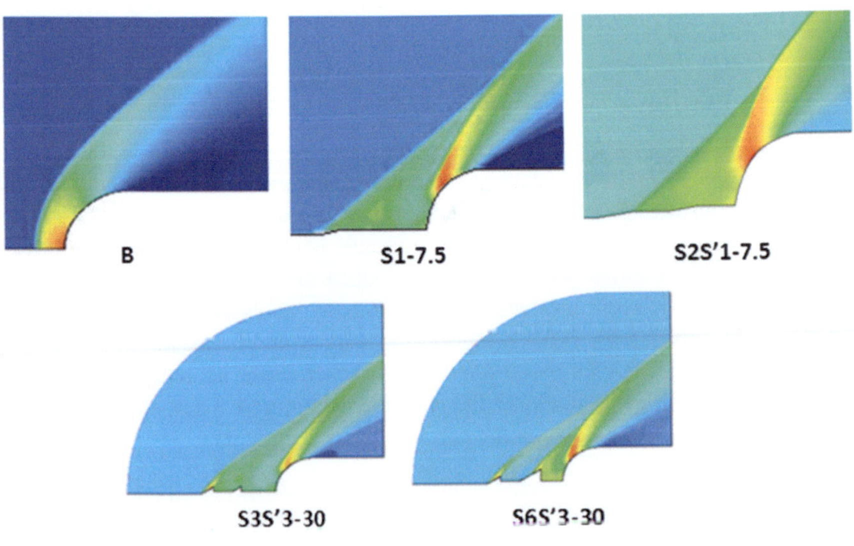

Fig. 3 Pressure contours on spiked body with change in spike shapes

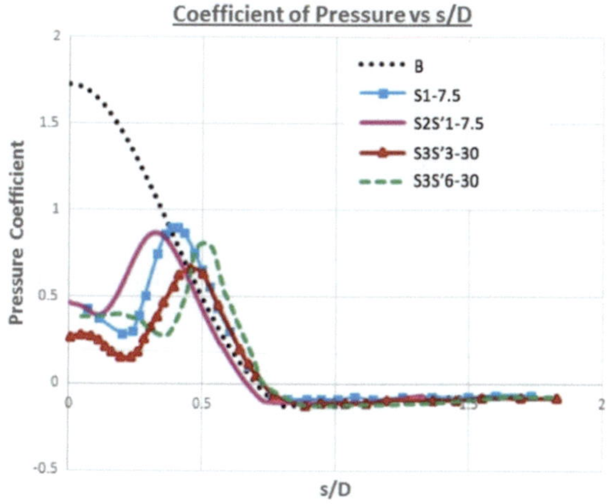

Fig. 4 C_p distribution on the axisymmetric model

Table 1 C_d obtained for different spike geometries

Geometry	C_d
B	0.8
S1-7.5	0.4
S2S'1-7.5	0.56
S3S'3-30	0.419
S6S'3-30	0.499

This shift in stagnation point was due to the formation of shear layer and recirculation zone. In cases of spike geometries, the weaker oblique shocks reattached themselves to the blunt fore body creating a high temperature and pressure region. The double-spike model gave interesting results where the pressure coefficient was even smaller but slightly higher drag coefficient. This is because of the intermixing of the two recirculation zones formed by the two spikes. There was also the formation of very weak oblique shock at the nose of the spike. This result may not be effective in missile bodies but has a great scope for reentry vehicles where optimum drag is needed for retardation during atmospheric reentry.

The net values of coefficient of drag hence obtained for all cases are given in Table 1.

The semi-cone angle vastly affects the recirculation zone formed around it. As the semi-cone angle increases in the case of S3S'3-30 and S6S'3-30, the recirculation zone weakens. This leads to decrease in pressure in the region which ultimately leads to lower drag generation. However, on comparing S3S'3-30 with S6S'3-30, we see that the C_d is more in the latter case. This is because of its design configuration. The diameter for the second spike in S6S'3-30 is more than that of S3S'3-30. Thus, the recirculation zone spans out more, shifting the reattachment point further downstream. This retards the flow field. Figure 4 gives the C_p distribution on the axisymmetric body with different types of spikes.

5 Conclusion

From the above results and discussions, the drag in the case of S1-7.5 has reduced by 50% and in the case of S2S'1-7.5 it has reduced by 30% as compared to the blunt body. The S2S'1-7.5 hence gives an intermediate value of C_d which helps in controlling the disadvantages caused by a body producing severely low drag. The intermediate value of C_d in the case of double spike is because pressure rise across shock takes place. This, in turn, shifts the reattachment point farther downstream. Reattachment of shock hugely depends upon the recirculation zone. The recirculation zone is of vital importance because if the pressure in this zone is high, reattachment point will shift toward the fore body. This can be quite harmful as it would completely nullify the point of attaching a spike to the geometry. Hence, an intelligent compromise has to be made while deciding the semi-cone angle as well as the length of the spike protruding out.

6 Future Scope

The future of this project lies in researching more for unsteady flow behavior for the geometries. Since unsteady flows are entirely time dependent, such a flow behavior is quite complex to solve. As all real cases, in fact, exist in unsteady flows, several different possibilities and problems get associated while dealing with these. Fluctuations in pressure of great magnitude are faced in such flows, making the computation a heavy task. Furthermore, an optimization study can be done for finding out fitting L/D ratios, which addresses the problem of aeroheating for each geometry. This means, for the same design, several length-to-diameter ratios can be modified. Acoustics is another theme which can be further studied for these optimized reentry vehicles. For this, vibration analysis has to be carried out which can be done using the concept of finite element methods. Material study can also be carried out after fetching desired computational results. For this, material sciences of alloys can be disserted which would further alleviate the requirement of having a robust thermal protection system.

References

1. Ahmed MYM, Qin N Recent advances in the aerothermodynamics of spiked hypersonic vehicles. In: Progress in aerospace studies, vol 47, no 6, pp 425–449
2. Elsamanoudy M, Ghorab A, Hendy M (2013) Drag reduction using aerodisks for hypersonic hemispherical bodies. In: Aerospace sciences & aviation technology, ASAT-15
3. Yadav R, Velidi G, Guven U (2014) Aerothermodynamics of generic re-entry vehicle with a series of aerospikes at nose. Acta Astronaut 96(1):1–10

4. Roveda R (2009) Benchmark CFD study of spiked blunt body configurations. In: AIAA paper, pp 1–16
5. Gilinsky M (2002) Spike-nosed bodies and forward injected jets in supersonic flow. In: 38th AIAA/ASME/SAE/ASEE joint propulsion conference and exhibit
6. Sahoo D et al (2013) Steady and unsteady flow over a spiked blunt body at supersonic speed (D), pp 310–314
7. Milicev SS, Pavloviv MD (2002) Influence of spike shape at supersonic flow past blunt-nosed bodies: experimental study. AIAA J 40(5):1018–1020
8. White JT (1993) AIAA 93-0968 Application of Navier-Stokes flowfield analysis to the aerothermodynamic design of an aerospike-configured missile Hughes missile systems company Pomona, California AIAAIAHS/ASEE aerospace design conference
9. Ma Y, Liu X, Ou P (2015) Numerical investigation of hypersonic unsteady flow around a spiked blunt-body. Proc Eng 126:163–168
10. Harish P, Rajagopal K (2014) Hypersonic winged re-entry vehicle using CFD 3(7):1177–1181

Effect of Leading-Edge Tubercle on Aerodynamic Performance of NACA 0021 Airfoil

Alok Mishra, Saravana Kumar Lakshmanan and Ashoke De

1 Introduction

Leading-edge tubercle on airfoil concept is motivated from the motion of the aquatic animal. Edel and Winn [1] were one of the first researchers who studied motion of the humpback whale and found that whale achieves the high degree of maneuverability through the high aspect ratio of their flippers. The flippers of humpback whale are having large rounded tubercles along the leading edge which act as passive-flow control devices. The leading-edge tubercles generate vortices which increase momentum exchange within the boundary layer and energize the flow over flipper. The additional momentum helps to counter the adverse pressure gradient, and the flow remains attached over the flipper which improves aerodynamic performance and maneuverability of humpback whale. Experimental analysis of the humpback whale flipper with and without tubercle modifications has demonstrated that the presence of tubercles produces a significant delay in the stall angle and increase in the total maximum lift coefficient. At post-stall region, the airfoil with tubercles shows a notable increase in the lift and decrease in the drag coefficient [2, 3].

The work done by Corsini et al. [4] was inspired by leading-edge tubercle for stall delay capability and applied the same for controlling the stall for axial fan blade. The three-dimensional numerical simulation was carried out over sinusoidal

A. Mishra (✉) · S.K. Lakshmanan · A. De
Department of Aerospace Engineering, Indian Institute of Technology Kanpur, Kanpur, India
e-mail: alokm@iitk.ac.in

S.K. Lakshmanan
e-mail: lsaravanakumar4791@gmail.com

A. De
e-mail: ashoke@iitk.ac.in

S. Singh et al. (eds.), *Proceedings of the International Conference on Modern Research in Aerospace Engineering*, Lecture Notes in Mechanical Engineering,
https://doi.org/10.1007/978-981-10-5849-3_17

leading edges on NACA-0015 and NACA-4415 airfoil with OpenFOAM using a nonlinear eddy viscosity k-ε closure model. They found that there is a significant increase in lift coefficient at post-stall region and drag coefficient shows very marginal difference as compared with the straight edge airfoil.

Although several numerical studies have been done on leading-edge tubercle with RANS models, very few studies exist in the literature which talks about three-dimensional effect and leading tubercle effect on NACA 0021 airfoil with SA [5] and k-kl-ω RANS models [6]. Efforts have been made to take a step ahead for deeper understanding of bifurcation effect of leading-edge tubercle over unmodified airfoil at post-stall region. The objective of this work is to study the three-dimensional and leading-edge tubercle effect on aerodynamics property of airfoil (unmodified NACA 0021 and modified (A2W7.5) NACA 0021 airfoil) operating at Re = 120000.

2 Computational Setup

2.1 Numerical Simulation Scheme

The numerical simulations are carried out in an open-source CFD toolbox OpenFOAM [7]. The standard incompressible solver pisoFoam is utilized for URANS computation. Second-order discretization schemes are used for discretizing all spatial and temporal terms in the governing equation. A variable time integration scheme is used to guarantee the local courant number less than 1 (CFL < 1). The convergence tolerance is set to 10^{-6} for all variables to get accurate solution at each time step. Averaging is done over 100 non-dimensional time units to obtain the mean flow quantities.

2.2 Grid Generation and Computational Domain

The computational grid for the airfoil is generated using ANSYS® ICEM-CFD [8]. The mesh is constructed using multi-block approach. The rectangular domain of size 30C × 30C is utilized in present simulations. The origin of the coordinate system is located at the leading edge of the airfoil. The flow inlet and outlet boundaries are located at 10C upstream and 20C downstream, respectively, from the leading edge of the airfoil. The top and bottom boundaries are located at a distance of 15C from the origin. The lateral distances are adequate enough to improve the stability of simulation and also to eliminate the far-field boundary effects. The chord length (C) for both the airfoils is 70 mm. The mesh for unmodified and modified (A2W7.5) NACA 0021 is shown in Fig. 1a, b, respectively. The total number of cells in the computational domain is roughly around 2.1 million. And the boundary conditions are shown in Fig. 1c.

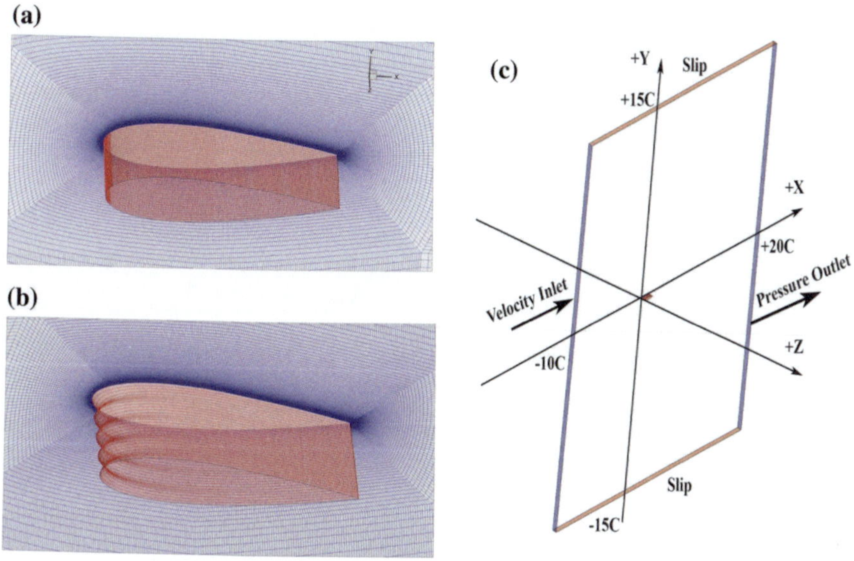

Fig. 1 **a** Unmodified NACA 0021 airfoil and **b** modified (A2W7.5) NACA 0021 airfoil and **c** geometry of the computational domain and boundary conditions

3 Results and Discussion

3.1 Lift and Drag Coefficient Results

3.1.1 Comparison Between 2D and 3D Results of Unmodified NACA 0021

The lift and drag coefficient variation of 3D numerical results is plotted with two different experimental results of Hansen et al. [9] and Jacobs [10] as shown in Fig. 2a, b. The 2D numerical results and XFOIL data of Lakshmanan et al. [11] of lift and drag coefficients are also shown in Fig. 2a, b. Between 8°–10° angle of attacks, the lift increases linearly. Unlike 2D simulation, the lift coefficient value matches well with the experimental results. In the stall region, 3D simulations are unable to capture the sudden stall as reported in experiments.

Drag coefficient versus angle of attack is plotted with experimental results as shown in Fig. 2b. Between 8°–12° angle of attacks, the SA model underpredicts the drag values. SA model is not able to predict the stall. But in post-stall region like 17° and 20°, the results are in good agreement with the experiments. The simulation results are also having good agreement with XFOIL [12] results at low angles.

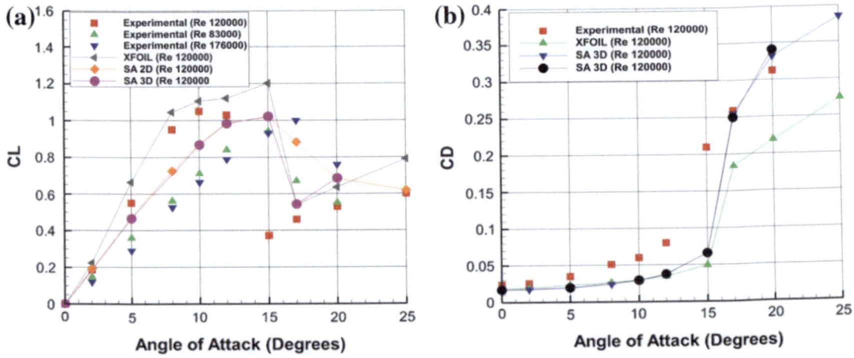

Fig. 2 **a**, **b** Lift and drag coefficients of unmodified NACA 0021 versus AOA plotted against experimental result of Hansen et al. [9], experimental result of Jacobs [10], 2D numerical and XFOIL results of Lakshmanan et al. [11]

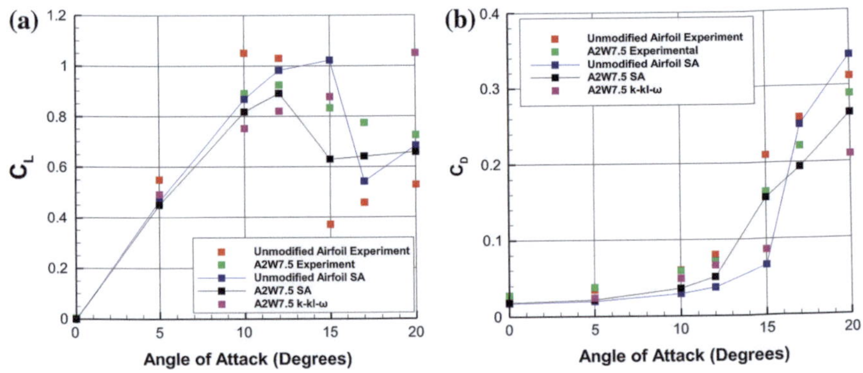

Fig. 3 **a**, **b** Lift and drag coefficients of modified NACA 0021 versus AOA plotted against experimental result of Hansen et al. [9] and experimental result of Jacobs [10]

3.1.2 Modified Airfoil (A2W7.5) Simulation Results

The lift coefficient of modified (A2W7.5) airfoils is plotted against experimental results of Hansen et al. [9] in Fig. 3a. From this plot, it is clear that at low angle of attacks, the simulation of lift coefficients with SA model yields good agreement with the experimental results. But in the stall region (12°–17°), the SA model is not able to predict it properly, whereas at 20°, results fairly match with the experimental results.

The predictions of k-kl-ω model are also included in Fig. 3a. The value of lift coefficient is underpredicted in pre-stall region. But this model is not able to capture the stall behavior and also does not predict stall up to 20° angle of attack. So it is appropriate to compare the SA simulation results of unmodified and modified (A2W7.5) airfoil.

The lift coefficient of simulation (SA model) for unmodified and modified (A2W7.5) airfoils is also shown in Fig. 3a. It is observed that there is only slight degradation in lift values up to 10° angle of attack. Between 12°–17°, the modified airfoil has less lift value than unmodified airfoil. But this is exactly the stall region, so it is not easy to conclude about the behavior of these two airfoils. On the other hand in post-stall region say at 20°, the lift coefficient value of modified (A2W7.5) and unmodified airfoil has same value.

The drag coefficient of modified airfoils (A2W7.5) is plotted against experimental results in Fig. 3b. From this plot, it is clearly evident that both models underpredict the drag values for almost all angles of attacks. For post-stall region, k-kl-ω model is not able to predict the drag coefficients.

The drag coefficient of simulation (SA model) for unmodified and modified (A2W7.5) airfoils is also shown in Fig. 3b. From this plot, it is clear that there exists a certain drag penalty at lower angles like 10° and 12°. Between 12°–17°, the behavior of SA model is peculiar in these locations. So at the post-stall angle (say 20°), it is clear that the drag value is less for the same amount of lift as shown in Fig. 3a, b. So it can be concluded that there is reduction in drag for modified airfoil (A2W7.5) for higher angle of attack.

3.2 *Surface Pressure Coefficient (C_P) Results*

3.2.1 Comparison Between 2D and 3D Results of Unmodified NACA 0021

The pressure coefficient C_P for 10° and 20° is plotted against normalized chordwise position as shown in the Fig. 4a, b. There exists a good agreement between 2D and 3D simulation results at almost all angles. At 20°, one can observe the difference in the values near the leading edge of the suction side.

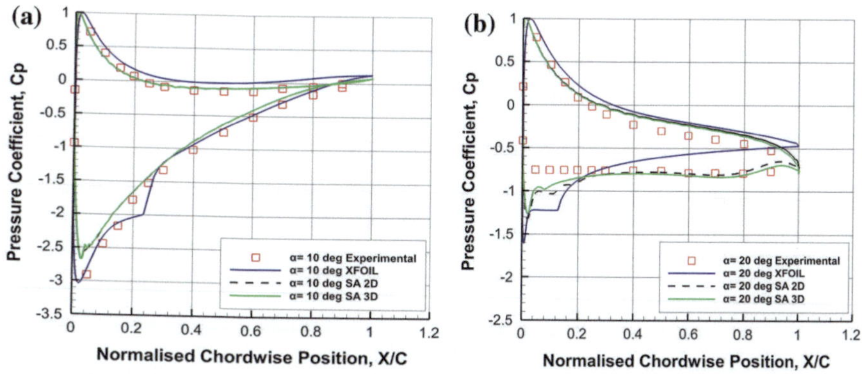

Fig. 4 a, b Normalized pressure distributions compared with experimental results of Hansen et al. [9] and XFOIL [11] results

At low angles of attacks, SA model underpredicts the maximum suction pressure on the upper airfoil. On the other hand, the pressure on the lower surface matches reasonably well at most of the angle of attacks. Similar to the 2D simulation, 3D simulation results also show poor performance in stall region.

3.2.2 Modified Airfoil (A2W7.5) Simulation Results

Simulation results of pressure coefficients are plotted for unmodified airfoil (SA model) and modified airfoil (SA and k-kl-ω models). Fig. 5a, b shows the C_P distribution of modified airfoil compared with unmodified airfoil at the same angles of attacks.

At 5°, the effect of tubercles starts appearing in the form of spanwise gradient. The maximum negative pressure is located at trough location. At mid-location, the values are same as the unmodified airfoil. At peak location, the negative pressure coefficient values are less (toward positive side). The effect of tubercles also appears at pressure side of an airfoil near tubercles. Figure 5a shows that k-kl-ω model is able to predict laminar separation bubble and SA model is failed to predict the same.

Figure 5b shows that prediction of SA model is better than k-kl-ω model at 20° angle of attack, which is also similar to lift and drag coefficient results. Due to this, further studies at 20° angle of attack are carried out with SA model.

3.3 *Results of Skin Friction Coefficient (C_f) at 20° Angle of Attack*

Figure 6a shows the variation of skin friction coefficient on the upper surface of an unmodified airfoil as a function of non-dimensional chordwise distance. It is clear

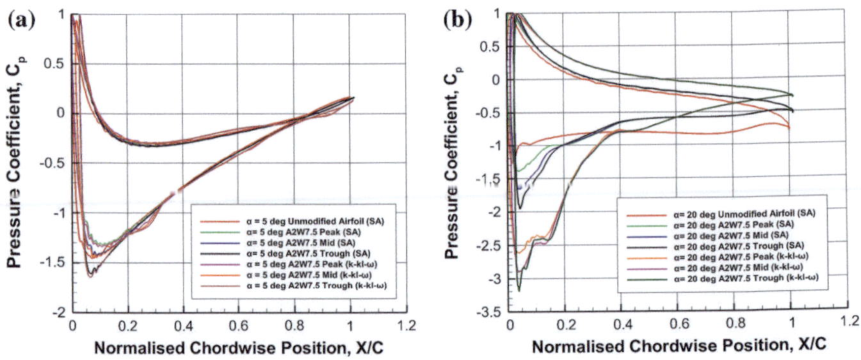

Fig. 5 **a, b** Normalized pressure distributions compared with experimental results of Hansen et al. [9] and XFOIL [11] results

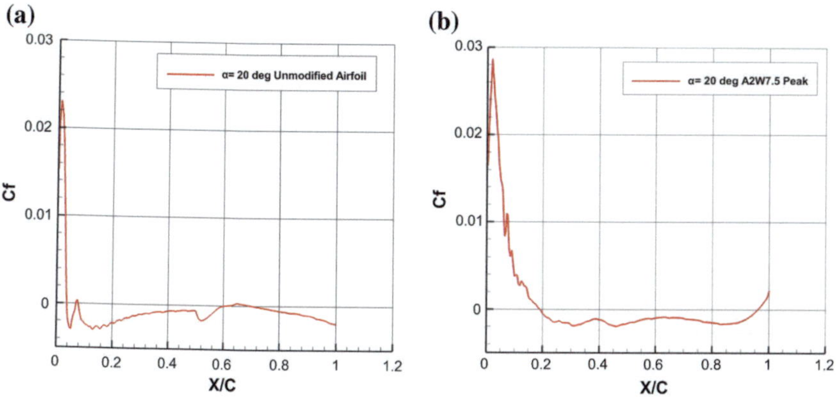

Fig. 6 **a**, **b** Skin friction coefficient (C_f) versus non-dimensional chordwise position at 20° angle of attack

that the flow has been separated near leading edge (sign change of the C_f value). The exact separation location is identified as 0.035 X/C. Skin friction coefficient of tubercles airfoil at peak location is plotted in Fig. 6b. The performance gain is clearly visible, and the flow separation is delayed. The exact separation location in this case is identified as 0.19 X/C.

4 Conclusion

The 3D simulations have been carried out for unmodified airfoil and modified airfoil (A2W7.5) NACA 0021. First of all, the unmodified 3D results are compared with the 2D simulation results obtained earlier. Except at 17°, the values are in good agreement with the 2D simulations. This can be attributed to the three-dimensional nature of the flow present at higher angles. Next, the simulations are compared with experiments as well as with the 3D results of modified airfoil. This is done for pressure coefficient values also. Finally, the modified airfoil simulation results of SA and k-kl-ω models are compared with unmodified 3D results. k-kl-ω model is able to predict laminar separation bubble and performs well in pre-stall region but unable to predict for stall and post-stall region. For SA model, the maximum lift value is less for modified airfoil when compared to unmodified one. But the stall nature of this airfoil is gradual. To obtain some quantitative information about the performance enhancement, a detailed analysis has been carried out at 20°. The result reveals that there exists a small gain due to the leading-edge tubercles. There is a shift in flow separation point due to the presence of tubercles at leading edge and reduction in drag at higher angles of attack.

Acknowledgements Simulations are carried out on the computers provided by the Indian Institute of Technology Kanpur (IITK) (www.iitk.ac.in/cc), and the manuscript preparation as well as data analysis has been carried out using the resources available at IITK. This support is gratefully acknowledged.

References

1. Edel RK, Winn HE (1978) Observation on underwater locomotion and flipper movement of humpback whale Megaptera novaeangliae. Marine Biol 48(3):279–287
2. Fish FE, Weber PW, Murray MM, Howle LE (2011) The tubercles on humpback whales' flippers: application of bio-inspired technology. Integr Comp Biol 51(1):203–213
3. Miklosovic DS, Murra MM, Howle LE, Fish FE (2004) Leading-edge tubercles delay stall on humpback whale (Megaptera novaeangliae) flippers. Phys Fluids 16:39–42
4. Corsini A, Delibra G, Sheard AG On the role of leading-edge bumps in the control of stall onset in axial fan blades. J Fluids Eng 135:081104-1
5. Spalart PR, Allmaras SR (1994) A one equation turbulence model for aerodynamic flows, 30th aerospace sciences meeting and exhibit. Reno, Nevada
6. Walters DK, Cokljat D (2008) A three-equation Eddy viscosity model for Reynolds-averaged Navier-Stokes simulations of transitional flow. J Fluids Eng 130:121401-1–121401-14
7. OpenFOAM Foundation Ltd. (2012) OpenFOAM user guide version 2.1.1
8. ANSYS Inc. (2012) ANSYS ICEM CFD 14.5 user manual, Canonsburg
9. Hansen KL, Kelso RM, Dally BB (2011) Performance variations of leading-edge tubercles for distinct airfoil profiles. AIAA J 49(1):185–194
10. Jacobs EN (1932) The aerodynamic characteristics of eight very thick airfoils from tests in the variable density wind tunnel. National advisory committee for aeronautics rept. 391
11. Lakshmanan SK, Mishra A, De A (2015) Assessment of different turbulence models for predicting laminar separation bubble over thick airfoils. In: Proceedings of ASME 2015 gas turbine india conference GTIndia 2015-1277, pp V001T01A012; 11
12. Drela M, Youngren H (2001) "XFOIL 6.94 user guide", aeronautical and astronautical engineering. MIT, Cambridge, MA

Estimation of Aerodynamic Coefficient for a Variable Span Wing and a C-Wing for Application in Personal Air Vehicles (PAV)

R. Naveen, B. Abinaya, D. Abinaya and K. Kalayarasi

1 Introduction

The main concerns of the Federal Aviation Administration (FAA) are the absence of enough air traffic control to hold greater number of flying vehicles and troubles such as smashed pilots and flying without a permit. The global society would also have to be of the same opinion on universal standards, the transformation of air miles to nautical miles, and so on. Above all, the FAA feared the accident of PAV (Personal Air Vehicles) in metropolitan areas, as oblivious built machines and also pilots' errors causing community nuisances. The researchers were trying to build a flying car more than a decade, but only a small amount of designs yet succeed in airborne through the atmosphere and driving on the conventional road. The American Defense Advanced Research Projects Agency has shown an awareness in the consideration with a sixty-five million dollar program called Transformer to develop a four place roadable aircraft by 2015. The vehicle is mandatory to take off vertically and has a 280-mile range. Terrafugia, AAI Corporation, and other Tex-tron companies have been awarded the contract. Flying cars fall into one of two styles: incorporated (all the pieces can be carried in the vehicle), or modular (the aeronautical sections are left at the airport when the vehicle is driven).

R. Naveen (✉) · B. Abinaya · D. Abinaya · K. Kalayarasi
Aeronautical Department, Bannari Amman Institute of Technology, Coimbatore, Tamilnadu, India
e-mail: aeror.naveen@gmail.com

B. Abinaya
e-mail: abinayab.ae13@bitsathy.ac.in

D. Abinaya
e-mail: abinayad.ae13@bitsathy.ac.in

K. Kalayarasi
e-mail: kalayarasik.ae13@bitsathy.ac.in

S. Singh et al. (eds.), *Proceedings of the International Conference on Modern Research in Aerospace Engineering*, Lecture Notes in Mechanical Engineering,
https://doi.org/10.1007/978-981-10-5849-3_18

The modern flying car inventions like the Terrafugia Transition which shows amazing promise. The Terrafugia Transition is a light sport, roadable airplane under development by Terrafugia since 2006 [1–3]. Even though flying car idea has been around since the early days of motoring, the optimized aerodynamic design of a roadable car is still a daunting task [1–7]. The journalism evaluation reveals the researchers who have tried to build a more than ten decades, but only a few designs ever succeeded in flying through the air and driving on the road. The first flying car or roadable aircraft came in 1917 via Wright Brothers rival Glenn Curtiss who has been beaten into the air designed the three-wing Curtiss Autoplane. The vehicle could only hop, but spawned an engineering race that, despite modern successes, has yet to come of age. The designer of a PAV has the major fear on the creation of upward force (lift) with low takeoff velocity also with minimum drag which united with better stability and control warranting promises. This gives result overall elevated performance both in road and air. Remind that the amount of upward force created by the flying car is depending mainly on two parameters like the shape (surface area, aspect ratio, and a cross section of the vehicle) and the device's orientation (or angle of attack). The objective of the paper is to design and predict the aerodynamic performance of a conceptual wing design theoretically and meet the flying car designs. Figure 1 shows the overall design process in this paper.

Fig. 1 Design Methodology

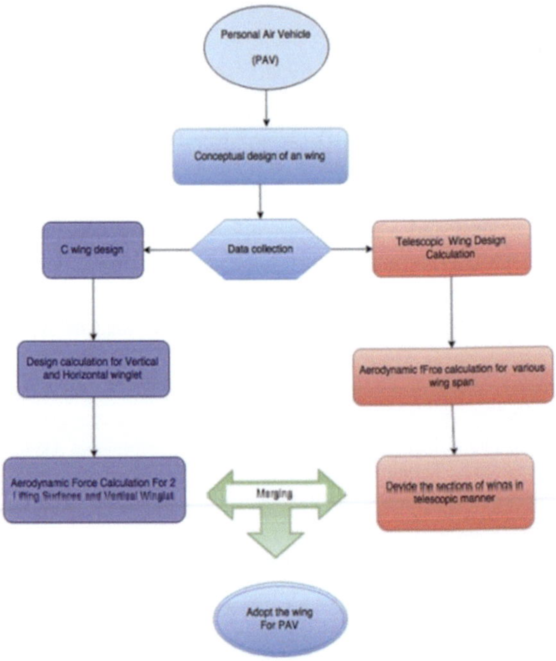

2 Design Methodologies

The wing was designed based on the lift requirement during various times of flight like takeoff, cruise, and landing conditions. The optimization of wing design parameters (wingspan, root, and tip chord) is based on the theoretical calculation of takeoff and landing velocities and wing length. Since the wing is variable span wing, the wingspan can be changed with respect to the amount of lift required during the flight.

The wing has total span of 14 m when extended and 6 m during retraction. A symmetrical airfoil NACA six series has been used for main wing box and vertical winglet. The horizontal winglet uses NACA 0012 airfoil.

While wing is extended, takeoff and landing lengths are reduced, by which endurance, range, and rate of climb are improved. When wings retracted the ice formation can be eliminated to $\frac{1}{3}$rd of the wingspan which helps in the reduction of drag.

The lift coefficient can be predicted for the vehicle by using theoretical estimation. The wing has been designed by combining telescopic wing and C-wing. The telescopic wing concept is extracted from the genesis Grever designs in which wing can retract and expand based on lift required. From the nonconventional aircraft design, C-wing design has been chosen and implied on flying car. C-wing consists of vertical and horizontal winglet which works like rudder and elevator of conventional aircraft. But there will be a structural problem by using this kind of combination and it has to be rectified. The designs are given in Figs. 2 and 3.

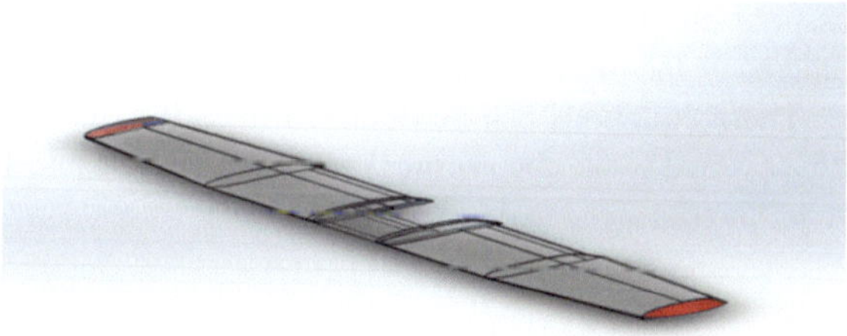

Fig. 2 CAITA model of telescopic wing

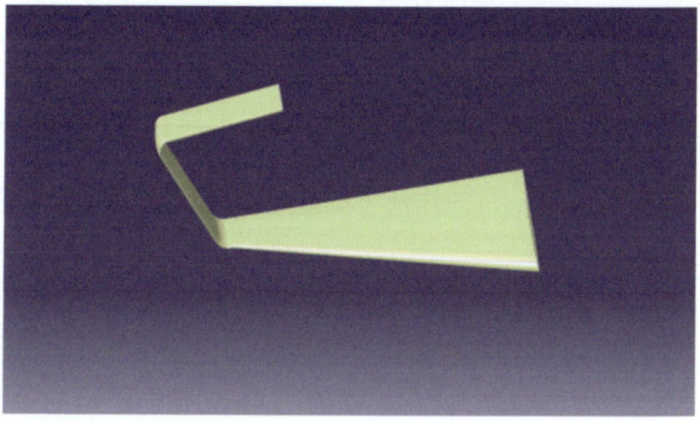

Fig. 3 Proposed wing design for PAV

3 Results and Discussion

This concept mainly deals with the wing design and its parameters like span, wing area, aspect ratio, wing loading, drag, and lift. Since the design combines the telescopic and C-wing configuration, the performance of wing design has been estimated using theoretically and analyzed by using ANSYS software.

3.1 *Theoretical Analysis*

3.1.1 Prandtl's Lifting Line Theory

Prandtl's lifting line theory describes the flow over the wing. It states that the low-pressure flow over a finite wing could force the air to roll over and around the tip, pushing the flow over the wing to move inboard and the same happens to the air under the wing to move outboard. It makes a difference in span-wise velocity which causes the air to roll up into a several stream-wise vortices influencing the lift force along the span.

For multiple lifting surface [8], the induced drag equation modifies to

$$D_i = \frac{L_1^2}{\pi q b_1^2} + \frac{L_2^2}{\pi q b_2^2} + \frac{2L_1L_2\sigma}{\pi q b_1 b_2} \tag{1}$$

where σ indicates the interference coefficient, and in the above expression we assume elliptical lift distribution over the wing, otherwise the effect of Oswald efficiency factor "e" is added in each term in the above expression.

These equations can be obtained from Fig. 4, where the x-axis is the span ratio b/b_{wing} and r is the vertical gap between each lifting surfaces expressed as $r = 2Gap/b_{wing}$.

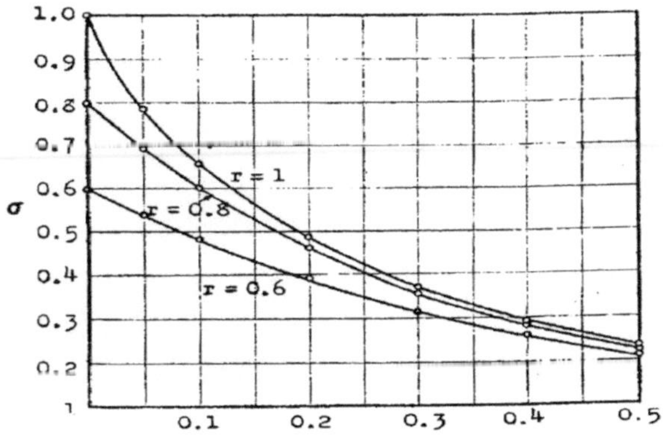

Fig. 4 Prandtl's interference factor σ

3.1.2 Kroo's Method

Dr. Kroo proposed a method to explain the minimum-induced drag span load over the wing by a secondary-loaded lifting surface [9, 10]. In order to label the downwash velocities induced by a secondary lifting surface on the wing, an analytical method given by von Kármán and Burgers was applied [11]

$$D_i = \frac{L_1^2\sigma^*}{\pi q b_1^2} + \frac{L_2^2}{\pi q b_2^2} + \frac{2L_1L_2\sigma}{\pi q b_1 b_2}, \quad (2)$$

where

$$\sigma = \frac{4I_{12(1)}}{\pi (b_1/b_2)^2} \quad (3)$$

$$\sigma^* = 1 - \frac{16}{\pi (b_1/b_2)^2} \sum_3^N \frac{I_{12(n)^2}}{n} \quad (4)$$

$$I_{12(n)} = \int_0^1 \left(\frac{\omega_{12}}{\omega_{01}}\right) \sin\,[\mathrm{n}(cos^{-1}y)]dy. \quad (5)$$

The σ factor obtained by Kroo is the same as Prandtl's interference factor which is given in Fig. 5.

By using Prandtl multiple lifting line theory, aerodynamic performance is predicted for proposed wing design for three cases (b = 6 m, 9 m, 14 m). When wing span is varied, total lift is changed but the lift on C-wing remains constant.

The graph is plotted for the value of C_L and C_D for the different angle of attacks. Three cases have been considered for plotting the graphs:

Casc-I—wing at 14 m span;

Casc-II—wing at 9 m wing span;

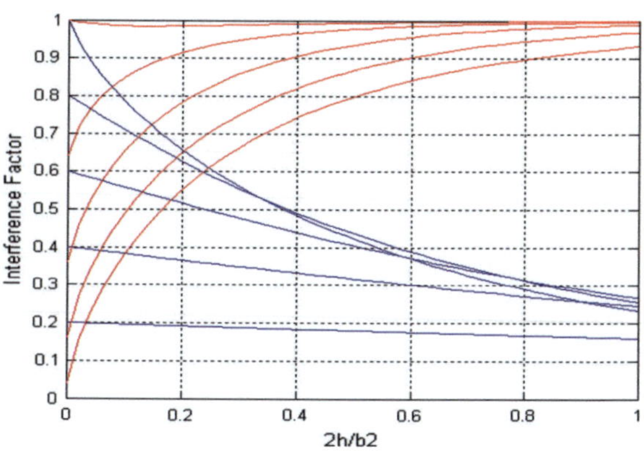

Fig. 5 Interference factor for I. Kroo method

Case-II—wing at 6 m wing span.

Aerodynamic forces have been calculated for the proposed wing design and without C-wing as shown in Figs. 6 and 7.

In Fig. 5, blue and violet curve indicates the primary and C-wing lifting surface, and it results in the similar lift and meets the required lift for the proposed design. Figure 8 indicates the induced drag coefficient for the coefficient of lift. By using C-wing approach, it has been drastically reduced because of Oswald efficiency. In the drag polar equation, the “e” is inversely proportional to induced drag and vice versa (Fig. 9).

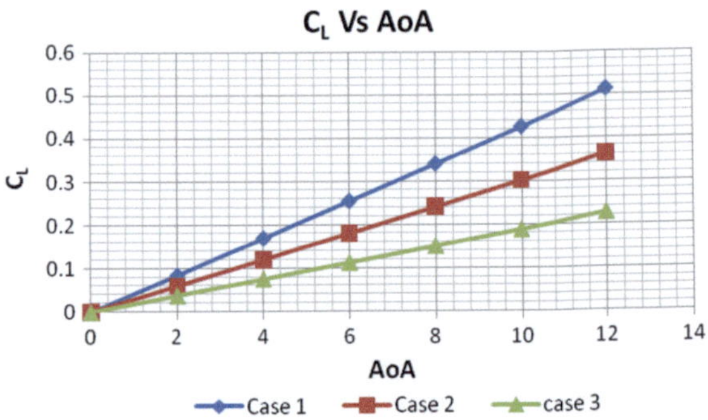

Fig. 6 C_L versus AoA for proposed wing design

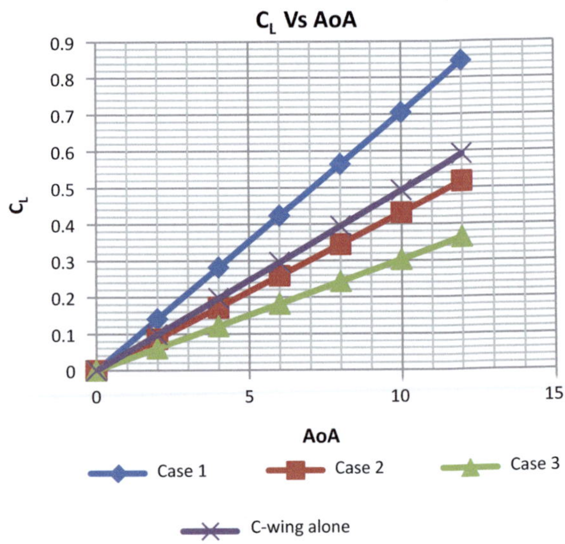

Fig. 7 C_L versus AoA for without C-wing

Fig. 8 C_L^2 versus $C_{D\ \text{induced}}$

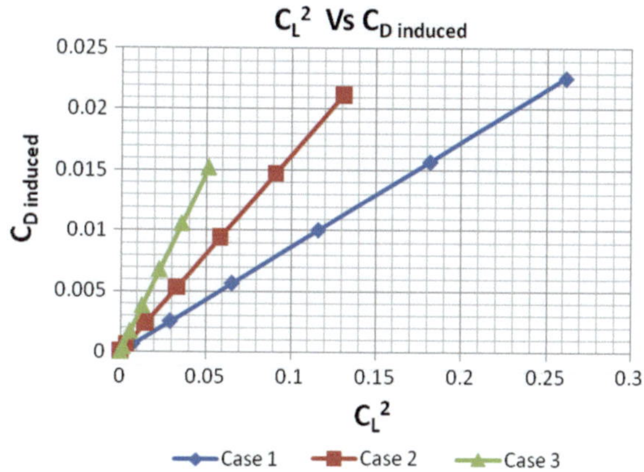

Fig. 9 C_L versus C_D graph

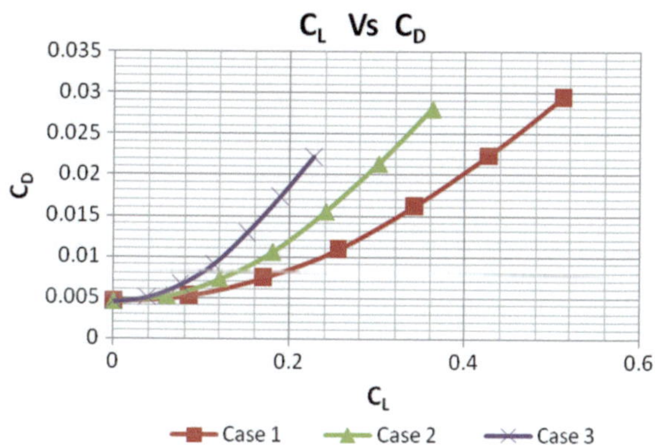

3.2 *Computational Analysis*

To simplify the analysis, the C-wing has been separated into the three sections as shown in Fig. 10. The suitable mesh is carried out by choosing domain across the airfoil which predicts boundary layer effects. The grid points are accelerated, and results are computed using multigrid technique.

Fig. 10 Splitting the sections of C-wing

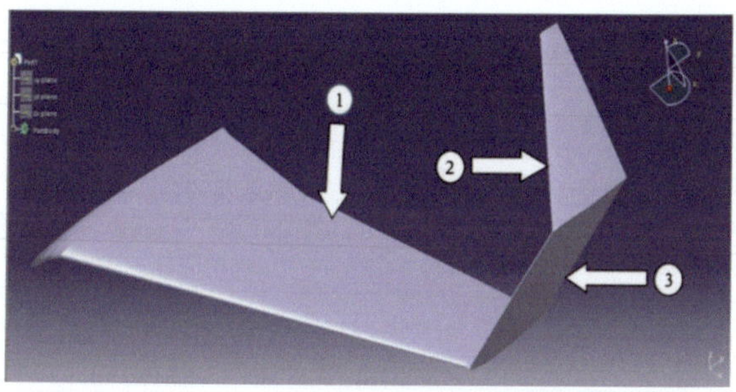

Computational analysis was performed for the wing at 1 and 2° angle of attack. The analysis is to get the knowledge of flow over the proposed wing. The detailed analysis has to be carried out in future for various angle of attacks (Fig. 11).

The graph from Fig. 12a,b shows that the pressure distribution for the various sections of the C-wing in terms of coefficient of pressure. For a symmetrical airfoil, there will be null lift at zero angle of attack, whereas Sect. 3 shows lift production, and this can be clearly understood from the above graph due to the pressure difference between upper and lower surface of the wing (Fig. 13).

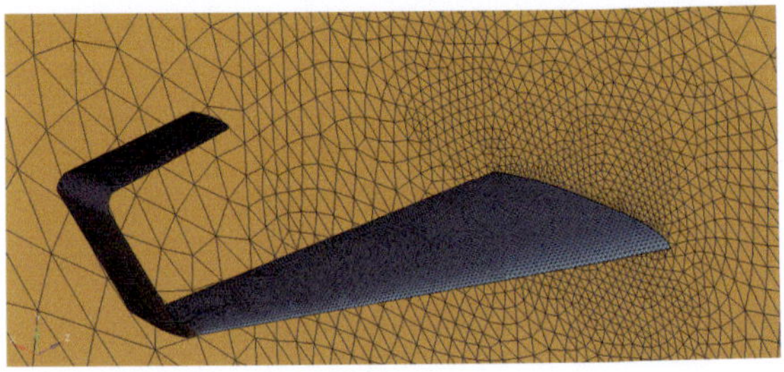

Fig. 11 Computational mesh of a proposed wing

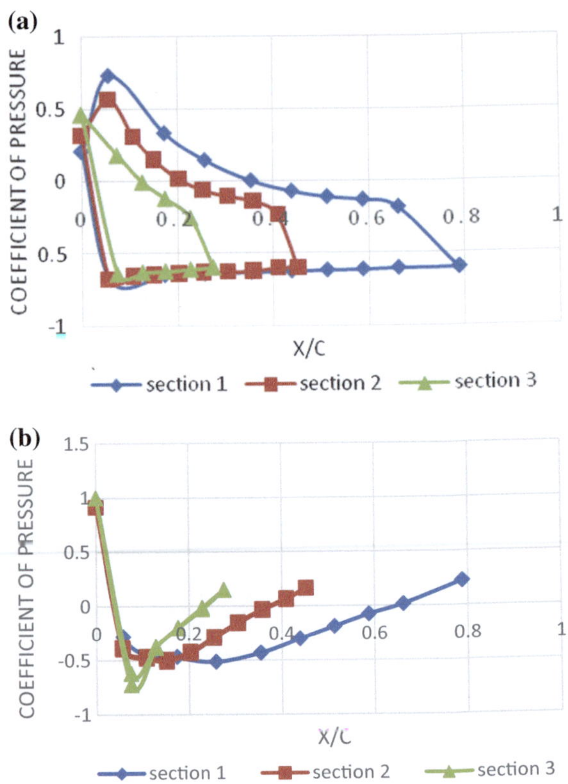

Fig. 12 Pressure distribution for various sections of the C-wing at **a** $\alpha = 0$ and **b** $\alpha = 15$

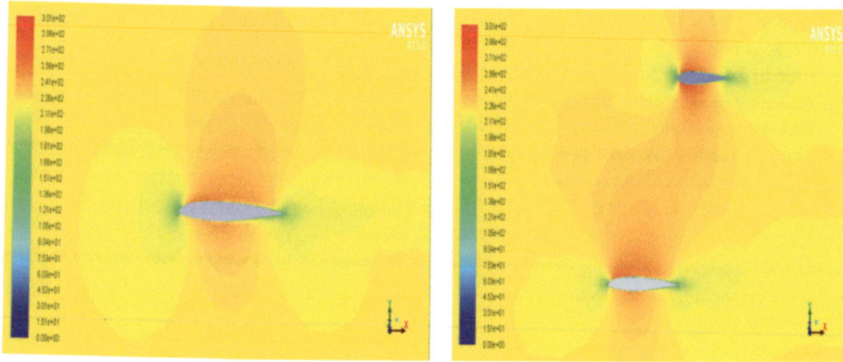

Fig. 13 Velocity magnitude for angle of attack at 3°

4 Conclusion

In this paper, the conceptual design of a PAV is developed and it is been found that it fulfills the basic requirements of preliminary design aspects. This design concept overcomes the aspect ratio problem by introducing the combination of C-wing and telescopic wing. From the aerodynamic analysis, we can state that the lift force of the preliminary variable span wing does not affect the C-wing performance. However, the detailed computational approach should be taken out for further improvements for whole vehicle. As the C-wing contains the horizontal and vertical stabilizer, the structural stability has to be studied in future. This type of flying car does not need any separate takeoff and landing requirements, and it can take off and land in both normal conventional roads and also in airport runway. So making an automobile to fly becomes very easy when using this design.

References

1. Sams GD, Gurunathan K, Selvan P (2012) Conceptual design and analysis of ferrari F430 flying car, Final year undergraduate project report, Department of Aeronautical Engineering, Kumaraguru College of Technology, Coimbatore 641 049, Tamil Nadu, India, April 2012
2. Haines TB (2009) First roadable airplane takes flight. Aircraft Owners and Pilots Association (AOPA). Retrieved 19 March, 2009
3. Durbin DA (2012) Flying car gets closer to reality with test flight. boston.com. Associated Press. Retrieved April 20, 2012
4. First flight for Terrafugia. Retrieved 2 April, 2012
5. Dietrich AM (2009) TransitionSpecs-FirstFlight-200. Terrafugia. Retrieved 2 April, 2009
6. North American Free Trade Agreement Land Transportation Standards Subcommittee, ìHighway safety performance criteria in support of vehicle weight and dimension regulations: candidate criteria & recommended. Thresholdsî Working Draft for Consultation, November 1999

7. Fielding John P (1999) Introduction to aircraft design. Cambridge University Press, New York, p 84
8. Prandtl L (1924) Induced drag of multiplanes. NAA TN 182, 22 p
9. Laurendeau E (1990) Traînée Induite de Configurations d'Avions. École Nationale Supérieure de l'Aéronautique et de l'Espace (ENSAE), Toulouse, France (September), 109 p
10. Kroo IM (1984) A general approach to multiple lifting surface design and analysis. AIAA paper 84-2507, October 1984
11. von Kármán T, Burgers JM (1963) General aerodynamic theory—perfect fluids. In: Durand WF (ed) Aerodynamic theory, vol 2, Dover edn. New York
12. Prandtl L (1921) Application of modern hydrodynamics to aeronautics. NACA TR 116, 59 p

Evaluation of the Structural Integrity of Full-Scale Wing of Aerial Vehicle

D. Venkata Sivakumar, M. Varadanam, K. Kishore Kumar and P.S.R. Anjaneyulu

1 Introduction

In the past 20 years, a significant use of aerial vehicles, i.e., manned or unmanned, has been increased in defense domain for the purpose of surveillance, target acquisition and tracking, battlefield monitoring, target assessment, signal intelligence, electronic warfare, and relay. At the same time, the economical aspects such as terrain and communication tracks monitoring, media, or energy transfer lines have raised interest in the application of aerial vehicles for civilian purpose. Composite structures are extensively used in the aerospace and defense areas due to various advantages such as high specific strength, low specific weight, ease of fabrication, and integral structures. The composite wings are extensively used for aerial vehicles to withstand the aerodynamics, inertia, flight maneuver and landing loads. This paper presents the results of an investigation for examining the strength and stiffness characteristics of a carbon/epoxy (CFRP) composite wing of aerial vehicle [1].

The criticality in the evaluation of structural properties is due to the use of composite materials in the fabrication of the wing. Unlike metallic structures, the structural behavior of composite wing is process dependent and hence not easily predictable. The process of qualifying, certifying composite materials and structures for strength, buckling and damage tolerance remains almost entirely empirical. The best way to estimate the structural properties such as local stiffness of the wing at various locations, overall stiffness of the full length wing, and buckling behavior of the wing for typical load is by experimental evaluation techniques [2, 3].

D. Venkata Sivakumar (✉) · M. Varadanam · K. Kishore Kumar · P.S.R. Anjaneyulu
Defence Research & Development Laboratory, Hyderabad 500058, India
e-mail: siva.doniparthi@gmail.com

S. Singh et al. (eds.), *Proceedings of the International Conference on Modern Research in Aerospace Engineering*, Lecture Notes in Mechanical Engineering,
https://doi.org/10.1007/978-981-10-5849-3_19

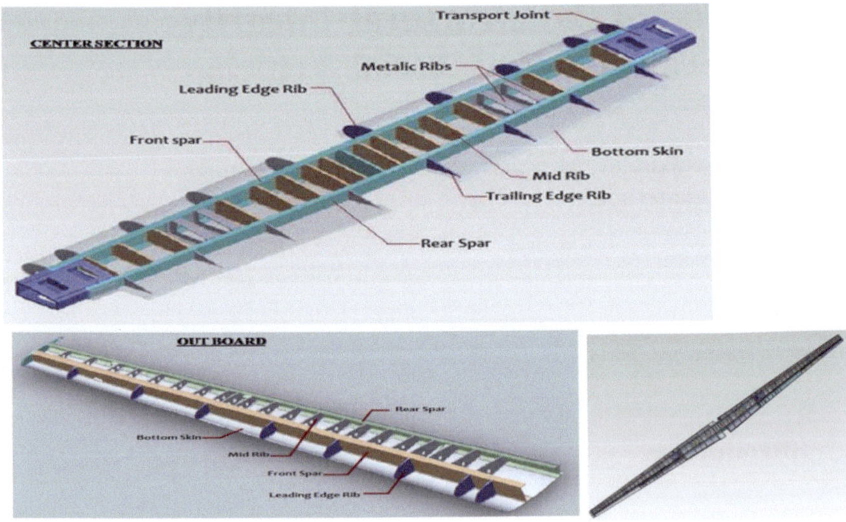

Fig. 1 Solid model of the wing assembly

2 Description of the Aerial Wing Structure

The total span of the composite wing was more than 20 m, with an aspect ratio of about 15. The solid model of the wing assembly as shown in Fig. 1. Wing was fabricated with three modules for ease of manufacturing and transportation, namely center section, outboard port side and outboard star side. The wing assembly consists of structural elements, namely CFRP foam sandwich composite skin, front and rear spar, composite ribs, metallic ribs. Metallic ribs are used for wing and engine frame interface for landing load diffusion. The engine frames were attached to the wing by using wing attachment brackets.

3 Experimental Test Setup and Procedure

A test rig with loading fixtures was designed and developed for supporting and testing the composite wing. The schematic test setup of flight loads on wing is shown in Fig. 2. An innovative testing methodology was introduced to load the full-scale wing to simulate different symmetrical maneuver flight (*Case-I*) and landing load conditions (*Case-II*). To simulate the exact boundary conditions, the wing was firmly fixed on test rig through test fixture at the wing mounting brackets, which is fabricated for simulating interface joint between fuselage body and wing mounting. The aerodynamic pressure loads on the surface are converted to statically equivalent loads. These loads are distributed from the front spar to the rear spar

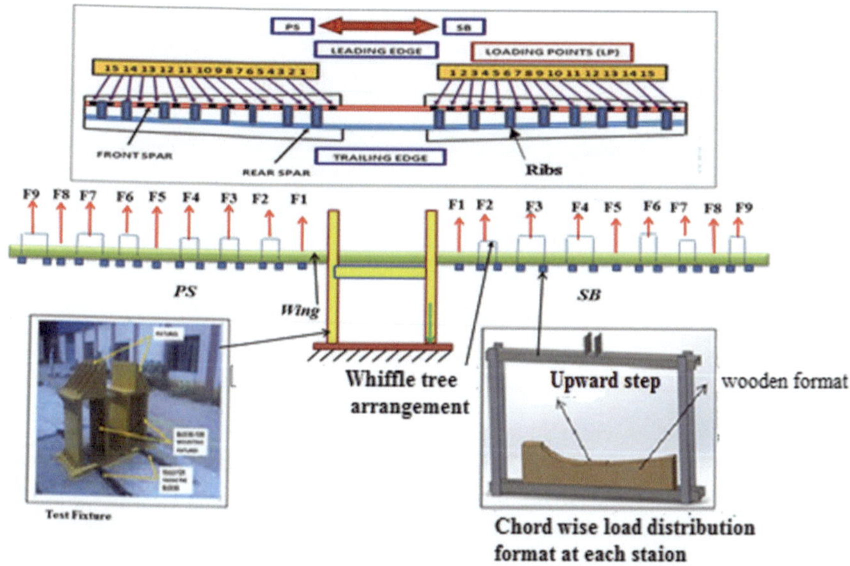

Fig. 2 Schematic test setup for flight loads on wing (Case-I: +2.5 g for flight loads)

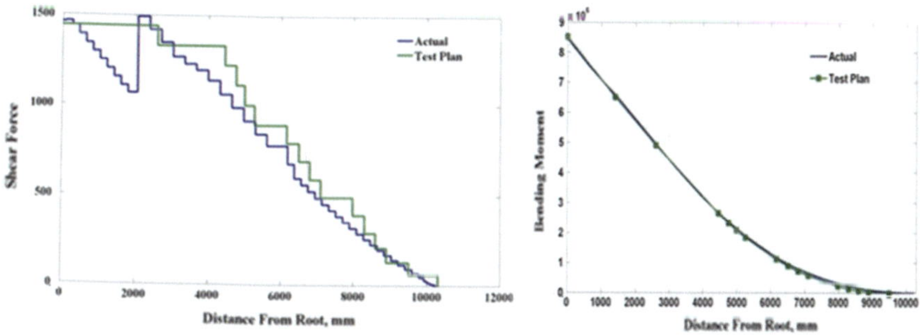

Fig. 3 Comparison of span-wise shear force and bending moment

through ribs along the span of the wing. This kind of approximation is covered by the St. Venant's principle, and the global response of the structure will not be altered. Major challenge lies in converting aerodynamic load to statically equivalent load where theoretical and calculated test loads are matched in terms of shear force and bending moment to capture the global behavior of the structure as shown in Fig. 3. The distributed loads were discretized and applied at 30 stations to simulate as close as to the flight bending moment envelope.

Structural loads were applied with the help of hydraulic actuators independently and in combination with whiffle load trees, to simulate the loads as experienced by the wing during flight load case (Case-I). Unlike metallic section and symmetrical airframe sections, loads should not be applied on the composite aerofoil wing using

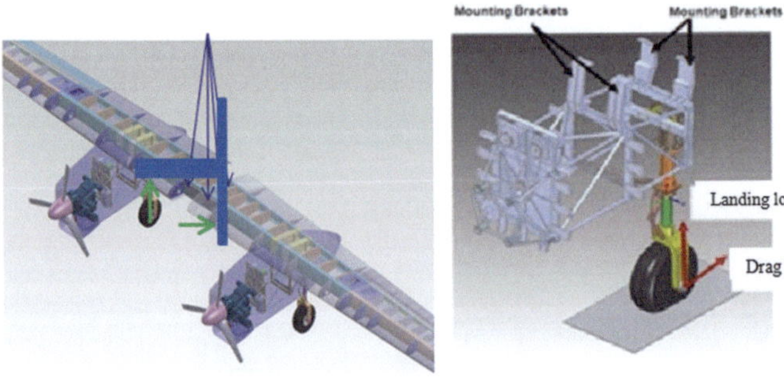

Fig. 4 Schematic test setup for landing gear loads on wing-engine mounting frame assembly (Case-II)

regular polyester/nylon slings. This difficulty was overcome by using aerofoil contour plywood formats having upward step (near the center of pressure location) with cushions which are made of felt material. The wooden formats (30 no.'s), one at each loading station along wing span, were designed in such a way to match the aerofoil contour. So that load was not distributed along the chord-wise direction and only resultant load applied at the center of pressure, or else it will give twisting and unnecessary moments and lead to improper load distribution with localized stress concentrations and may also even cause catastrophic failure of the structure. The design of the whiffletree to distribute the loads in the desired fashion is one of the key parameters in this testing process. Synchronization and coupling of the load application was maintained during the testing of the composite wing [4–6].

To simulate the landing gear loads (Case-II) separately [7], wing was fixed to metallic test fixture at the wing-attached brackets. The schematic test setup of landing gear loads on wing-engine mounting frame assembly is shown in Fig. 4. The wing was firmly fixed on test rig through test fixture, and engine mounting frame was assembled to the wing, by using wing-attached brackets for simulating interface joint between wing-engine mounting frames. The landing gear loads were applied on the engine mounting frame with the help of hydraulic actuators independently.

4 Instrumentation

Hydraulic actuation system instrumented with load cells was used to control and apply various loads at different locations on the wing. Selection of strain gage location for the composite was a challenging task. Selection of location depends on the layer orientation, layer drop, and joints, and it should be free from manufacturing defects. Locations of strain gages on wing assembly are shown in Fig. 5. Twenty-three numbers of biaxial (on composite section) and triaxial strain gages (on metallic portion) were used to measure the static response of the wing assembly at

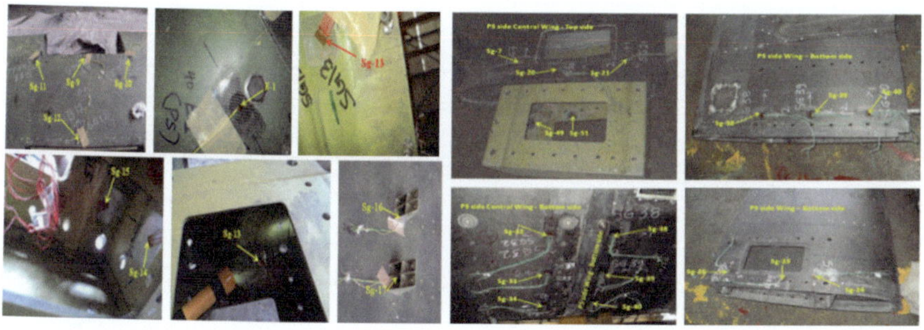

Fig. 5 Location of strain gages on wing assembly

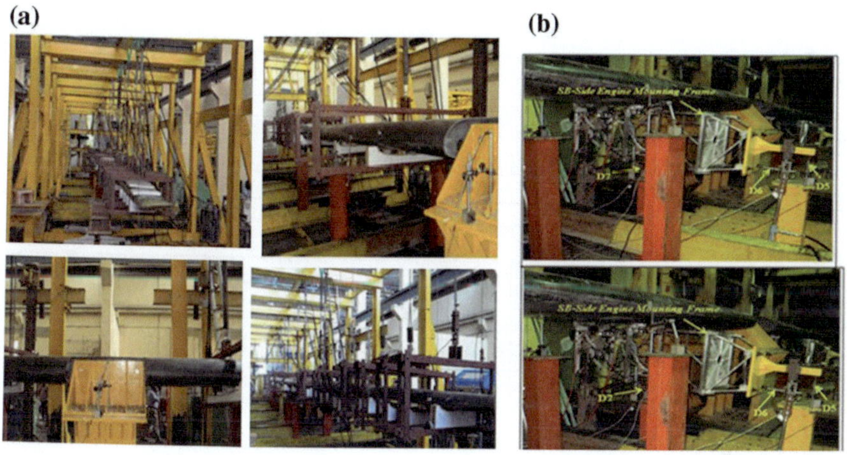

Fig. 6 a Experimental setup of wing for flight loads (Case-I). b Experimental setup of wing for landing loads (Case-II)

different longitudinal locations, and the wing is loaded incrementally up to limit loads (100% load) for Case-I and Case-II. Eighteen numbers of linear variable displacement transducers (LVDT) have been effectively used for the measurement of deflection. Data acquisition system was used for measuring strains and deflections [8]. Photograph of experimental test setup which includes hydraulics and instrumentation setup of the wing for flight and landing gear loads is shown in Fig. 6.

5 Results and Discussions

The strains and deflections measured during the test are shown in Figs. 7, 8, 9, and 10. The measured strains were compared with the failure strains of the composite material. Principal strains and corresponding principal stresses were obtained from

Fig. 7 Limit load versus strain

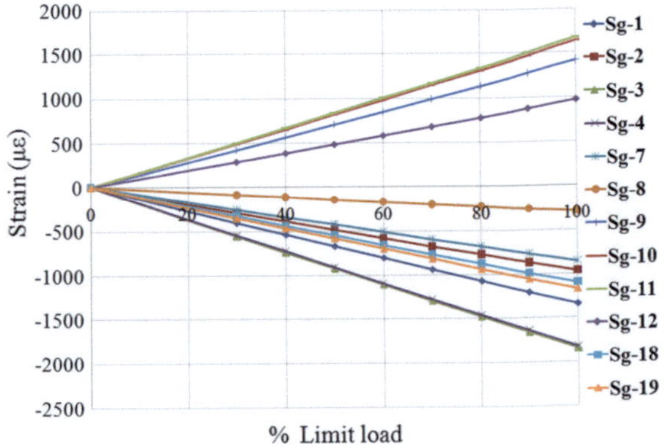

Fig. 8 Limit load versus deflection

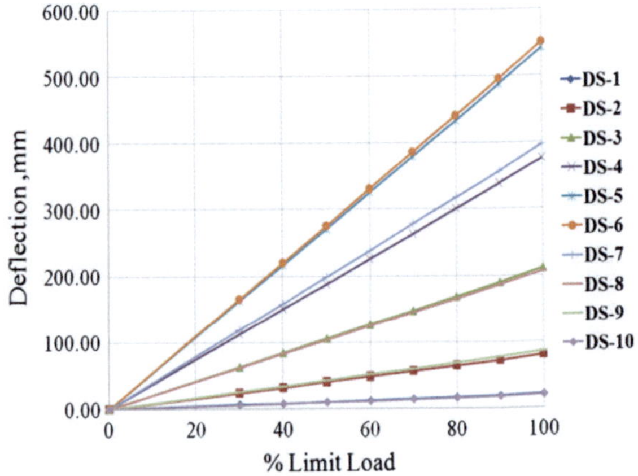

Fig. 9 Limit load versus principal stress

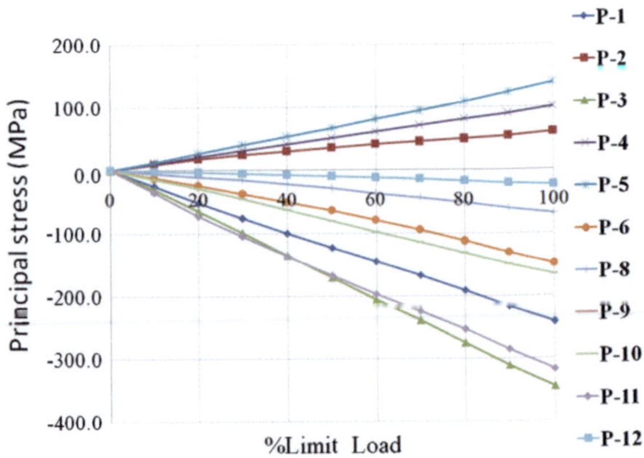

Fig. 10 Limit load versus deflection

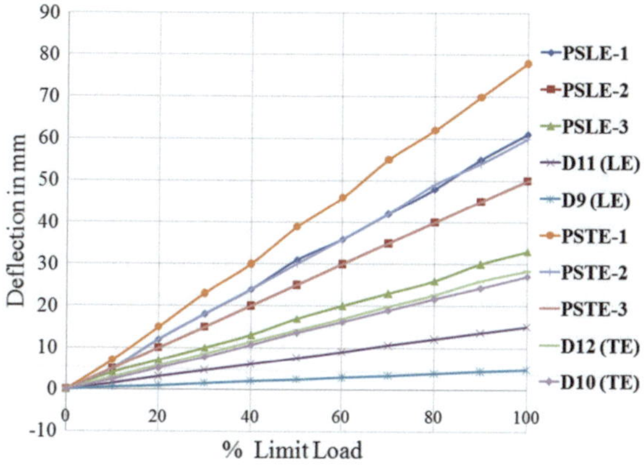

measured strain data on metallic portions. The principal stresses are compared with yield strength of the material. The maximum strain recorded on composite wing, during the test against the flight maneuver loads, was −1842 με (***Case-I***). This is well below the failure strain (9000 με) of the wing structure material. Maximum deflection measured was 550 mm, during flight maneuver load case, whereas predicted deflection was 510 mm. Few strain gages (Sg-3 and Sg-4) were bonded at critical locations of expected buckling zone as predicted from the FE analysis. The presence of nonlinearity in the strain data is a measure of buckling phenomenon on the zone. These strain gages behaved linearly as seen from Fig. 7. Therefore, the wing has not shown any signs of buckling up to limit load during the test. Due to landing gear loads (***Case-II***), the maximum principal stress measured on wing-engine mounting bracket of metallic section was 323 MPa (compressive) at the limit load, which is less than the yield strength of metallic material. The linearity in strain and deflection variation denotes that the carbon/epoxy (CFRP) composite wing structure has no abnormalities or defects like delamination, debonding, and withstood for the applied loads. Similarly, engine frame assembly was not yielded for the applied loads.

In addition to structural load test, the natural frequency of the full-scale composite wing was also measured before and after the structural load test to ensure the stiffness and structural integrity of the full-scale wing, and no deviation was observed in the measured frequency [9]. The objectives of the structural load test have been met for the various mentioned load cases on the full-scale wing. The wing deflection and strains have been verified to be within specified values and confirmed there is no permanent set on release of the limit loads. However, after successful completion of structural test followed by visual inspection, the wing inspected with NDT techniques and found that the structure has no defects.

Case-I: Strains and deflections measured due to flight loads (+2.5 g case) on the composite wing.

Case-II: Principal stresses and deflections on the wing-engine mounting frame due to landing gear loads (+3.0 g case).

6 Conclusion

The test methodology to execute qualification of a full-scale composite wing for structural loads has been successfully established. Moreover, the challenges faced and the innovative execution of the structural test were discussed. The experimental test results as an input to refine the finite element model for further analysis. Critical measurements of strains and deflections have aided in evaluating the structural integrity of the full-scale wing. Maximum strains are within the failure strain of the respective materials. This paper includes the results obtained from various tests conducted on a full-scale composite wing and established the procedure for evaluation of the structural integrity of the wing. A complete overview on structural evaluation has been discussed, and structural integrity was proven by applying flight and landing gear loads.

Acknowledgements This research was supported by structural test facility, DRDL, and we are thankful to testing team for assistance in all aspects. The authors are thankful to Director, DRDL for pursuing this research paper.

References

1. Tsai SW, Wu EM (1980) A general theory of strength for anisotropic materials. J Compos Mater 5:58–80
2. Frulla G, Cestino E (2008) Design, manufacturing and testing of a HALE-UAV structural demonstrator. Compos Struct 83:143–153
3. Gupta SG, Ghonge MM, Dr. Jawandhiya PM (2013) Review of unmanned aircraft system (UAS). Int J Adv Res Comput Eng Technol (IJARCET) 2(4)
4. Sullivan RW, Hwang Y, Rais-Rohani M, Lacy T (2009) Structural analysis and testing of an ultra-light unmanned-aerial-vehicle carbon-composite wing. J Aircr 46(3)
5. Maksimovic S, Ilic I (2013) Structural analysis and static strength testing of a tactical unmanned aerial vehicle. Sci Tech Rev 63(2):58–62
6. Sullivan R, Rais-Rohani M, Lacy T, Alday N (2006) Structural testing of an ultra-light UAV composite wing. Presented at the 47th AIAA/ASME/ASCE/AHS/ASC structures, structural dynamics, and material conference, Newport, Rhode Island, 1–4 May 2006
7. Pritchard JI (1976) An overview of landing gear dynamics, NASA/TM-1999-209143-TR-1976
8. Dally JW, Riley WF (1991) Experimental stress analysis. McGraw- Hill Inc
9. Simsiriwong J, Ullivan RW (2010) Vibration testing of a carbon composite fuselage. Int J Veh Noise Vib

Finite Element Analysis of Hyperbolic Paraboloid Composite Shells for Static Analysis Under Uniform Pressure

Anil Garhwal, Yogesh Kaushik, Sabita Madhvi Singh and Divya

Notations

a, b	Length and width of the shell in plan
c	Rise of the hyperbolic paraboloid shell
$\{d\}$	Global displacement vector
$\{d_e\}$	Element displacement vector
E_1, E_2, E_3	Moduli of elasticity
G_{12}, G_{23}, G_{13}	Shear moduli of a lamina
$\mu_{12}, \mu_{23}, \mu_{13}$	Poison's ratio
RSR	Rise-to-span ratio
CCCC	All sides clamped
SSSS	All sides simply supported
CSCS	Alternately clamped and simply supported
CP	Cross-ply laminate
AP	Angle-ply laminate
u, v, w	Translational degrees of freedom at each node
α, β	Rotational degrees of freedom at each node

A. Garhwal · Y. Kaushik (✉) · S.M. Singh
Department of Civil Engineering, Amity University,
Noida 201313, Uttar Pradesh, India
e-mail: kaushik.yogesh2@gmail.com

A. Garhwal
e-mail: garhwalanil@gmail.com

Divya
Department of Civil Engineering, South Point Technical Campus,
Sonepat 131001, India
e-mail: divyav30730@gmail.com

S. Singh et al. (eds.), *Proceedings of the International Conference on Modern Research in Aerospace Engineering*, Lecture Notes in Mechanical Engineering,
https://doi.org/10.1007/978-981-10-5849-3_20

1 Introduction

Composites are the materials combined from two different materials, cumulative properties of these materials are higher than each material individually. In fibre reinforced composites, fibres are the primary load carriers enclosed in matrix to keep them in position. Composites have high-strength-to-weight ratio. In the present study orthotropic defined shell element is considered. Orientation of fibres with respect to reference axis defines the lamination as cross-ply or angle-ply. Hyperbolic paraboloids are doubly curved non-developable surfaces, when used in form of shell elements can sustain high stresses. Hyperbolic paraboloid shells have broad applications in civil engineering as they can cover large obstruction free areas. These shells are anticlastic shells and get their strength from their shape as the stresses are distributed in three dimensions.

Deflections, forces, moments and free vibration of composite stiffened shell roofs under concentrated load, vibration characteristics of composite hypar shell with various edge supports was explored by Sahoo and Chakravorty [6, 16, 17]. Static characteristics of delaminated composite conoidal shell subjected to point load was studied by Kumari and Chakravorty [1]. Dynamic instability behavior of laminated hypar and conoidal shells using a higher-order shear deformation theory was presented by Pradyumna and Bandyopadhyay [5]. Bending of delaminated composite conoidal shells under uniformly distributed load was examined by Kumari and Chakravorty [7]. Impact response of simply supported skewed hypar shell roofs by finite element was studied by Pradyumna and Bandyopadhyay [13].

Studies have been done on various parameters for hyperbolic paraboloid shells but it's evident to extract a relation between rise and span for best suitability hence there is a need of obtaining a relation between rise of hyperbolic paraboloid with maximum stresses and deflection, therefore the present investigation aims to study finite element analysis of hyperbolic paraboloid under uniform pressure with change in rise to span ratio with different laminations and boundary conditions to conclude the most favorable RSR of hyperbolic paraboloid to be used.

2 Mathematical Formulation

An eight nodded quadratic iso-parametric shell element is used for the present study by using finite element method. The coordinates and displacements at any point within element are represented by coordinates & displacements of the nodes of the element and the shape functions. Hyperbolic paraboloid shell of radius of cross curvature Rxy is considered. By differentiating the surface equation of the shell in the form $z = f(x, y)$ the radius of curvature may be evaluated. The five degrees of

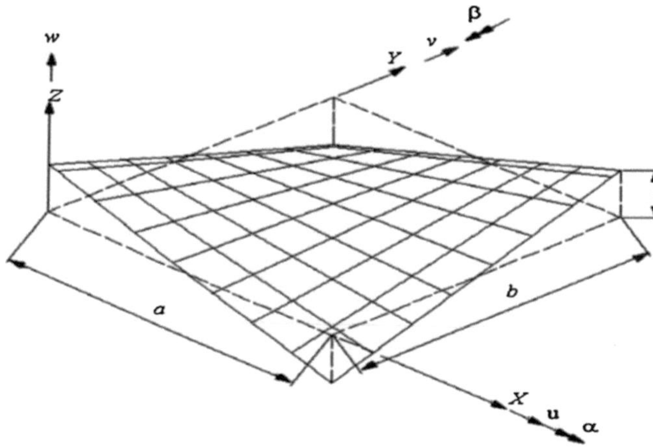

Fig. 1 Surface of hyperbolic paraboloid and degrees of freedom

freedom considered for each node are u, v, w, α, and β (Fig. 1). The interpolation polynomials derived for these are a function of ξ and η having form [9]:

$$u(\xi,\eta) = A_0 + A_1\xi + A_2\eta + A_3\xi^2 + A_4\xi\eta + A_5\eta^2 + A_6\xi^2\eta + A_7\xi\eta^2$$

The shape function derived as:

$$N_i = \frac{1}{4}(1+\xi\xi_i)(1+\eta\eta_i)(\xi\xi_i+\eta\eta_i-1) \quad i=1,2,3,4$$

$$N_i = \frac{1}{4}(1+\xi\xi_i)\left(1+\eta^2\right) \quad i=5,7$$

$$N_i = \frac{1}{4}(1+\eta\eta_i)\left(1+\xi^2\right) \quad i=6,8$$

Here Ni represents the shape function for ith node (Fig. 2).

A four layered hyperbolic paraboloid laminate with antisymmetric stacking for equal thickness of layers with simply supported end conditions, clamped and alternately simply supported and clamped on four sides under transverse uniform pressure is considered for analysis:

The material properties used for the plate are:

$E_1 = 39 \times 10^3$ N/mm^2, $E_2 = E_3 = 8.6 \times 10^3$ N/mm^2
$G_{12} = G_{23} = G_{13} = 3.8 \times 10^3$ N/mm^2
$\mu_{12} = \mu_{23} = \mu_{13} = 0.28$
The uniform pressure applied is 0.05 N/mm^2.

In the present problem, the effect of rise to span ratio (c/a) varying from 0 to 0.5 with linear increment of 0.05 is studied on maximum stresses (σ_x, σ_y, σ_z, τ_{xy}, τ_{yz}, τ_{xz}) and maximum deflection (U_z). The geometric dimensions of the present problem are:

Length, a = breadth, b = 1000 mm
Rise, c = 0 to 500 mm
Thickness, h = 8 mm

Laminations of composite layers used are cross-ply (0°/90°/0°/90°) and angle-ply (45°/−45°/45°/−45°). ANSYS 15.0 is used for the modeling and analysis of the problem. The eight nodded shell element, [SHELL281] is used in the present problem with five degrees of freedom.

2.1 *Formulating Static Problem*

The expression for static equilibrium can be represented as,

$$[K]\{d\} = \{Q\}$$

The above equation is solved by the Gauss elimination technique and from the global nodal displacement vector {d} thus obtained, the element displacement vector {de} is known (Fig. 3) [14].

3 Results and Discussions

The various cases of simply supported, clamped and alternately simply supported and clamped with antisymmetric stacking are discussed (Fig. 4). Change in stresses and deflection due to the increase in rise to span ratio (c/a) from 0 to 0.5. are discussed. A four-layered shell is considered for present study with equal thickness

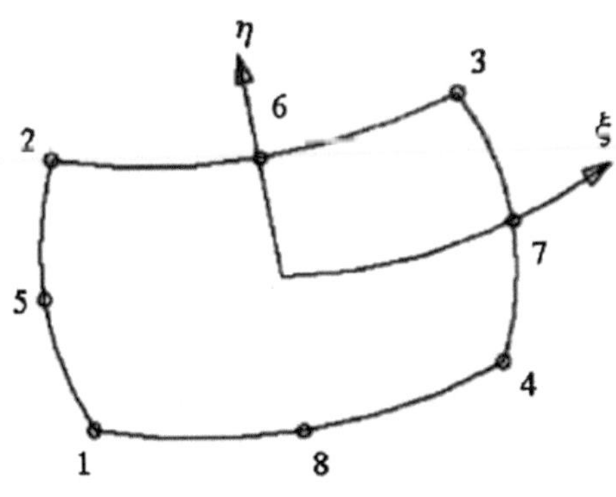

Fig. 2 Eight nodded quadratic element

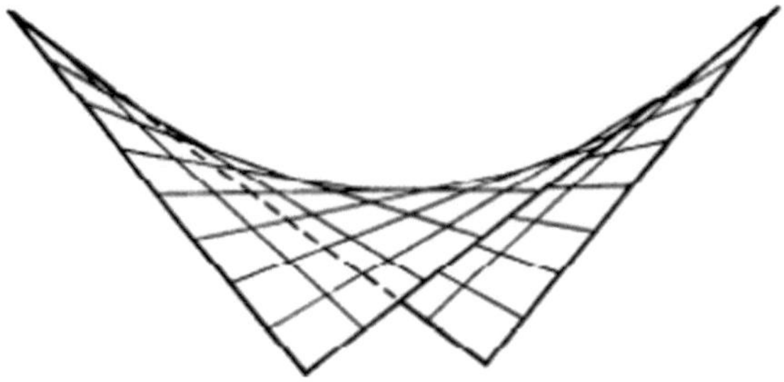

Fig. 3 Hyperbolic paraboloid shell

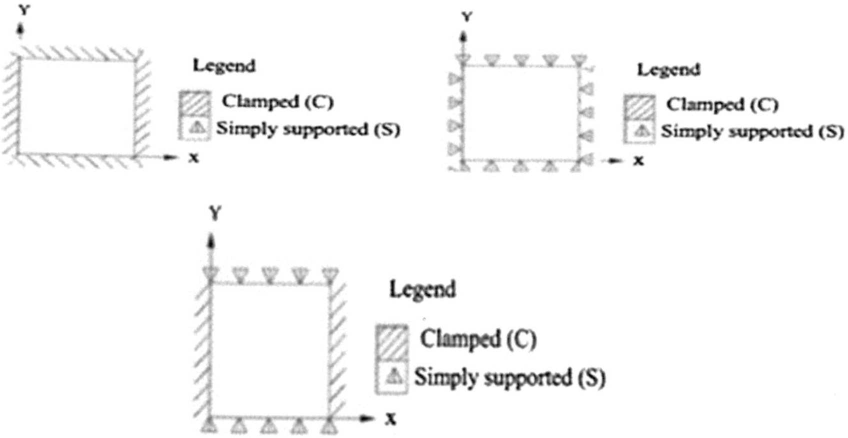

Fig. 4 Boundary conditions for hyperbolic paraboloid shell

of all layers. Uniform transverse pressure of 0.05 N/mm^2 is applied. Based on the present study following cases has been discussed as described below:

$$\text{SURFACE EQUATION, } Z = \frac{4c}{ab}(x - a/2)(y - b/2)$$

3.1 *Clamped Hyperbolic Paraboloid with Cross-Ply (0°/90°/0°/90°) Lamination*

This case projects the relation between the ratio c/a and the maximum stresses and deflection for clamped cross-ply lamination stacked antisymmetric. Change of normal stresses, shear stresses and deflection with corresponding change in rise-to-span-ratio is shown in Table 1. It can be noticed that with increase in RSR normal stresses decrease considerably initially and for the case of transverse shear stresses the case is vice-versa i.e. stresses increase with increase in RSR. Clamped hyperbolic paraboloid with angle-ply (45°/–45°/45°/–45°) lamination.

Table 1 Stresses and deflections for all cases considered with change in rise to span ratio

RSR	Normal stress σ_x (N/mm^2)	Normal stress σ_y (N/mm^2)	Normal stress σ_z (N/mm^2)	Shear stress τ_{xy} (N/mm^2)	Shear stress τ_{yz} (N/mm^2)	Shear stress τ_{xz} (N/mm^2)	Deflection ion U_z (mm)
CCCC CP (0°/90°/0°/90°)							
0	378.229	159.627	0	22.41	0.463e-13	0.326e-14	76.387
0.05	113.087	33.6205	0.454635	17.3813	5.95132	5.89486	14.7323
0.1	56.8276	18.8762	1.18634	9.20	7.3187	7.08839	3.45112
0.15	39.6024	13.4736	1.98812	5.78757	8.09189	7.57225	1.64911
0.2	30.8757	10.5167	2.80732	4.34104	8.64264	7.72763	1.05881
0.25	25.6758	8.44009	3.59938	3.89536	8.97641	7.61842	0.779356
0.3	22.0239	6.91201	4.32837	4.30241	9.11329	7.3001	0.626799
0.35	19.1429	6.02	4.96441	4.77196	9.04353	6.87384	0.533957
0.4	16.7098	5.9071	5.46245	5.21033	8.77023	6.34192	0.47015
0.45	14.581	5.92244	5.79947	5.58233	8.37036	5.74154	0.423738
0.5	12.6878	6.08455	5.95294	5.86894	7.91732	5.10983	0.387934
CCCC AP (45°/−45°/45°/−45°)							
0	218.574	218.172	0	81.5333	0.267E-13	0.997E-14	78.0372
0.05	41.4098	41.048	0.627686	21.7707	4.36334	4.38925	6.30562
0.1	21.1366	20.3584	1.41549	10.8649	4.70274	4.85569	1.40215
0.15	14.4737	13.4026	2.17714	7.33401	4.66064	5.03431	0.717475
0.2	11.2281	10.3188	2.86717	6.34416	4.41508	5.07992	0.467316
0.25	9.38374	9.91916	3.45455	6.02143	4.05778	5.07033	0.349537
0.3	8.22429	9.93015	3.92467	5.97889	3.63452	5.0227	0.290296
0.35	7.51804	10.169	4.27071	6.09006	3.18545	4.94188	0.255266
0.4	7.3717	10.4729	4.50075	6.22201	2.754	4.83174	0.235348
0.45	7.27864	10.7694	4.68129	6.34186	2.63954	4.69654	0.223237
0.5	7.20615	11.0577	4.78488	6.4315	2.75936	4.52604	0.214209
SSSS CP (0°/90°/0°/90°)							
0	425.13	121.13	0	107.21	0.149E-13	0.246E-13	283.564
0.05	45.363	13.198	0.32393	19.311	3.3011	3.2981	14.7489
.1	23.965	7.1652	0.80823	8.8279	3.9499	3.9098	3.46162
0.15	16.4638	4.89399	1.3073	5.79188	4.25936	4.1289	1.72438
0.2	12.525	3.7909	1.7533	4.6514	4.4497	4.1609	1.10135
0.25	9.8315	3.1495	2.1028	4.179	4.4928	4.0239	0.816466
0.3	8.1774	2.7111	2.3142	4.0427	4.3879	3.7304	0.657457
0.35	7.9657	2.6424	2.3764	4.0558	4.157	3.3239	0.56241
0.4	7.9079	2.7886	2.3143	4.1509	3.8367	2.8731	0.494127
0.45	7.8681	2.9313	2.1712	4.2759	3.4702	2.3833	0.441041
0.5	7.7769	3.6799	1.9943	4.3939	3.0899	2.4821	0.398412

(continued)

Table 1 (continued)

RSR	Normal stress σ_x (N/mm^2)	Normal stress σ_y (N/mm^2)	Normal stress σ_z (N/mm^2)	Shear stress τ_{xy} (N/mm^2)	Shear stress τ_{yz} (N/mm^2)	Shear stress τ_{xz} (N/mm^2)	Deflection U_z (mm)
SSSS AP (45°/−45°/45°/−45°)							
0	182.081	182.5	0	227.312	0.2E-13	0.2E-13	206.883
0.05	23.0756	23.3682	0.93245	30.5687	4.9153	4.9478	5.21194
0.1	11.7961	12.3589	1.70136	14.281	4.4355	4.5572	1.43359
0.15	8.08189	8.89111	2.31601	8.92567	3.9528	4.2047	0.724032
0.2	6.30948	7.3488	2.79924	6.32624	3.4870	3.8951	0.476702
0.25	5.27739	6.52235	3.15028	4.79868	3.0316	3.6058	0.365668
0.3	4.58238	5.99269	3.3662	5.34574	2.5849	3.3170	0.311242
0.35	4.06046	5.58493	3.44849	4.8688	2.1550	3.0201	0.279348
0.4	3.69478	5.22271	3.40856	4.46278	1.7543	2.7154	0.259667
0.45	3.4702	4.87493	3.2636	4.09549	1.3955	2.4099	0.247229
0.5	3.50534	4.76719	3.04663	3.62904	1.3140	2.1052	0.239278
CSCS CP (0°/90°/0°/90°)							
0	255.43	52.203	118.81	42.8579	0.685E-15	0.677E-14	118.817
0.05	33.187	13.358	0.27641	5.04185	2.72174	7.04926	14.7862
0.1	17.586	7.5236	0.63049	2.40724	3.54775	7.94738	3.45244
0.15	12.275	5.2167	1.0650	5.78389	3.96772	8.18033	1.67949
0.2	9.2864	3.9758	1.4668	1.97483	4.30841	8.11689	1.07183
0.25	7.3027	3.2351	1.7758	2.74924	4.49017	7.79212	0.786245
0.3	6.5104	2.7718	1.9619	4.76157	4.52356	7.30648	0.630351
0.35	6.3689	2.4294	2.0214	3.67465	4.42337	6.69924	0.533003
0.4	6.4195	2.5567	1.9713	5.63228	4.21449	6.00017	0.466898
0.45	6.5283	2.7261	1.8354	3.90071	3.92502	5.44322	0.417419
0.5	6.6188	2.9048	1.6669	3.83693	3.59081	5.23421	0.376276
CSCS AP (45°/−45°/45°/−45°)							
0	280.96	144.42	113.67	114.424	113.678	113.678	113.678
0.05	42.554	24.636	0.7696	13.4964	3.88404	5.07131	5.78402
0.1	21.7	11.926	1.6319	11.3094	3.91864	5.35198	1.43768
0.15	14.845	7.4868	2.4436	4.82458	3.73513	5.39763	0.729881
0.2	11.567	6.3727	3.1651	6.24479	3.45027	5.41279	0.479979
0.25	9.6863	5.9615	3.7785	5.75231	3.10786	5.36938	0.367297
0.3	8.481	5.8090	4.2638	3.24792	2.73388	5.27241	0.311395
0.35	7.8884	5.7630	4.6125	5.49337	2.34602	5.16344	0.279427
0.4	7.7102	5.7585	4.8256	2.83469	1.96278	5.04137	0.260366
0.45	7.5754	5.7624	5.0014	2.65224	1.73843	4.89842	0.248564
0.5	7.4649	5.7588	5.1046	2.47534	1.85134	4.72937	0.24131

The stresses are nearly equal for normal stress in x and y-direction as shown in Table 1. Normal stresses decrease with increase in RSR while in-plane shear stress increases. Transverse shear stresses initially increases and then decreases after RSR of 0.3 value and deflection decreases considerably with in RSR from 0 to 0.2 and then decrement is continuous but with very minute changes.

3.2 Simply Supported Hyperbolic Paraboloid with Cross-Ply (0°/90°/0°/90°) Lamination

The variation of stresses and deflection with change in rise to span ratio is shown in Table 1. The normal stresses decrease with increase in RSR rapidly in the initial stage and then gradually but decrement is continuous with increase of RSR. As compared to clamped cases, normal stresses are high in case of simply supported cross-ply lamination. As RSR increases the normal stresses are reduced in such a way that the stresses are lower than the clamped case.

3.3 Simply Supported Hyperbolic Paraboloid with Angle-Ply (45°/−45°/45°/−45°) Lamination

The variation of stresses and deflection with change in RSR is shown in Table 1. The stresses in this case are lowest of the above all cases, considering cross-ply or angle-ply with clamped edges.

3.4 Alternately Simply Supported and Clamped with Cross-Ply (0°/90°/0°/90°) Lamination

This case projects the relation between c/a ratio and stresses and deflections as define by Table 1. The in-plane shear stresses are fluctuating in this case i.e. increasing and decreasing with the increase in rise to span ratio showing mysterious behavior. The lowest stress is noticed at 0.2 RSR and the normal stresses decreases up to RSR of 0.3 value, then there is approximately no change with increase in RSR.

3.5 Alternately Simply Supported and Clamped with Angle-Ply (45°/−45°/45°/−45°) Lamination

The variation of stresses and deflection with change in RSR is shown in Table 1. The normal stresses are decreasing with increase in RSR initially. The in-plane

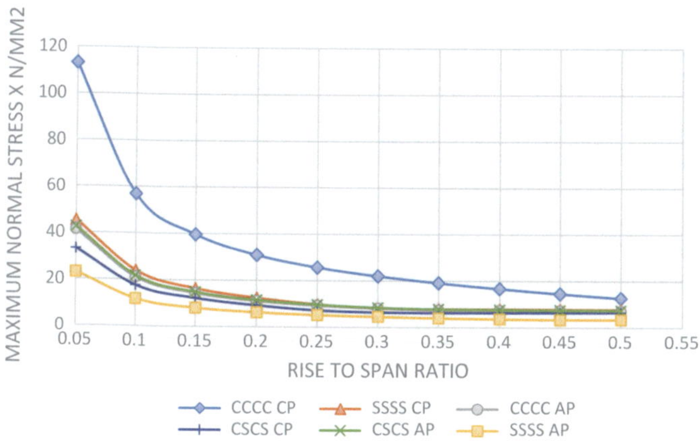

Fig. 5 Comparison of maximum normal stress σ_x with change in rise to span ratio

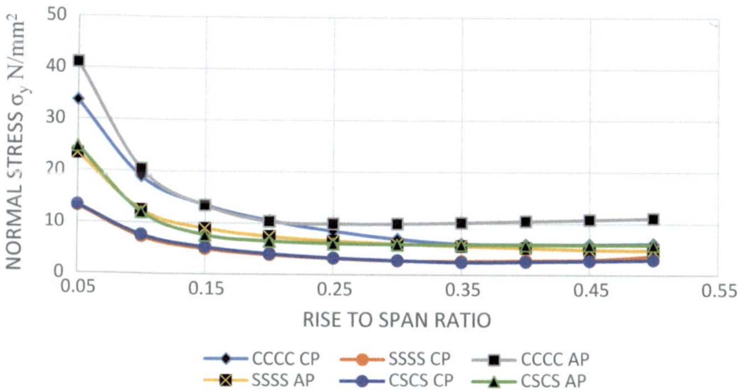

Fig. 6 Comparison of maximum normal stress σ_y with change in rise to span ratio

shear stress fluctuates at a lower rate as compared to the case of CSCS CP. Deflection is lower than other cases for cross-ply laminations.

Hence the stresses and deflection are compared for all cases by analyzing them graphically as below:

- Normal stress σ_x
  The stresses in (CCCC CP) are maximum from 113.087 to 12.68, all other cases are arranged in decreasing order as SSSS CP, CSCS AP, CSCS CP, and minimum in SSSS AP. As shown in Fig. 5.
- Normal Stress σ_y
  The stresses are initially high in case of CCCC AP and decrease up to RSR 0.2 value but remains constant after that, stresses in cases of CSCS CP and SSSS CP are nearly equal. The maximum stress at RSR 0.5 is least in all cases except for the case of CCCC AP. As shown in Fig. 6.

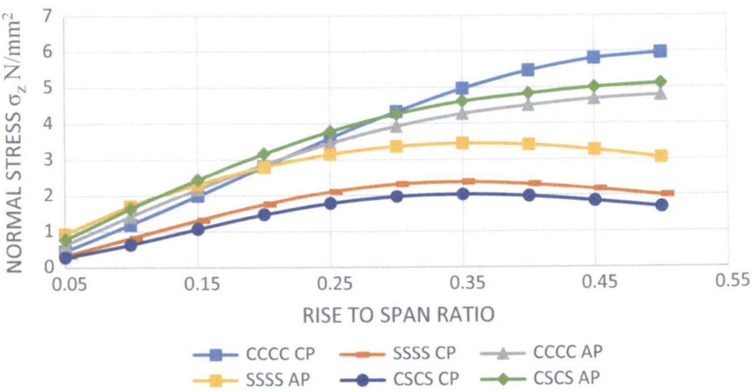

Fig. 7 Comparison of maximum normal stress σ_z with change in rise to span ratio

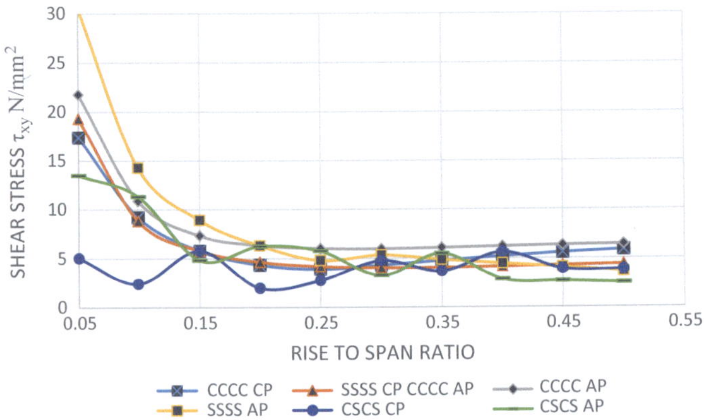

Fig. 8 Comparison of in-plane shear stresses τ_{xy} with change in rise to span ratio

- Normal Stress σ_z
  The stresses are negligible initially at RSR 0.05 but keeps on increasing with increase in RSR. Maximum stress increases more in cases of CCCC CP, CSCS AP, CCCC AP which shows that except for the case of clamped edges, minimum stresses are observed in case of cross-ply laminate. As shown in Fig. 7.
- In-plane shear stress τ_{xy}
  The stresses in cases of CSCS boundary conditions are fluctuating with increase in rise to span ratio, in all other cases it decreases greatly up to RSR 0.2 and thereafter little change is noticed. As shown in Fig. 8.
- Transverse shear stress τ_{yz}
  The stresses for all cross-ply laminates increases initially with increase in RSR up to 0.3 and then decreases, while continues decrease of stresses is noticed in

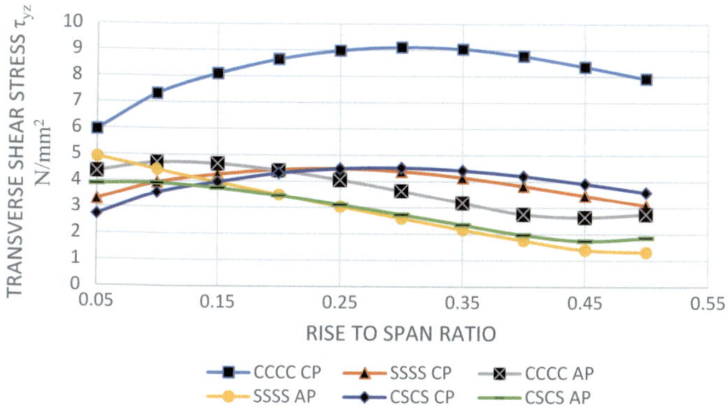

Fig. 9 Comparison of maximum transverse shear stresses τ_{yz} with change in rise to span ratio

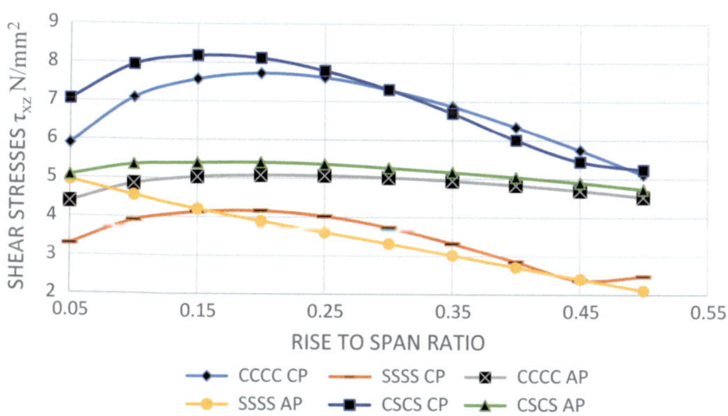

Fig. 10 Comparison of maximum transverse shear stresses τ_{xz} with change in rise to span ratio

the cases of angle-ply laminates. Maximum stress is in the cases of CCCC CP. As shown in Fig. 9.

- Transverse shear stress τ_{xz}
  The stresses are maximum for case of CSCS CP. Change of stresses in case of CSCS AP and CCCC AP is very small. Gradual decrease of stresses is noticed in case of SSSS AP. As shown in Fig. 10.
- Deflection U_z
  The deflection is maximum initially up to RSR value 0.3 for all cross-ply cases as compared to angle-ply. Deflection decreases considerably up to RSR 0.25 for most of the cases. Deflection at RSR 0.5 is nearly equal for all cases but not depends on boundary conditions or type of laminations. As shown in Fig. 11.

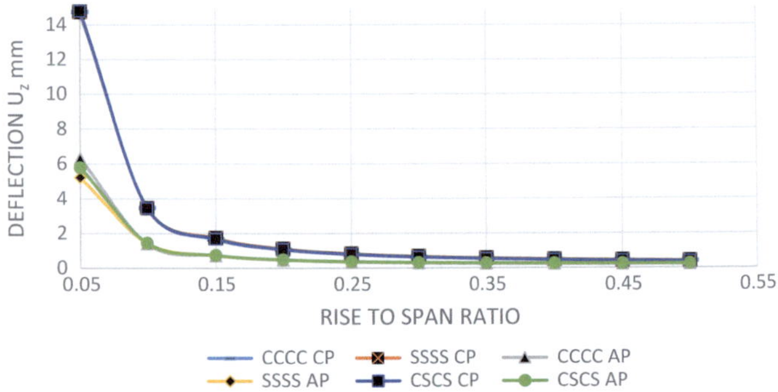

Fig. 11 Comparison of deflection U_z with the change in rise to span ratio

4 Conclusions

In the present study an attempt has been made to analyze the most favorable RSR for a hyperbolic paraboloid considering the static characteristics and different boundary conditions under uniform pressure. Following conclusions have been made based on the analysis of present problem:

- Angle ply performs better than cross ply laminates in all cases for static deflections. Deflection decrease in all cases, initially decreasing with large difference when rise to span ratio is increased from 0.05 to 0.3 in all cases studies.
- As compared to a plate of same geometry (RSR = 0), boundary conditions, and loadings, to hyperbolic paraboloid a small rise of edges decreases normal stresses and deflection to a large extent in all cases discussed.
- Rise to span ratio shows considerable changes in normal and shear stresses up to RSR 0.3 in most of the cases, beyond which increase of RSR shows little change of stresses.
- Maximum stresses are noticed in the case of all sides' clamped boundary conditions. For case of CSCS, angle-ply laminates show higher stresses than cross-ply laminates.
- As per the results, derived from present study the most favorable Rise to Span Ratio for all cases considered is from 0.2 to 0.3, as the further increase of rise to span ratio shows negligible change of stresses.

References

1. Bolster ED, Cuypers H, Itterbeeck PV, Wastiels J, Wilde WPD (2008) Use of hypar-shell structure with textile reinforced cement matrix composites in lightweight constructions. Compos Sci Technol Elsevier
2. Bernasconi J (2012) Buckling studies into large scale hyperbolic paraboloid shell and lattice structures. PhD thesis, University of Queensland
3. Bakshi K, Chakravorty D (2013) Static and dynamic characteristics of composite conoidal shell roofs. Hindawi Publishing Corporation, advances in acoustic and vibration
4. Chen D, Hao J, Wang Z (2011) Study on shape formation of post-tensioned and shaped hypar space truss. Adv Mater Res 243–249
5. Das HS, Chakravorty D (2009) A finite element application in the analysis and design of point-supported composite conoidal shell roofs: suggesting selection guidelines. JSA582
6. Kim JW, Kim JJ, Rhew HJ (2005) Analysis and experiment for the formation and ultimate load testing of a hypar space truss. J Constr Steel Res Elsevier
7. Kumari S, Chakravorty D (2010) Study of static characteristics of delaminated composite conoidal shell subjected to point load. Adv Mater Res
8. Kumari S, Chakravorty D (2011) Bending of delaminated composite conoidal shells under uniformly distributed load. Am Soc Civil Eng
9. Kumar A, Bhargava P, Chakrabarti A (2012) Vibration of laminated composite skew hypar shells using higher order theory. Thin Wall Struct Elsevier
10. Kumar A, Bhargava P, Chakrabarti A (2012) Natural frequencies and mode shapes of laminated composite skew hypar shells with complicated boundary conditions using finite element method. Adv Mater Res 585
11. Laursen PT, Saliklis EP (2012) Finite-element limit analysis of the tucker high school gymnasium roof failure. J Arch Eng 18, ©ASCE
12. Neogi SD, Karmakar A, Chakravorty D (2011) Impact response of simply supported skewed hypar shell roofs by finite element. J Reinf Plast Compos
13. Pradyumna S, Bandyopadhyay JN (2010) Dynamic instability behaviour of laminated hypar and conoidal shells using a higher-order shear deformation theory. Thin Wall Struct Elsevier
14. Pradhan N, Jena J (2012) Static characteristics of stiffened conoidal shell roofs under concentrated load. Int J Adv Eng Technol
15. Rha CS, Kim SN, Yu E (2015) Behaviour of gabled hyperbolic paraboloid shells. J Asian Arch Build Eng
16. Sahoo S, Chakravorty D (2005) Deflections, forces, and moments of composite stiffened hypar shell roofs under concentrated load. JSA117 IMechE
17. Sahoo S, Chakravorty D (2005) Finite element vibration characteristics of composite hypar shallow shells with various edge supports. J Vib Control Sage
18. Sahoo S, Chakravorty D (2006) Stiffened composite hypar shell roofs under free vibration: behaviour and optimization aids. J Sound Vib Elsevier
19. Sahoo S (2012) Design and selection guidelines for composite hypar shell roofs with cutouts. Acad Res J IJACEA

Behavior of Rectangular Laminated Composite Plate with Central Hole Under Transverse Loading

Sumit Rathi, Yogesh Kaushik and Sabita Madhvi Singh

Notations

E_1, E_2, E_3	moduli of elasticity of lamina x, y, and z directions
G_{12}, G_{23}, G_{13}	shear moduli of a lamina in xy, yz, and xz directions
μ_{12}, μ_{23}, μ_{13}	Poison's ratio in xy, yz, and xz directions
1, 2, 3	x, y, and z directions
D/H	hole diameter to width of the plate

1 Introduction

Composite materials consist of a combination of materials that are mixed together to achieve specific structural properties. The properties of the composite material are superior to the properties of the individual materials from which it is constructed. The composite plates consist of layers, and it can be single layer or multilayer. The composite materials are also based on the symmetric; they can be symmetric or antisymmetric. In this study, the layer study has applied on the rectangular plate with central circular hole with different orientations of layers. The study includes the variation of ratio, but the material remains same. The analysis is carried out to identify the behavior of laminate composite plate with central circular hole with the help of ANSYS mechanical APDL 15.0 software.

S. Rathi · Y. Kaushik (✉) · S.M. Singh
Department of Civil Engineering, Amity University,
Noida 201313, Uttar Pradesh, India
e-mail: kaushik.yogesh2@gmail.com

S. Rathi
e-mail: er.sumit.rathi@gmail.com

S.M. Singh
e-mail: smsingh1@amity.edu

S. Singh et al. (eds.), *Proceedings of the International Conference on Modern Research in Aerospace Engineering*, Lecture Notes in Mechanical Engineering,
https://doi.org/10.1007/978-981-10-5849-3_21

Composite materials have been used in aircraft industry for the lightweight, high-temperature-resistant, and high-performance structure which bend in one direction but not in another. The aerospace industry including military and commercial aircrafts uses composite materials for the high stiffness, high strength, and high performance properties of fiber-reinforced composite materials.

The comparison of stress and deflection on the isometric and orthotropic plates with central circular hole under tension load was calculated by Abdullah [1]. It was done with the help of finite element method carried by the ANSYS. The stress analysis of composite plates with different types of cutouts was carried out by Riyah and Ahmed [7]. In the normal strain at the circular and square holes with the different numbers of layers and types of composite materials with the help of strain gauges under tensile loads. Various orientation angles were used for the analysis. The stress and load–displacement analysis of fiber-reinforced composite laminates with a circular hole under compressive load was done by Manoharan and Jeevanantham [5] in which the 3-D orientation of the fibers has been analyzed with the help of finite element method carried by ANSYS.

The finite element analysis of the shear deformable laminated composite plate was reported by Kam and Chang [4] in which the Mindlin plate theory was used for the finite element formulation and shear correction factors. Harn C. Chen obtained the stress analysis of laminates with hole by special finite element method [2] in which the complex variable technique has been used to formulate the problem and to find the solution. The stress around square and rectangular cutouts in symmetric laminates was calculated by Nageswara Rao et al. [6] in which the complex variable technique and mapping function have been used to formulate the problem for finding the solution. A general solution for stress resultants and moments around holes in unsymmetric laminated was given by Ukadgoanker and Rao [9] in which the complex variable method of Lekhnitskii and Savin has been applied in the paper.

The elastic–plastic stress analysis of simply supported and clamped metal–matrix-laminated plates with a hole was achieved by Sayman and Aksoy [8] in which the classical lamination theory was applied to formulate the problem. Dheeraj Gunwant and J.P Singh represented the stress and displacement analysis of the rectangular plate with central elliptical hole [3], in which the finite element method was carried out with the help of ANSYS and von Mises stresses, and deflection of the plate was calculated.

This study reveals the effect of orientation angle with different ratio of the diameter of hole to plate width on the axial stresses, shear stresses, and deflections in the end conditions of clamped edges. The dimensions of the plate remain same throughout all cases except the diameter of hole. The angle orientation is also imposed for the symmetric and antisymmetric. The stress and deflection changes show the impact of the angles and diameter of hole.

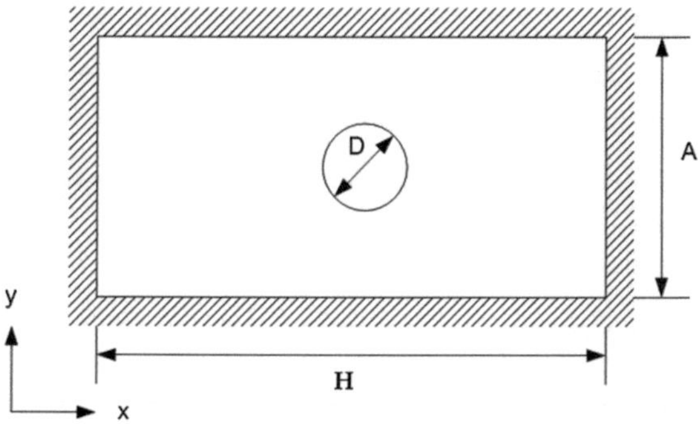

Fig. 1 Rectangular laminate with clamped boundary conditions on all four sides

2 Mathematical Formulation

A four-layer rectangular laminate in symmetric (60°/-60°/-60°/60°) and antisymmetric (60°/-60°/60°/-60°) stacking with central circular hole with equal thickness of layer with end conditions of fixed on all four sides (shown in Fig. 1) under transverse uniform pressure load is analyzed by ANSYS mechanical APDL 15.0. Also, the comparison has been done with the four-layer rectangular laminate in same symmetric (60°/-60°/-60°/60°) and antisymmetric (60°/-60°/60°/-60°) stacking without hole. The material properties used for the plate are $E_1 = 39 \times 10^3$ N/mm^2, $E_2 = E_3 = 8.6 \times 10^3$ N/mm^2, $G_{12} = G_{23} = G_{13} = 3.8 \times 10^3$ N/mm^2, and $\mu_{12} = \mu_{23} = \mu_{13} = 0.28$. The uniform pressure is taken as 0.005 N/mm^2 for analysis. In the present study, the effect of D/H ratio varying from 0.1 to 0.5 with increment of 0.1 has been studied on maximum stresses (σ_x, σ_y, τ_{xy}) and maximum deflection (Uz) for all layers. Length and width of the plate are taken as 1000 mm and 500 mm (x and y directions), respectively. The laminate considered for the study has four layers with 2 mm thickness (z direction) of each layer

$$
\begin{aligned}
u = & \frac{1}{4}(u_I(1-s)(1-t)(-s-t-1) + u_J(1+s)(1-t)(s-t-1) \\
& + u_K(1+s)(1+t)(s+t-1) + u_L(1-s)(1+t)(-s+t-1)) \\
& + \frac{1}{2}(u_M(1-s^2)(1-t) + u_N(1+s)(1-t^2) \\
& + u_O(1-s^2)(1+t) + u_P(1-s)(1-t^2)).
\end{aligned}
\tag{1}
$$

ANSYS 15.0 is used for the finite element modeling of the plate. The eight noded shell elements Shell 281 with element length of 10 mm is selected for the plate modeling. Each node has six degrees of freedom. The shape function used for the shell 281 is shown in Eq. 1. The model is created for which the stress and deflection variations are shown in Figs. 2 and 3. The effect of change of D/H ratio from 0.1 to 0.5 on the stresses and deflections is shown in the form of graphs and tables (under Results and Discussions phase). The symmetric (60°/-60°/-60°/60°) and antisymmetric (60°/-60°/60°/-60°) sequences for 60° have been discussed.

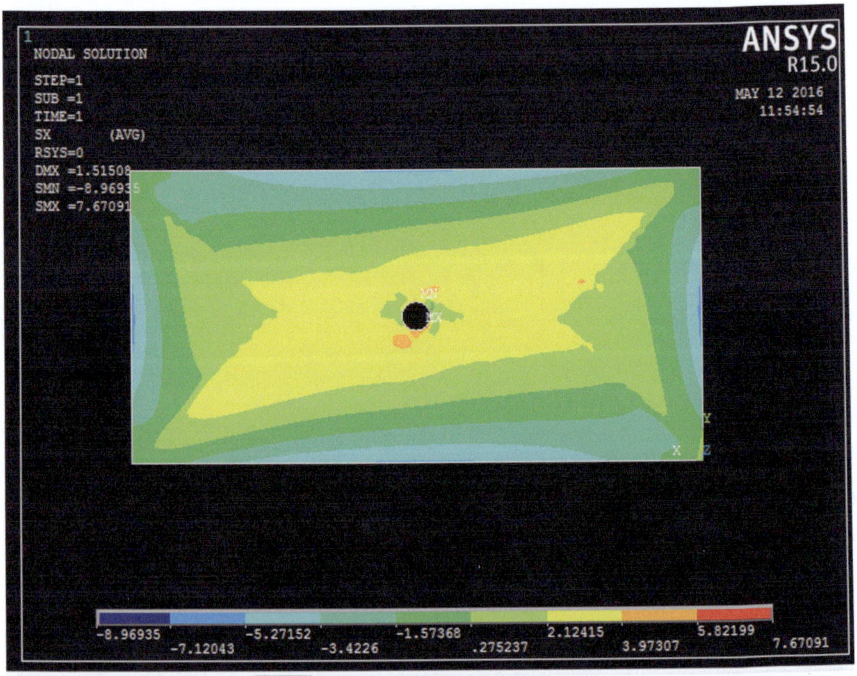

Fig. 2 Stresses in x direction clamped case 60° orientation angle

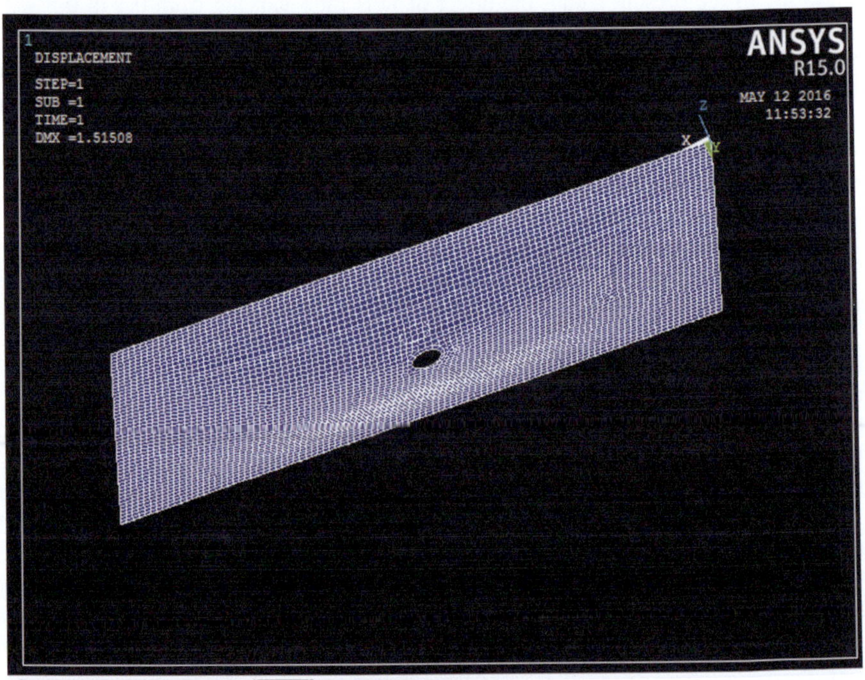

Fig. 3 Deflection clamped case 60° orientation angle

3 Results and Discussions

The case of clamped boundary conditions with symmetric and antisymmetric stacking has been studied in the present study in which the stress and deflection due to the increase in the ratio of D/H from 0.1 to 0.5. The stress and deflection changes have been observed according to the orientation and D/H ratio and the following observation is made to meet the desired results as discussed below.

3.1 *Comparison of Symmetric (60°/-60°/-60°/60°) and Antisymmetric (60°/-60°/60°/-60°) with Hole*

1. In Fig. 4, the stresses in x direction are shown. The stresses in symmetric case are varying from 6.3 to 6.5 Mpa and for antisymmetric case from 7.17 to 5.5 Mpa. The stresses in x direction vary in a manner so that the stresses in symmetric case are less only for 0.1, 0.4 ratio, and for other values of ratio, it shows larger stresses for the antisymmetric case.
2. In Fig. 5, the stresses in y direction are shown. The stresses in symmetric case are varying from 11.08 to 11 Mpa and for antisymmetric case from 13.01 to 10.12 Mpa. The stresses in y direction vary in a manner so that the stresses in symmetric case are low for 0.1, 0.4 ratio, and for other values of ratio, stresses in antisymmetric are more.
3. In Fig. 6, the shear stresses in xy direction are shown. The stresses in symmetric case are varying from 3.95 to 3.52 Mpa and for antisymmetric case from 5.44 to 3.6 Mpa. The shear stresses in xy direction are varying in a manner so that the stresses in symmetric case are lower than the stresses in antisymmetric.
4. In Fig. 7, deflections in z direction are shown. The deflections in symmetric case are varying from 1.14 to 1.04 mm and for antisymmetric case from 1.23 to 1.09 mm. The deflections in z direction vary in a manner so that the deflection in symmetric case is less than the deflection in antisymmetric.

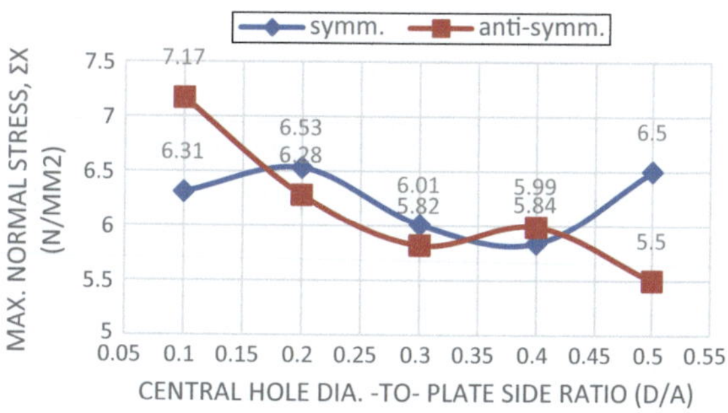

Fig. 4 Comparison of stresses in x direction for 60° angle

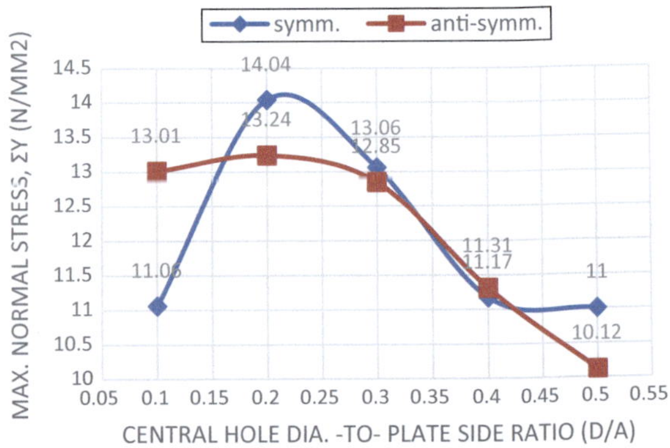

Fig. 5 Comparison of stresses in y direction for 60° angle

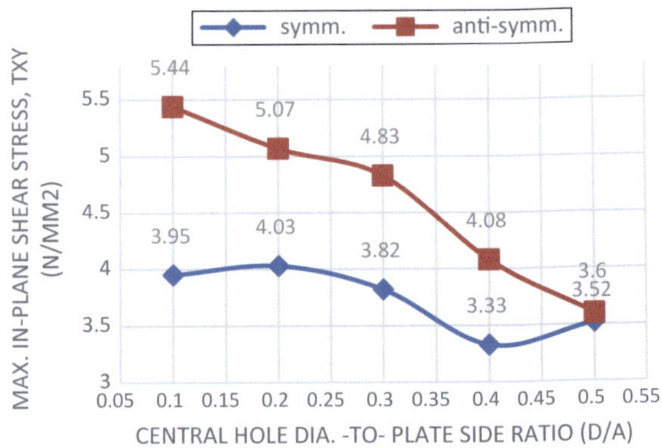

Fig. 6 Comparison of shear stresses in xy direction for 60° angle

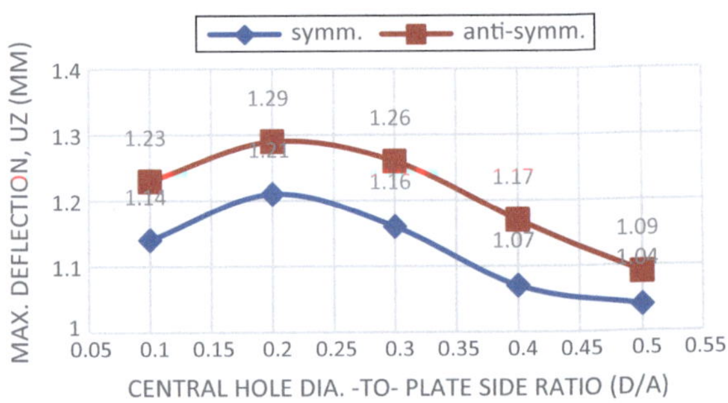

Fig. 7 Comparison of deflections in z direction for 60° angle

3.2 *Comparison of Symmetric (60°/-60°/-60°/60°) and Antisymmetric (60°/-60°/60°/-60°) Without Hole*

1. The stress in x direction for the symmetric case is 4.3534 N/mm^2 which increases for the antisymmetric case at 4.39 N/mm^2.

Table 1 Maximum stresses and deflection in laminates

	Stacking	Max. normal stress σ_x (N/mm^2)	Max. normal stress σ_y (N/mm^2)	Max. in-plane shear stress τ_{xy} (N/mm^2)	Max. deflection U_z (mm)
With hole	Symmetric	6.5 at top and bottom	14.04 at top and bottom	3.95 at top and bottom	1.21
	Antisymmetric	7.17 at top and bottom	13.24 at top and bottom	5.44 at top and bottom	1.29
Without hole	Symmetric	4.35 at top boundaries at width	9.75 at top boundaries at length	3.50 at top boundaries at length	0.717
	Antisymmetric	4.39 at top boundaries at width	9.42 at top boundaries at length	1.53 at top throughout the plate	0.761

2. The stress in y direction for the symmetric case is 9.75 N/mm^2 which decreases for the antisymmetric case at 9.42 N/mm^2.
3. The shear stress in xy direction for the symmetric case is 3.50 N/mm^2 which decreases for the antisymmetric case at 1.53 N/mm^2.
4. The deflection for the symmetric case is 0.717 mm which increases for the antisymmetric case at 0.761 mm.

Table 1 depicts the maximum stresses and deflections in the laminates as defined below.

4 Conclusion

The present study points out the stresses and deflections in the rectangular laminated plate with central circular hole under transverse loading for symmetric and unsymmetric orientation. The following conclusions are drawn:

- In laminates with hole, the stress variation through the thickness of lamina is discontinuous, but due to stacking orientation in symmetric (60°/-60°/-60°/60°) and antisymmetric (60°/-60°/60°/-60°), maximum stresses for x, y, and xy directions appears at the top and the bottom layers of the plate at the boundaries of hole not in the middle layer.
- In laminates with hole due to the stacking sequence difference in first and last layers, the stresses in x direction and shear stresses in xy direction increase in antisymmetric case.
- Due to symmetric, D/H does not produce large variation and specific order in stresses and deflection, but in antisymmetric case (60°/-60°/60°/-60°), the D/H ratio increases normal stresses and shear stresses, and deflections decrease,

because as the diameter of hole goes small, the area of the plate increases and stresses get distributed on larger area.

- The stresses and deflections in antisymmetric found to be in directly proportional.
- The maximum stresses and deflections are less in case of laminates without hole than laminates with hole, because insertion of hole in the plate increases stresses and deflections.
- The maximum stresses in laminate without hole are distributed on the sides at the top area unlike the laminates with hole, because with hole insertion only the boundaries of hole will have maximum stresses which can be at the top area or the bottom area of the plate.

References

1. Ahmed MA (2011) Comparison the stresses and deflections of an isotropic and orthotropic rectangular plates with central circular hole under tension load. AL-Rafidain Eng J 20(4): 137–153
2. Chen CH (1995) Stress analysis of laminates with holes by special finite element. Compos Struct 31:99–106
3. Gunwant D, Singh JP (2013) Stress and displacement analysis of a rectangular plate with central elliptical hole. Int J Eng Innov Technol 3(3):387–392
4. Kam TY, Chang RR (1993) Finite element analysis of shear deformable laminated composite plates. J Energy Res Technol 115:41–48
5. Manoharan R, Jeevanantham AK (2011) Stress and load-displacement analysis of fiber reinforced composite laminates with a circular hole under compressive load. ARPN J Eng Appl Sci 6(4):64–74
6. Nageswara Rao DK, Babu MR, Reddy KRN, Sunil D (2010) Stress around square and rectangular cutouts in symmetric laminates. Compos Struct 92:2845–2859
7. Riyah NK, Ahmed NE (2009) Stress analysis of composite plates with different types of cutouts. Anbar J Eng Sci 2(1):11–29
8. Sayman O, Aksoy S (2001) Elastic-plastic stress analysis of simply supported and clamped aluminum metal-matrix laminated plates with a hole. Compos Struct 53:355–364
9. Ukadgaonker VG, Rao DKN (2000) A general solution for stress resultants and moments around holes in unsymmetric laminates. Compos Struct 49:27–39

Effect of Aspect Ratio Change on Nonlinear Resonance of Rectangular Aircraft Panel

Nisar Ali and M.S. Mahesh

1 Introduction

Normally, airframe surroundings in the flight are rich in vortex motion of air. The vortices are appearing naturally through viscous effects due to boundary layer separation from aircraft surface or three-dimensional geometrical configuration of the airframe. Today, vibration of airframe panels continues to be a difficulty, generating undesirable cabin noise and causing catastrophic failure in some cases. The initial case of problems related to panel flutter is believed to be the failure of German V2 missiles during World War II [1]. The cause-and-effect associations are not well understood, but limit cycle oscillations often seem to be associated with the event of stimuli such as vortices in subsonic flow and shock waves in supersonic flow [2]. In this paper, the effects of aspect ratio change on nonlinear resonance of a thin rectangular aircraft panel clamped on all edges with forcing frequency near to the frequency of the first linear mode of the rectangular aircraft panel are studied. In this study of aircraft panel oscillations all edges clamped rectangular plates are considered and it is believed to be representative of typical aircraft panels, which are generally attached to structural members of the airframe.

When the plate is forced at frequencies near to the natural frequencies of the linear mode, a nonlinear resonance phenomenon arises leading to a complicated limit cycle behaviour. Limit cycle oscillations are complicated oscillations of aircraft surfaces for which the cause-and-effect relationships are poorly understood [3]. Nonlinearities in the structure or the aerodynamic forces lead to limiting the amplitude of oscillations of an unstable system [4]. It is observed that very complicated vibration

N. Ali (✉) · M.S. Mahesh
Department of Aerospace Engineering, Defense Institute of Advanced Technology (DU), Girinagar, Pune, Maharashtra, India
e-mail: nisar.ada2008@gmail.com

M.S. Mahesh
e-mail: maheshms7@gmail.com

S. Singh et al. (eds.), *Proceedings of the International Conference on Modern Research in Aerospace Engineering*, Lecture Notes in Mechanical Engineering,
https://doi.org/10.1007/978-981-10-5849-3_22

motions can occur with bigger aircraft panels and, basically, disturbed motion can develop. Linear and nonlinear aeroelastic responses of panels with fixed supports on all four sides are called panel flutter and are generally of interest with respect to specific, local skin panel oscillations on a wing or fuselage. Limit cycle oscillations (LCOs) of the order of the plate thickness are possible by nonlinear, aeroelastic effect [5]. An aircraft can experience LCOs even though the oscillations are in the linear structural and aerodynamic lift ranges.

Aircraft panels are subjected to unsteady loads in operation due to vortex motion and turbulence of air. In this study, vibrations of clamped beam and plate, which are subjected to forcing near a linear resonance frequency by a convective pressure wave of the form [6]

$$p = P_0 \sin(\alpha x - \omega t), \tag{1}$$

will be considered. Here, p denotes the pressure difference in the fluid above, and below the plate, t is time, α is the wave number, ω is the forcing frequency, and the forcing amplitude P_0 is assumed constant. When the solution for w is written in expression 17 and orthogonality is applied, the net result is that the equation for each mode contains a forcing function involving $\cos(\omega t)$ and $\sin(\omega t)$. An equivalent formulation is to take the forcing pressure to be

$$\begin{aligned} \text{for beam,} \quad & p = P_0 \sin(\omega t) X_1(x) \\ \text{for plate,} \quad & p = P_0 \sin(\omega t) \phi_1(x, y) \end{aligned} \tag{2}$$

where $X_1(x)$ and $\phi_1(x, y)$ are the first eigenfunctions of the linear problem for beam and plate. Other modes come into play, but the dominant response is associated with $\phi_1(x, y)$ for ω near ω_1. Here, the interest is in forcing frequencies near the first linear frequency, and therefore, let $\omega = (1 + \epsilon_0)\omega_1$, where ϵ_0 is assumed small and ω_1 is the first eigenvalue of the linear problem for beam and plate.

2 Governing Equations

The classical theory of large deformations of elastic plates was developed by Kirchhoff [7] and later extended by Von Karmann and Herrmann [8] is used in this study. Large amplitude deflections of plates are explained by a formulation normally credited to Von Karmann [9]. Analytical solution for rectangular plate clamped at four edges is not known. In recent times, Gorman [10] presented analytical series solutions for different boundary conditions. Various approximation methods were later developed and reviewed [11]. The governing partial differential equations were firstly derived through consideration of a free body diagram of a deformed plate element. The formulation was refined by Herrmann [8] to account for unsteady deformations. Consider a plate of thickness h and dimensions a and b in each of the $(\hat{x}, \hat{y})$ directions, respectively, along two edges of the plate as shown in Fig. 1.

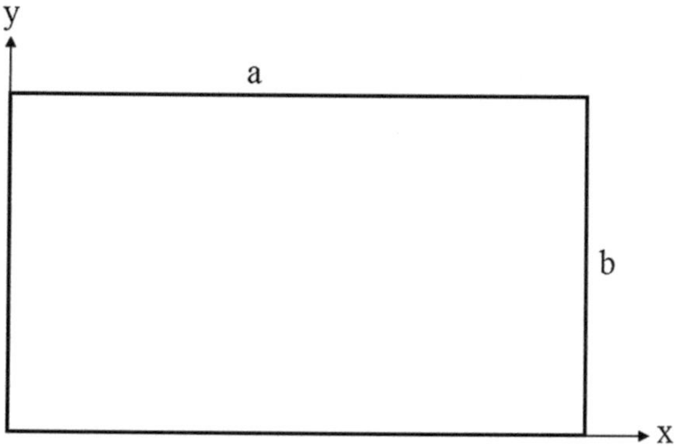

Fig. 1 Rectangular aircraft panel

Displacements in the $(\hat{x}, \hat{y}, \hat{z})$ directions are denoted by $(\hat{u_x}, \hat{v_y}, \hat{w})$, respectively, where $\hat{z}$ measures distance normal to the equilibrium position of the plate centre. It is assumed that

1. The aspect ratio, $r = \frac{b}{a}$, is $O(1)$, and plate is thin means that $\hat{e} = \frac{h}{a} \ll 1$.
2. w is $O(\hat{e})$, and the gradients $\frac{\partial \hat{w}}{\partial \hat{x}}, \frac{\partial \hat{w}}{\partial \hat{y}}$ are small.
3. All strain components are small, and Hookes law holds.
4. There is no applied force in the $\hat{x}$ and $\hat{y}$ directions.

Kirchhoff's hypothesis holds. The dimensionless plate equations are: [12]

$$\frac{\partial N_x}{\partial x} + r\frac{\partial N_{xy}}{\partial y} = 0 \tag{3}$$

$$\frac{\partial N_{xy}}{\partial x} + r\frac{\partial N_y}{\partial y} = 0 \tag{4}$$

$$\frac{\partial^4 w}{\partial x^4} + 2r^2\frac{\partial^4 w}{\partial x^2 \partial y^2} + r^4\frac{\partial^4 w}{\partial y^4} + \frac{\partial^2 w}{\partial t^2} = 12\left(N_x\frac{\partial^2 w}{\partial x^2} + 2rN_{xy}\frac{\partial^2 w}{\partial x \partial y} + r^2 N_y\frac{\partial^2 w}{\partial y^2}\right) - p \tag{5}$$

where p is the forcing function and dimensionless stress resultants are:

$$N_x = \frac{\partial u}{\partial x} + vr\frac{\partial v}{\partial y} + \frac{1}{2}\left(\frac{\partial w}{\partial x}\right)^2 + \frac{1}{2}vr^2\left(\frac{\partial w}{\partial y}\right)^2 \tag{6}$$

$$N_y = r\frac{\partial v}{\partial y} + v\frac{\partial u}{\partial x} + \frac{1}{2}r^2\left(\frac{\partial w}{\partial y}\right)^2 + \frac{1}{2}v\left(\frac{\partial w}{\partial x}\right)^2 \tag{7}$$

$$N_{xy} = \frac{1}{2}(1 - v)\left[r\frac{\partial u}{\partial y} + \frac{\partial v}{\partial x} + r\frac{\partial w}{\partial x}\frac{\partial w}{\partial y}\right] \tag{8}$$

For beam, aspect ratio $r = \frac{b}{a} \longrightarrow 0$ and from Eq. 5

$$\frac{\partial^4 w}{\partial x^4} + \frac{\partial^2 w}{\partial t^2} - 6\left[\int_0^1 \left(\frac{\partial w}{\partial x}\right)^2 dx\right]\frac{\partial^2 w}{\partial x^2} = p \tag{9}$$

This is beam equation with nonlinear and unsteady term. The boundary conditions for a clamped plate are used in this study:

$$u = v = w = \frac{\partial w}{\partial x} = 0 \; at \; x = 0, 1, \quad u = v = w = \frac{\partial w}{\partial y} = 0 \; at \; y = 0, 1 \tag{10}$$

Assume that the solution is

$$w(x) = \sum_{i=1}^{N} c_i(t) X_i(x) \tag{11}$$

where X_i clamped–clamped beam eigenfunctions. The clamped–clamped beam eigenfunctions are [6]

$$X_i(x) = (\cosh(\gamma_i x) - \cos(\gamma_i x)) - \frac{(\cosh(\gamma_i) - \cos(\gamma_i))}{(\sinh(\gamma_i) - \sin(\gamma_i))}(\sinh(\gamma_i x) - \sin(\gamma_i x)) \tag{12}$$

where $\gamma = \omega^{1/2}$ represents the eigenvalues. Substituting the expression of w in Eq. 9 and multiplying by X_n and using the eigenfunctions orthogonality, it can be easily shown that

$$\frac{d^2 c_n(t)}{dt^2} + \omega_n^2 c_n(t) + 6\left(\sum_{j=1}^{N}\sum_{k=1}^{N} c_j(t) c_k(t) I_{jk}\right)\sum_{i=1}^{N} c_i(t) I_{in} = \int_0^1 p(x,t) X_n(x) dx \tag{13}$$

where $n = 1, 2, 3, \ldots$ and $I_{ij} = \int_0^1 X_i'(x) X_j'(x) dx$ [13].

The differential equations governing the large amplitude oscillations of rectangular plates, when posed in terms of deflections u, v and w, are given by Eqs. 3 to 8. The boundary conditions for a clamped plate are given in 10. A general set of ordinary differential equations is obtained using the following basis functions:

$$u(x,y,t) = \sum_{i=1}^{K^2} a_i(t)\psi_i(x,y), \quad v(x,y,t) = \sum_{i=1}^{K^2} b_i(t)\psi_i(x,y)$$

$$w(x,y,t) = \sum_{i=1}^{M} c_i(t)\phi_i(x,y), \quad \psi_i(x,y) = \sin(m\pi x)\sin(n\pi y)$$

where $\phi_i(x, y)$ are the eigenfunctions of the linear problem ordered in an array corresponding to increasing natural frequencies:

$$\phi = \sum_{i=1}^{N} \sum_{j=1}^{N} a_{ij} X_i(x) X_j(y) \tag{14}$$

where $X_i(x)$ and $X_j(y)$ are the beam eigenfunctions Eq. 12, a_{ij} are coefficients were determined separately, and N is the total number of modes retained in each direction.

Substituting the expansions of u, v and w into Eqs. 3 and 4, multiplying the resulting equations by ψ_k, integrating over the plate surface and using the orthogonal properties of the sine functions, it can be shown that

$$\alpha_0 r \sum_{i=1}^{K^2} F_{ik} b_i + \frac{1}{4} a_k \pi^2 \{(m(k))^2 + \beta_0 r^2 (n(k))^2\} = \sum_{i=1}^{M} \sum_{j=1}^{M} M_{ijk} c_i c_j \tag{15}$$

$$\alpha_0 r \sum_{i=1}^{K^2} F_{ik} a_i + \frac{1}{4} b_k \pi^2 \{\beta_0 (m(k))^2 + r^2 (n(k))^2\} = \sum_{i=1}^{M} \sum_{j=1}^{M} N_{ijk} c_i c_j \tag{16}$$

where $\alpha_0 = \frac{1}{2}(1 + \nu)$ and $\beta_0 = \frac{1}{2}(1 - \nu)$. Consequently, once an estimate of c is known, the coefficients a and b for u and v can be evaluated directly from Eqs. 15 and 16. For the c, the expression given in Eq. 14 is substituted into Eq. 5, which is then multiplied by ϕ_n and integrated over the plate surface. Using the orthogonal properties of the basis function leads to [6]

$$\frac{d^2 c_n}{dt^2} + \omega_n^2 c_n = \int_0^1 \int_0^1 p\phi_n dxdy + 12\left\{ \sum_{i=1}^{M} \sum_{j=1}^{K^2} U_{ijn} a_j c_i + \sum_{i=1}^{M} \sum_{j=1}^{K^2} V_{ijn} b_j c_i \right. \tag{17}$$
$$\left. + \sum_{i-1}^{M} \sum_{j-1}^{M} \sum_{k=1}^{M} W_{ijkn} c_i c_j c_k \right\} \quad \text{where } n = 1, 2, 3, \ldots$$

3 Results and Discussions

In this study, Runge–Kutta fourth-order method with fixed time step 0.01 is used to solve nonlinear beam equation 13 and plate equation 17 by using MATLAB.

The values of I are evaluated separately and stored. The nonlinear plate equation is more challenging to solve because of the complexity of the nonlinear term in Eq. 17. The quantities F_{ik}, M_{ijk}, N_{ijk}, U_{ijn}, V_{ijn} and W_{ijkn} were calculated precisely for each value of aspect ratio r, with Simpsons $\frac{1}{3}$ rule of double definite integral. 251 integration points are considered to evaluate the definite integral by using MATLAB in each of the coordinate directions, before initiation of a specific time-dependent integration for a given physical case. The value of a and b is evaluated by inverting the matrix on the right-hand side of Eqs. 15 and 16. $M = 12$, $K = 9$ and number of

Table 1 Effect of aspect ratio change on fundamental frequencies

$r = \frac{b}{a}$	0.00	0.25	0.50	0.75	1.00
ω_1	22.373	22.799	24.578	28.701	35.986
ω_2	61.673	24.148	31.837	48.175	73.395
ω_3	120.903	26.571	44.767	67.539	73.412
ω_4	199.859	30.256	63.337	80.184	108.267

modes in each direction are used in this numerical solution for convergence of results. The fundamental frequency of the clamped panel is increasing, while the aspect ratio r is increasing as shown in Table 1.

The shape of the panel normally corresponds to the first linear mode. The panel is initially at rest, and motion is activated only by the forcing function. The plotted figures represent the time history of the centre point of the panel. Results are plotted for different values of r, ϵ_0 and $P_0 = 20$ as shown in Figs. 2, 3, 4, 5 and 6. The vibrational motion shows two frequencies: one is faster and another is slower beating frequency. The amplitude of vibration increases and reaches to a maximum value when the forcing frequency approaches the linear resonance frequency and the beating frequency gradually decreases. The pattern of the beating frequency is distorted and considerably different from a simple sine wave as $\epsilon_0 = 0$ is reaches its critical value. The

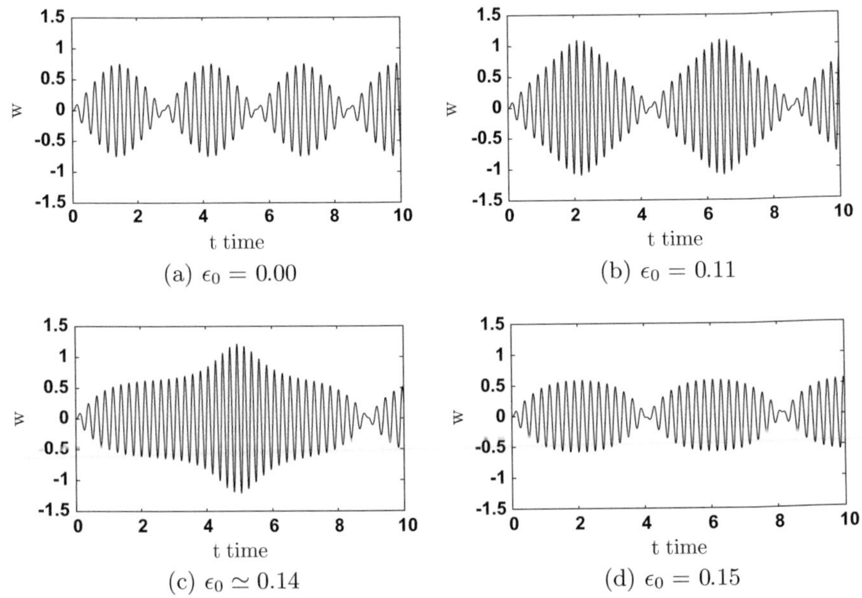

Fig. 2 Centre point displacement of beam at $r = 0.00$

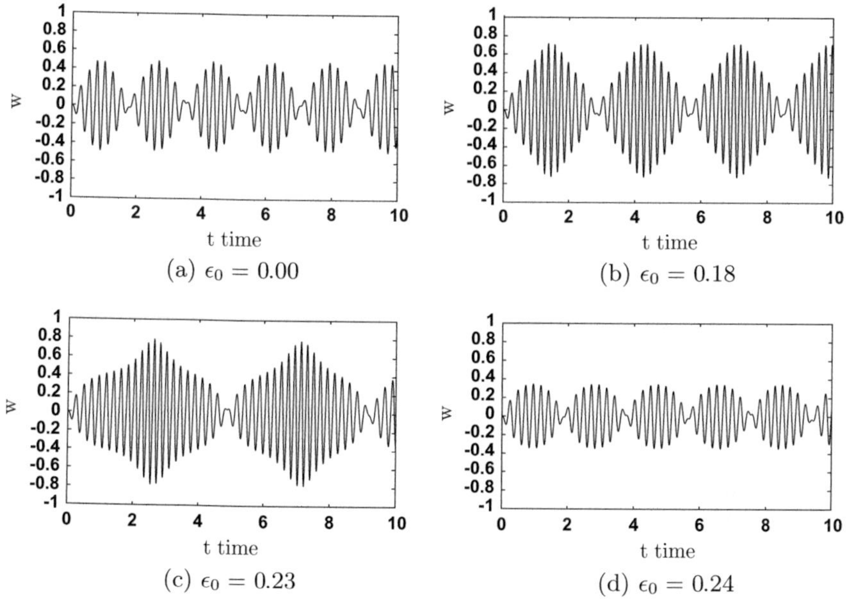

Fig. 3 Centre point displacement of beam at $r = 0.25$

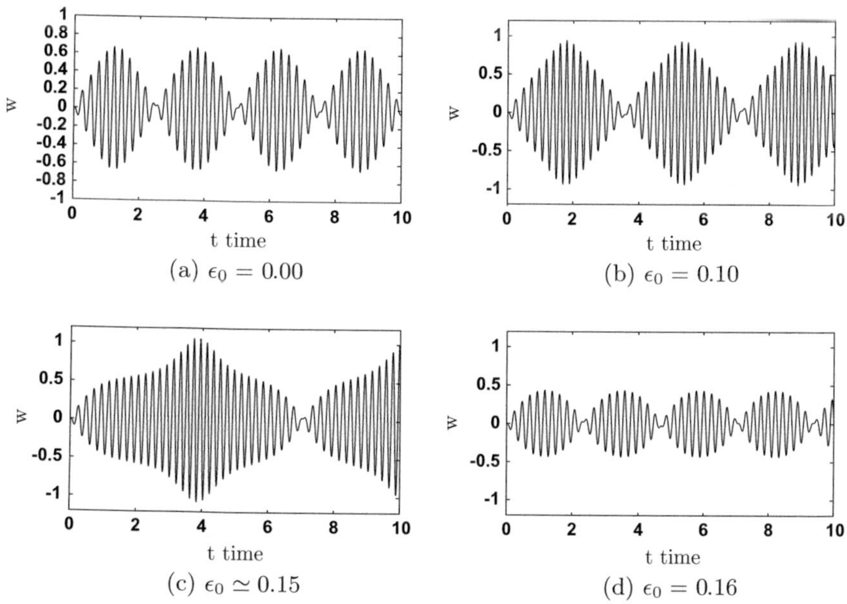

Fig. 4 Centre point displacement of panel at $r = 0.50$

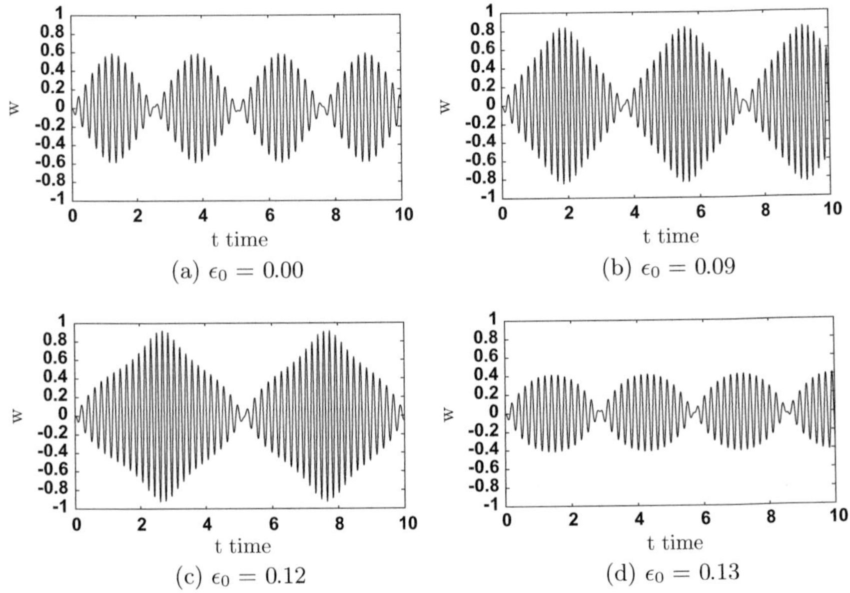

(a) $\epsilon_0 = 0.00$ (b) $\epsilon_0 = 0.09$

(c) $\epsilon_0 = 0.12$ (d) $\epsilon_0 = 0.13$

Fig. 5 Centre point displacement of beam at $r = 0.75$

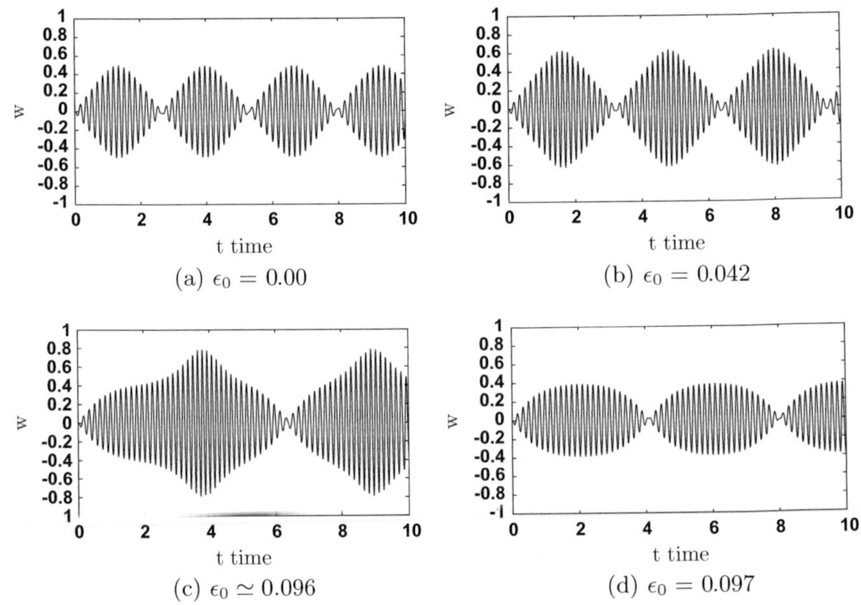

(a) $\epsilon_0 = 0.00$ (b) $\epsilon_0 = 0.042$

(c) $\epsilon_0 \simeq 0.096$ (d) $\epsilon_0 = 0.097$

Fig. 6 Centre point displacement of panel at $r = 1.00$

maximum amplitude achieved suddenly drops, and the beating frequency increases as the critical value of $\epsilon_0 = 0$ is passed. The pattern is same for all values of aspect ratio r. Increasing aspect ratio of the aircraft panel acts to increase the fundamental frequency and decreases beating frequency, amplitude and the value of ϵ_0.

4 Conclusions

In this paper, the oscillations of a thin clamped rectangular aircraft panels, which are excited by a convective pressure wave, have been considered. It has been observed that change in aspect ratio plays an important role in the vibration of plate. Increasing aspect ratio of the rectangular aircraft panel basically acts to increase the fundamental frequency of the clamped panel. A phenomenon of nonlinear resonance has been shown to occur when the forcing frequency is close to the frequency of the first linear mode of the panel. Increasing aspect ratio of the aircraft panel fundamentally acts to stimulate higher modes of vibration and has the overall effect of increasing the frequency of the plate oscillation. Increasing aspect ratio of the aircraft panel acts to increase the fundamental frequency and decreases beating frequency, amplitude and the value of ϵ_0. The vibration is categorized by a high-frequency vibration linked with the forcing and a beating frequency. The present results may be useful in explaining the complex limit cycle oscillations witnessed on aircraft panels in flight.

References

1. Jordan PF (1956) The physical nature of panel flutter. Aero Digest 72(2):34–38
2. Cunningham Jr AM, Geurts EG (1997) Analysis of limit cycle oscillation/transonic high alpha flow visualization. In: LOCKHEED MARTIN CORP FT WORTH TX TACTICAL AIRCRAFT SYSTEMS
3. Bunton RW, Denegri CM (2000) Limit cycle oscillation characteristics of fighter aircraft. J Aircraft 37(5):916–918
4. Patil MJ (2002) Limit cycle oscillations of aircraft due to flutter-induced drag. In: Proceedings of the 43rd structures, structural dynamics and materials conference, Denver, Colorado
5. Tang D, Dowell EH, Hall KC (1999) Limit cycle oscillations of a cantilevered wing in low subsonic flow. AIAA J 37(3):364–371
6. Geveci B, Walker JDA (2001) Nonlinear resonance of rectangular plates. In: Proceedings of the royal society of London A: mathematical, physical and engineering sciences, vol 457, no 2009. The Royal Society, pp 1215–1240
7. Bauchau OA, Craig JI (2009) Kirchhoff plate theory. In: Structural analysis. Springer Netherlands, pp 819–914
8. Herrmann G (1956) Influence of large amplitudes on flexural motions of elastic plates. In: NACA Technical Note 3578, national advisory committee for aeronautics, Washington
9. Fung YC (1965) Foundations of solid mechanics. Englewood Cliffs, NJ, Prentice-Hal
10. Gorman DJ (1982) Free vibration analysis of rectangular plates. New York and Amsterdam: Elsevier
11. Leissa AW (1982) The free vibration analysis of rectangular plates. J Sound Vibr 31:257

12. Geveci B (1999) Flow induced nonlinear vibrations of rectangular plates. PhD thesis, Lehigh University, Bethlehem, PA, USA
13. Walker JDA, Zhikharev CN, Delph TJ (1997) Induced oscillations of a finite plate. J Guidance Control Dyn 20(6):1172–1180

Experimental Validation of the Structural Integrity of a CFRP Aircraft Fuselage

R. Gopikrishna, R.S. Rawat, R. Girish, K. Kishore Kumar and P.S.R. Anjaneyulu

1 Introduction

The fuselage represents the main body of any aerial vehicle. With the advancements in the design and manufacturing of composite materials, the design material of any fuselage is slowly being replaced from metallic to composite material. In recent times, the cross section of an aircraft fuselage ranges from the conventional circle up to a wide variety of other shapes. Carbon fibre-reinforced plastic (CFRP) and glass fibre-reinforced plastic (GFRP) are two most common composite materials used in current times [1]. The fuselage shell is chiefly made of CFRP material with aluminium bulkheads. GFRP is also used locally for sections like hatch covers, cockpit windows and so on. The testing of any design involving composite materials becomes mandatory, as the properties of composites show a wide amount of variation from the sample and coupon level results. Unlike testing of metallic sections, the structural testing of composite fuselage poses a real challenge. The structural integrity of the composite fuselage design will nevertheless require experimental validation of the design. The fuselage design needs to be validated for a variety of load cases, where it experiences as a part of its flight envelope. Furthermore, loads due to landing, loads from empennage and local loads from fuel tank and stores also need to be simulated during the experimental validation process. The details of the structural integrity validation test carried out on a composite fuselage of an aircraft are presented in this work.

R. Gopikrishna (✉) · R.S. Rawat · K. Kishore Kumar · P.S.R. Anjaneyulu
Defence Research Development Laboratory, Hyderabad, India
e-mail: tisipog557@gmail.com

R. Girish
Aeronautical Development Establishment, Bangalore, India

S. Singh et al. (eds.), *Proceedings of the International Conference on Modern Research in Aerospace Engineering*, Lecture Notes in Mechanical Engineering,
https://doi.org/10.1007/978-981-10-5849-3_23

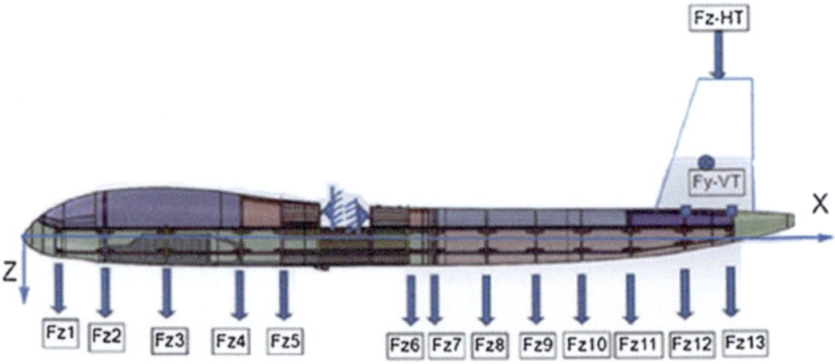

Fig. 1 Schematic of the fuselage test set-up

2 Test Loads

The manoeuvring envelope dictates the critical testing cases for the composite fuselage. These cases may occur during take-off, landing, cruise, climb or turn of the aircraft. The acceleration levels along pitch, yaw or roll axis during any of these flight cases lead to distribution of forces along the fuselage. The continuous force distribution results in a bending moment profile along the length of the fuselage for each load case. The simulation of the bending moment profile at a ground test facility is achieved by applying discrete loads along the length of the fuselage. Figure 1 shows the distributed loads along the fuselage and on the empennage simulator for one of the test load cases. The test load case shown here corresponds to landing case of the aircraft with acceleration of 2.5 g about the yaw axis. Thirteen loads ranging from 50 kgf to 1000 kgf are applied on the fuselage to simulate the bending moment profile experienced during this load case.

3 Experimental Set-up

The main challenges of testing include simulation of the actual boundary conditions and actual test loads. Unlike metallic section testing, loads could not be applied on the composite fuselage using polyester slings. This is because there are chances of local failure due to non-uniform load distribution from the polyester slings. This hurdle was overcome by using wooden formats with cushions made of felt material. The formats were custom-made to match the contour and cross section of the fuselage at each load point. The aerodynamic force acting at the centre of pressure of the horizontal and vertical tails is usually transferred to the fuselage through the attachments. These loads were also simulated and applied during the test. Figure 1 shows a schematic of the test set-up with the various loads applied. The aircraft

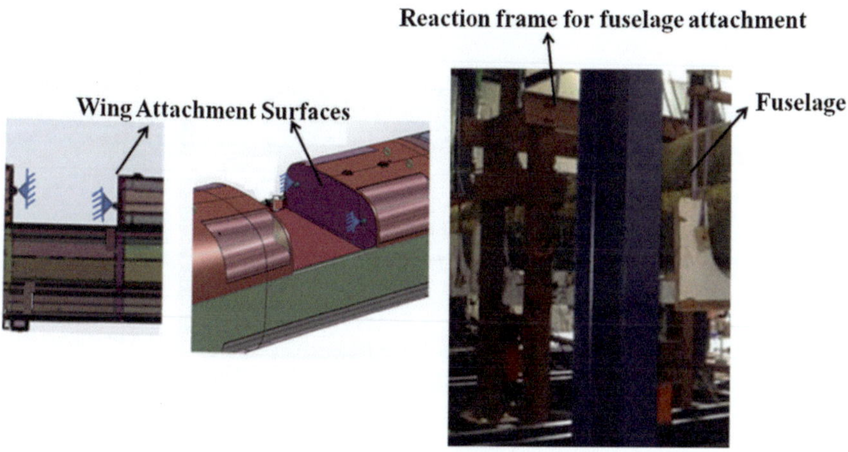

Fig. 2 Boundary condition details

fuselage was fixed at the wing attachment surfaces. Figure 2 shows the details of boundary conditions. The entire fuselage was suspended at these surfaces; therefore, the entire reaction loads occurred at these attachment surfaces.

4 Instrumentation Involved

Any experiment is futile without the aid of proper instrumentation. The input parameters need to be controlled, and the output parameters need to be recorded during the test. Load cells and pressure transducers were used to control the input parameters. Hydraulic actuation systems were applied the necessary loads on the fuselage. Strain gages and linear variable displacement transducers (LVDTs) were used to measure and record the output strains and deflections during the experiment. Fifteen rectangular rosette strain gages were bonded on both the port (PS) and starboard (SB) sides of the fuselage body. Strain gages used for composite materials are different from those that are used for metallic sections [2]. Composite materials are bad conductors of heat; as a result, the strain gage bonded area gets heated up locally causing errors in the strain measurement. This problem was overcome by using self-temperature compensation gages with higher gage length and also low excitation voltage as input to the gage [3]. Figure 3 shows few strain gages bonded on the composite section during the test. Twenty LVDTs were used along the length of the fuselage to capture the entire fuselage deflection for the applied loads.

Fig. 3 Strain gages bonded on the fuselage

5 Necessary Precautions for Testing of Composite Section

1. The loading polyester slings are to be replaced with wooden formats to ensure uniform distribution of applied load
2. Non-destructive evaluation prior to the load test to detect zones of delamination and other defects during manufacturing of the section
3. Surface preparation of composite material without causing fibre damage
4. Alignment and bonding of strain gage channels along and across fibre directions
5. Low excitation voltage as input for the strain gage—about 2VDC [4]
6. Flatness and finish of the surface at the points of LVDT contact to avoid picking up of local manufacturing defects
7. Load cycling of 0-10-20%-10-0 twice prior to the actual test to relieve slackness in the test set-up and to ensure functioning of the instrumentation.

6 Results and Discussion

The fuselage was subjected to three load cases as a part of the structural qualification tests. The natural frequency of the fuselage was measured and compared before and after each test case. The variation in natural frequency between the data before and after the test was observed to be less than 1 Hz, for each case. This denotes that the composite fuselage did not undergo any major change with respect to its mass and stiffness [5]. Strains and deflections were monitored during each load case with a sample rate of 5 samples per second. Hottinger Baldwin Messtechnik (HBM) MGCplus was used as data acquisition module for all strain gage based and displacement measurements [6]. The maximum test loads applied during each load case corresponded to 80% of the design load of that particular

Fig. 4 Experimental test set-up of fuselage

case. The photograph of the test set-up for load case corresponding to +2.5 g during landing is shown in Fig. 4.

The failure strains of carbon composite material are about 9000 με, while that of glass fibre composite is 7000 με. The overall maximum strain measured during +2.5 g load case during landing was 866 με. The corresponding maximum deflection was 29.73 mm at rear tip of the fuselage, along the yaw axis. During the +2 g case corresponding to cruise flight, an overall maximum of 1415 με was measured at the same location as earlier, with a maximum deflection of 41.74 mm along the pitch axis. The third case corresponding to −1.5 g during cruise was recorded an overall maximum of 322 με with a maximum deflection of 23.42 mm along the yaw axis. The qualification of the fuselage structure not only demands that the maximum measured strains be less than the failure strains of the material but also requires that the behaviour of strain gages is linear throughout the load application cycle. This was observed from the strain variation plots, one of which is shown in Fig. 5 corresponding to +2.5 g during landing. Furthermore, permanent strains measured after releasing the loads during each case were within the allowable 10% of maximum measured strain limit.

The necessary conditions for qualifying the fuselage structure were met for all the three load cases. The design and fabrication of the composite fuselage was validated for different load cases encountered during its life cycle. A non-destructive evaluation (NDE) of the fuselage after the completion of all the load cases was carried out to verify the health of the fuselage for the presence of defects, before approving its design and fabrication.

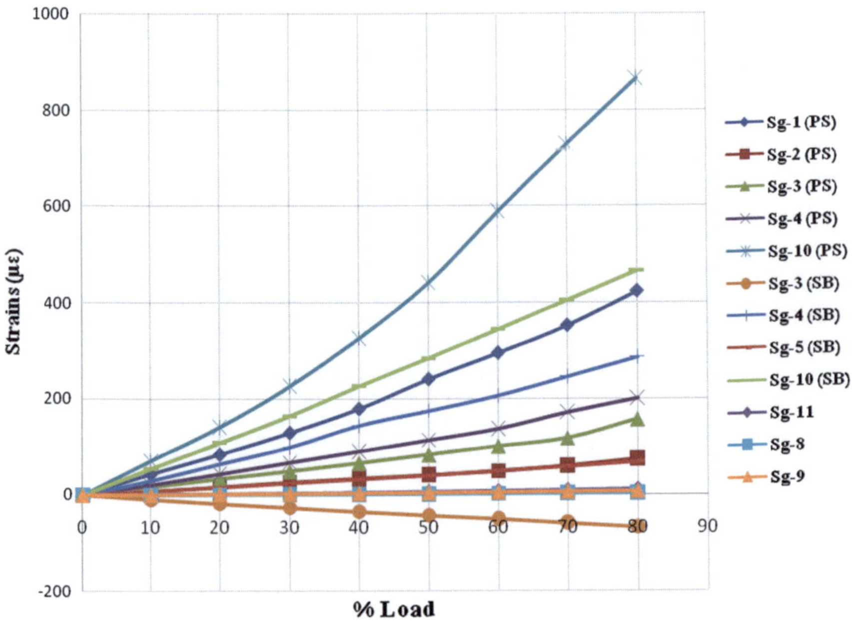

Fig. 5 Variation of strains with applied load

7 Conclusion

Structural testing of any fuselage is as important as the design of the fuselage itself. The simulation of different load cases was successfully achieved, taking necessary precautions for testing of composite structures. The fuselage has very high factors over the failure strain of the composite material. The fuselage may seem over-designed, but owing to the uncertainties in the fabrication of any composite structure, it is acceptable to maintain high design factors. The deflections recorded were a measure of the flexibility of the fuselage as a whole. The simulation of the bending moment profile using as many discrete loads as possible was helped in reproducing the loading conditions corresponding to the FE analysis. Strain measurement at many locations paved the way to judge the health of the fuselage structure during the application of the loads. The maximum of the strains of all the load cases on the composite material was less than the failure strain of the material, thereby ensuring that the fuselage has withstood the landing and flight loads without any visual failure.

References

1. Baker AA (2004) Composite materials for aircraft structures, 2nd edn. AIAA Education Series
2. Tuttle ME, Brinson HF (1984) Resistance-foil strain-gage technology as applied to composite materials. J Exp Mech 24:54–65
3. Dally JW, Riley WF (2005) Experimental stress analysis, 4th edn. College Home Enterprises
4. Scheider, Hugh N. Electrical instruments and testing: How to use the voltmeter, ammeter, galvanometer, potentiometer, ohmmeter and the wheatstone bridge. Nabu Press
5. Link M, Weiland M, Hahn Th (2008) Structural health monitoring of the Gaertnerplatz bridge over the Fulda river in Kassel. Proceedings of the 2nd international symposium on ultra high performance concrete, Structural materials and engineering series, H 10, No 10, Kassel University Press, March 2008
6. Hoffman K (1987) An introduction to stress analysis & transducer design, HBM

Effect of Natural Frequency of Slender Structure on VIV at Higher Reynolds Numbers

Rahul Jadhav, P.R. Maiti and Sabita Madhvi Singh

1 Introduction

Vortex-induced vibration arises in many engineering applications such as civil, mechanical, aerospace, ocean engineering. Flow-induced vibrations are complex and complicated involving coupling between structural and fluid dynamics. This vibration has a substantial influence on the fatigue life of structures and may even have disastrous consequences. Interest in predicting this vibration is rapidly growing. If we know the formation of the vortices which usually occupies a compact region behind structure, we can determine the entire flow in the case of an incompressible fluid. The alternate shedding of vortex from structure causes it to vibrate. When frequency of vortex shedding is close to natural frequency of structure, structure oscillates at higher amplitude. The drag and lift forces are fluctuating when vortex-induced vibration is observed.

Many researchers have carried out experimental as well as numerical studies on vortex induced vibrations of bluff bodies over the past few decades in this prospect. Belloli et al. [1] carried out experiments in wind tunnel on circular cylinder at Reynolds numbers ranging from 1.5×10^5 to 6×10^5 to compare characteristics of VIV in post critical flow regime with subcritical regime. Stappenbelt and Johnstone [8] investigated circular pivoted cylinder in flow channel to understand the influence of Reynolds number on VIV. An experiment on VIV of two cylinders

R. Jadhav · P.R. Maiti
Department of Civil Engineering, IIT (BHU), Varanasi 221005, Uttar Pradesh, India
e-mail: jadhav.rahul.civ14@itbhu.ac.in

P.R. Maiti
e-mail: prmaiti.civ@itbhu.ac.in

S.M. Singh (✉)
Department of Civil Engineering, Amity University, Noida, Uttar Pradesh, India
e-mail: smsingh1@amity.edu

S. Singh et al. (eds.), *Proceedings of the International Conference on Modern Research in Aerospace Engineering*, Lecture Notes in Mechanical Engineering,
https://doi.org/10.1007/978-981-10-5849-3_24

in tandem arrangement was conducted by Huera-Huarte and Bearman [6]. Their results show that at higher reduced velocities response of cylinders depends on the aspect ratio.

Han et al. [5] carried out numerical simulation of circular cylinder to study the effect of natural frequency of structure on VIV of cylinder. Mittal and Kumar [4] used space–time finite element method to investigate vortex-induced vibration of cylinder at Re in range of 10^3–10^4. Cui et al. [2] carried out numerical simulation on force and free oscillating cylinder in cross-flow at Re = 100. Result was compared with different numerical techniques, and response of cylinder was investigated over wide range of reduce velocities. Pan et al. [7] simulated VIV of cylinder at low mass-damping ratio numerically using SST k-ω turbulence model of RANS code. Author discussed about disturbances affecting observation and appearance of upper branch in experiments.

Present study is based on simulation of circular cylinder in cross-flow carried out at higher Reynolds numbers in unsteady flow. In this study, an attempt has been made to get an insight into the effect of natural frequency of structure VIV cylindrical structure in two-dimensional unsteady flow. For this purpose, k-ε model based on RANS equations is used in Fluent.

2 Governing Equations of Fluid Domain

2.1 Conservation of Momentum

The momentum equation for the incompressible fluid flow is

$$\frac{\partial u}{\partial t} + (u \cdot \nabla)u = -\frac{1}{\rho}\nabla P + \nu \nabla^2 u. \tag{1}$$

2.2 Continuity Equation

The mass conservation theory states that the mass will remain constant over time in a closed system.

$$\frac{\partial \rho}{\partial t} + \frac{\partial(\rho u)}{\partial x} + \frac{\partial(\rho v)}{\partial y} + \frac{\partial(\rho w)}{\partial z} = S_m, \tag{2}$$

where the source S_m is the mass added to the system and any user-defined sources. The density of the fluid is ρ, and the flow of mass in x, y and z direction is u, v and w.

2.3 RANS Equation

Turbulent model is needed to solve modified set of transport equation by introducing average and fluctuating components. This is Reynolds average Navier–Stokes (RANS) equation.

$$\frac{\partial \rho}{\partial t} + \frac{\partial \rho u_j}{\partial x_j} = 0, \tag{5}$$

$$\frac{\partial \rho u_j}{\partial t} + \frac{\partial(\rho U_i U_j)}{\partial x_j} = -\frac{\partial p}{\partial x_i} + \frac{\partial(\tau_{ij} - \rho \overline{U_i U_j})}{\partial x_j} + S_m, \tag{6}$$

where τ = molecular stress tensor (including both normal and shear components of the stress).

RANS equations are the time-averaged equations of motion of fluid flow. Two of the most popular models of the RANS are the k-ε model and k-ω model. The realizable k-ε model is used for present study.

2.4 k-ε Model

This two-equation model is based on model transport equation for turbulent kinetic energy (k) and its dissipation rate (ε). In k-ε model, it is assumed that the flow is fully turbulent and the effects of molecular viscosity are negligible. The modelled transport equations for k and ε in the realizable k-ε model are

$$\frac{\partial \rho k}{\partial t} + \frac{\partial(\rho k u_j)}{\partial x_j} = \frac{\partial}{\partial x_j}\left[\left(\mu + \frac{\mu_t}{\sigma k}\right)\frac{\partial k}{\partial x_j}\right] + G_k + G_b - \sigma\varepsilon - Y_M + S_k \tag{7}$$

and

$$\frac{\partial(\rho\varepsilon)}{\partial t} + \frac{\partial(\rho\varepsilon u_j)}{\partial x_j} = \frac{\partial}{\partial x_j}\left[\left(\mu + \frac{\mu_t}{\sigma\varepsilon}\right)\frac{\partial \varepsilon}{\partial x_j}\right] + \rho C_1 S_\varepsilon - \sigma C_2 \frac{\varepsilon^2}{k + \sqrt{\nu\varepsilon}} + C_{1\varepsilon}\frac{\varepsilon}{k} C_{3\varepsilon} G_b + S_\varepsilon, \tag{8}$$

where

$$C_1 = \max\left[0.43, \frac{\eta}{\eta + 5}\right]; \quad \eta = S\frac{k}{\varepsilon}; \quad S = \sqrt{2 S_{ij} S_{ji}}. \tag{9}$$

G_k generation of turbulence kinetic energy due to the mean velocity gradients,

G_b generation of turbulence kinetic energy due to buoyancy,
Y_M contribution of the fluctuating dilatation in compressible turbulence to the overall dissipation rate.
C_2 and $C_{1\varepsilon}$ constant where $C_{1\varepsilon} = 1.44$ and $C_2 = 1.9$.
σ_k and σ_ε turbulent Prandtl numbers ($\sigma_k = 1$ and $\sigma_\varepsilon = 1.2$).
S_k and S_ε user-defined source terms.

2.5 *Governing Equation of Motion of Structure*

Response of cylinder subjected to VIV is calculated by solving equation of motion as,

$$m\frac{\partial^2 y}{\partial t^2} + Ky = F_y, \tag{10}$$

where

y amplitude of cylinder in transverse direction.
K structural stiffness of cylinder.
F_y lift force on cylinder.

The unsteady flow field is solved using Fluent (CFD code) based on a finite volume method (FVM). Cylinder oscillation is simulated by a user-define function (UDF). Pressure-velocity coupled equations are solved by SIMPLE algorithm. The implicit first-order scheme is used for transient terms. Second-order upwind schemes for k and ε transport equation are used. Second-order upwind scheme is used for convective terms in momentum equation.

3 Computational Domain and Grid Formation

The entire computational domain of fluid of size 35 D × 20 D is used for simulation where the diameter of cylinder is D as shown in Fig. 1. Only single degree of freedom is allowed in VIV of cylinder. Finer mesh near cylinder region as shown in Fig. 2 is used to obtain more accurate result. Remaining flow domain is divided into uniform size mesh comparatively larger size to reduce computational time.

Boundary conditions used in Fluent for VIV problem are Inlet, outlet, wall and symmetry. Inlet boundary condition: (1) allows fluid to enter in domain at specified velocity at inlet. Outlet boundary condition: (2) allows exit of fluid out of the domain. Wall boundary condition: (3) confines fluid flow in domain i.e. no slip boundary condition. Symmetry boundary condition: (4) is use to specify that no

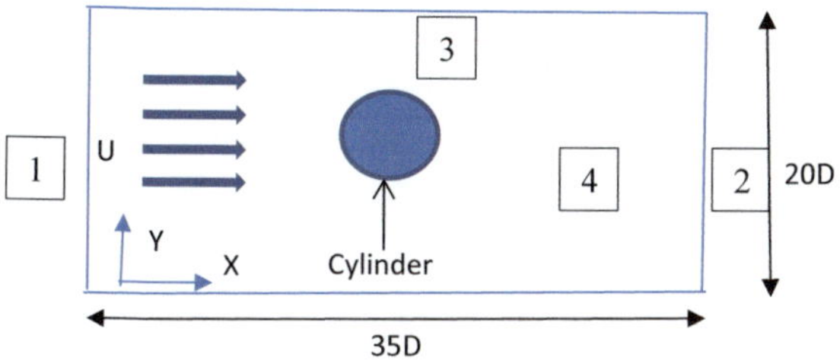

Fig. 1 Numerical set-up (two-dimensional flow domain)

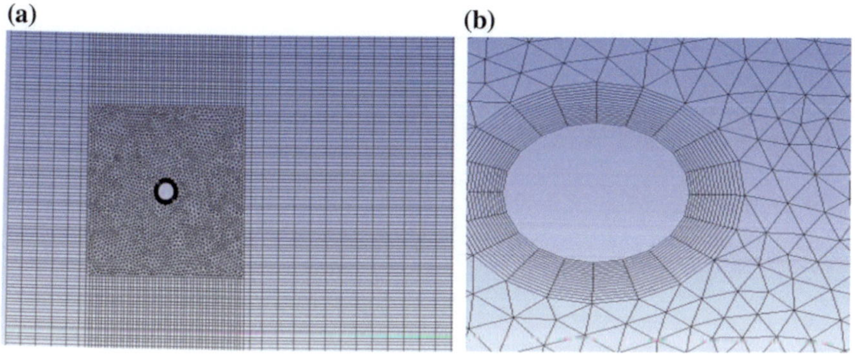

Fig. 2 **a** Computational mesh used for numerical simulation **b** Mesh near cylinder

flow occurs across the plane. Rigid motion of cylinder and deforming grid is simulated using user-defined function (UDF) and smoothening and re-meshing.

4 Vortex-Induced Vibration of Circular Cylinder

In this part, effect of natural frequency of structure on vortex-induced vibration is studied. For this purpose, a single cylinder is simulated in cross-flow and natural frequency is varied from the 0.05 Hz to 8 Hz. Simulation is carried out at Re = 5×10^4, 15×10^4 and 50×10^4. Result is obtained and presented in Table 1. Response of C_D and C_L with respect to time is given in Fig. 3.

It can be seen from Fig. 3 that behaviour C_D and C_L of cylinder varies with respect to natural frequency. The Strouhal frequency response for different natural frequency of structure is plotted in Fig. 4.

It can be seen from Fig. 4 that when cylinder of higher structural natural frequency is subjected to cross-flow in unsteady condition at upper subcritical regime; the vortex shedding frequency f_{st} is less affected and shows constant behaviour. Whereas at Reynolds number in critical regime, there is steady increment f_{st}

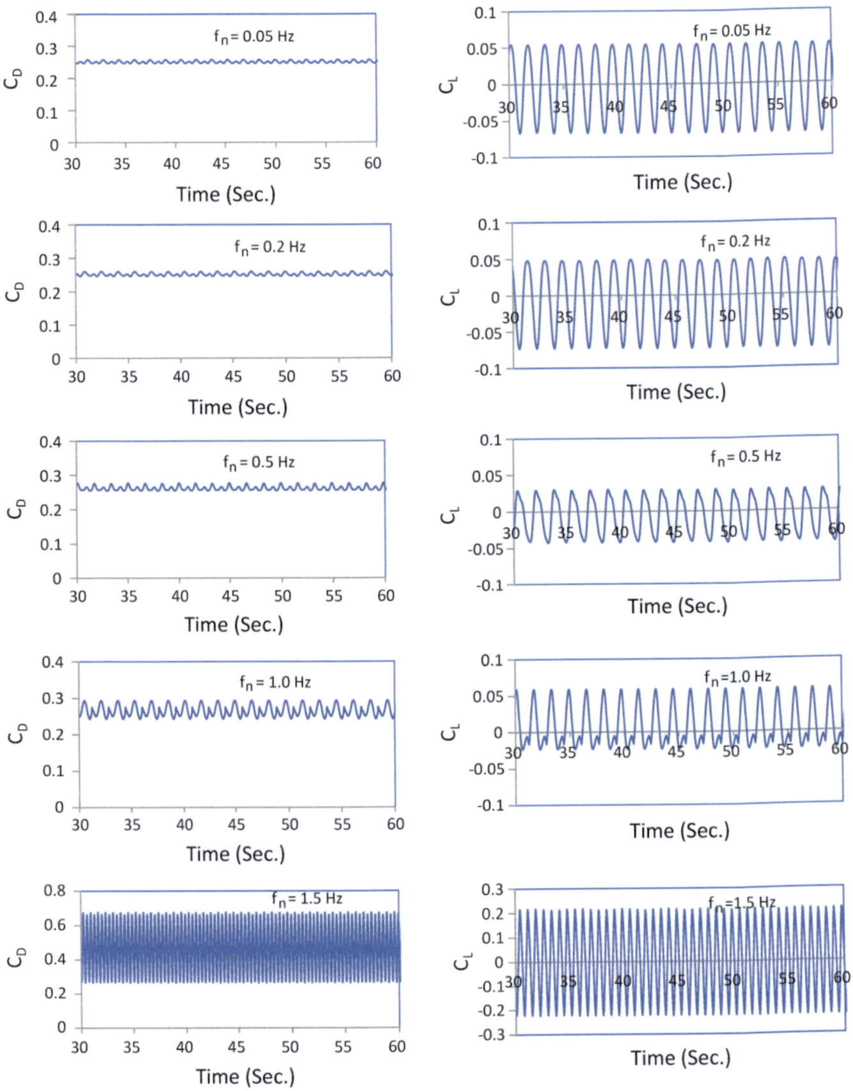

Fig. 3 Drag coefficient C_D and lift coefficient C_L versus time

observed. The above graph is useful for finding frequency ratio f^* as it signifies lock-in region, i.e. condition at which response of cylinder in flow-induced vibration's drag force and lift force are maximum on given bluff body.

Frequency ratio and C_D, C_L

After calculating frequency ratio f^*, drag force and lift force coefficients are plotted against f^* and presented in Fig. 5.

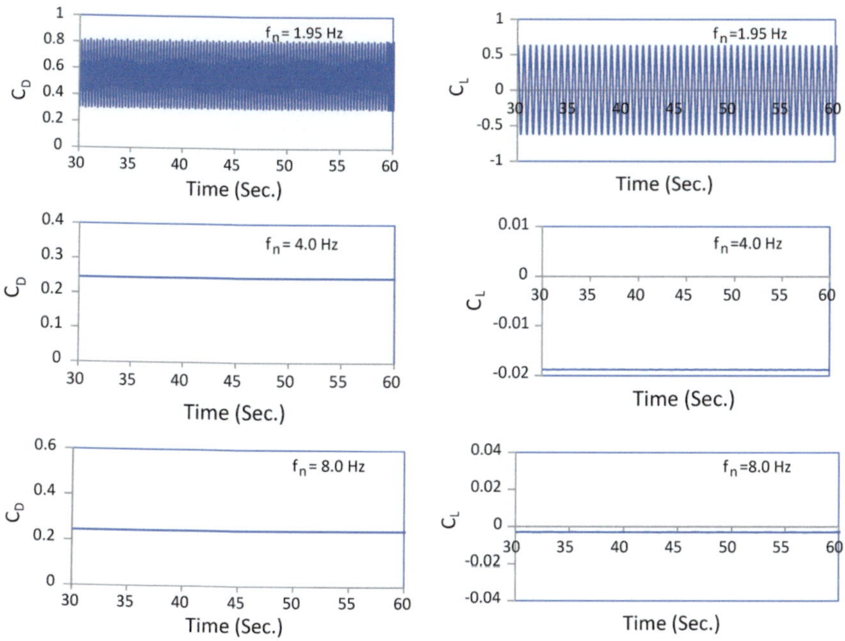

Fig. 3 (continued)

It is seen from Fig. 5 that drag and lift force coefficients are higher when f^* is near to unity. Cross-oscillation of the cylinder is maximum range f^* varies from 0.88 to 0.98 indicating condition at which Strouhal frequency f_{st} and natural frequency f_n are nearly equal (Fig. 6).

Frequency ratio f^* and reduced velocity U_r

U_r is plotted against f*, and reduce velocity range found out to be varying from 4 to 8. Results obtained are in good agreement with results from experiment conducted by Feng.

Wake formation behind cylinder

Figures 7 and 8 show vortices shed by the cylinder in cross-fluid flow. The figure shows transition in vortex shedding by changing natural frequency of structure.

It can be seen from Fig. 7 that when frequency ratio f^* is not close to unity, vortex street formed has two separate distinguishing tails of nearly equal strength in vorticity preventing from alternative shedding of vortices. Thus, drag and lift responses for the lower frequencies are less. When structure enters in lock-in zone, well-defined alternating vortices are observed in Figs. 7 and 8. In these conditions, C_D and C_L on circular cylinder surface and amplitude of oscillation are higher. At higher frequency f_n, lock-in zone is observed in critical flow regime and can be seen in Fig. 8. Depending on natural frequency of cylinder, length at which alternating shedding of vortices takes place varies.

Table 1 Vibration response of cylinder for different natural frequencies

Natural frequency f_n	At Re = 5 × 10^4				At Re = 15 × 10^4				At Re = 50 × 10^4			
	C_D	C_L	F_{st}	S_t	C_D	C_L	F_{st}	S_t	C_D	C_L	F_{st}	S_t
0.05	0.300	0.107	0.397	0.132	0.254	0.066	2.564	0.285	0.225	0.004	0.699	0.023
0.2	0.317	0.179	0.139	0.046	0.256	0.073	2.503	0.278	0.227	0.010	0.374	0.012
0.5	0.434	0.017	0.246	0.082	0.271	0.042	2.411	0.268	0.228	0.010	0.940	0.031
1	0.313	0.141	0.664	0.221	0.280	0.058	0.974	0.108	0.225	0.010	1.189	0.040
1.5	0.313	0.128	0.662	0.221	0.674	0.213	1.399	0.155	0.225	0.012	2.051	0.068
1.95	0.309	0.125	0.641	0.214	0.815	0.620	1.705	0.189	0.226	0.015	2.564	0.085
4	0.032	0.008	0.363	0.121	0.240	0.020	1.119	0.124	0.460	0.110	3.793	0.126
8	0.285	0.040	0.363	0.121	0.240	0.005	1.118	0.124	1.000	1.200	6.777	0.226

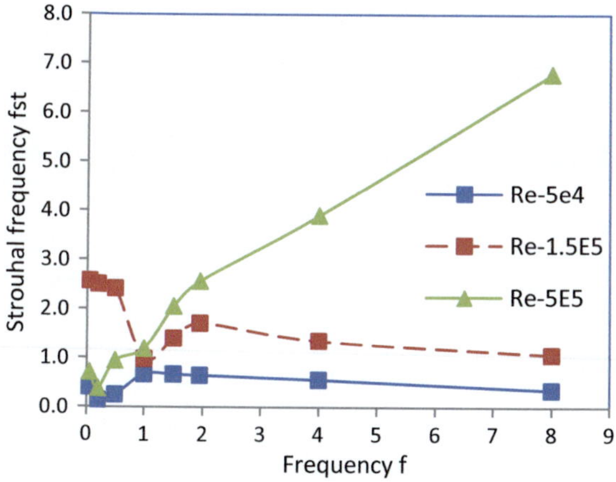

Fig. 4 Strouhal frequency f_{st} versus natural frequency f_n

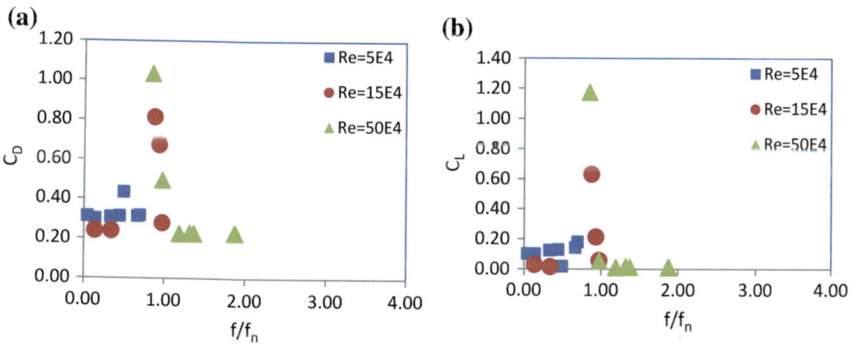

Fig. 5 C_D and C_L versus frequency ratio f^* **a** Drag coefficient **b** Lift coefficient

Fig. 6 Frequency ratio f^* versus reduced velocity Ur

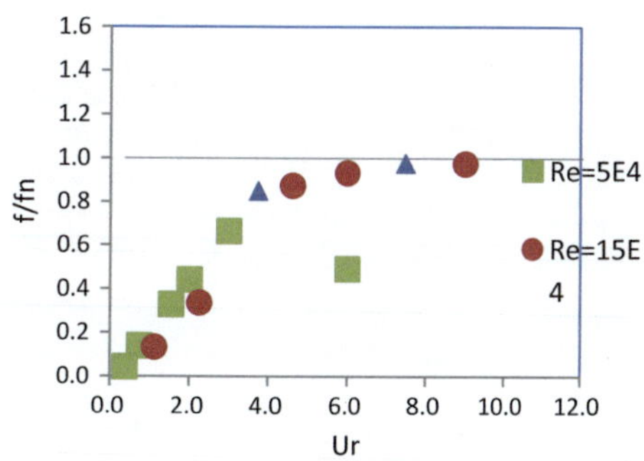

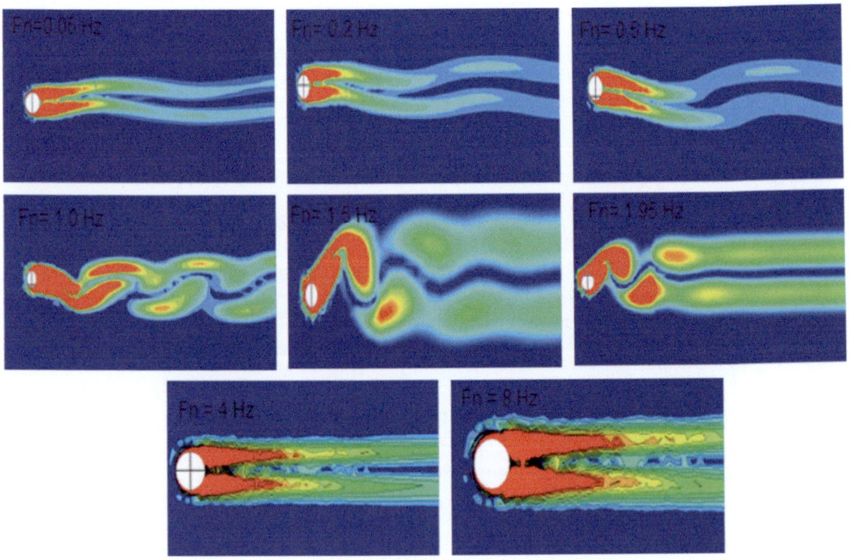

Fig. 7 Vorticity behind cylinder for different natural frequency at Re = 15×10^4

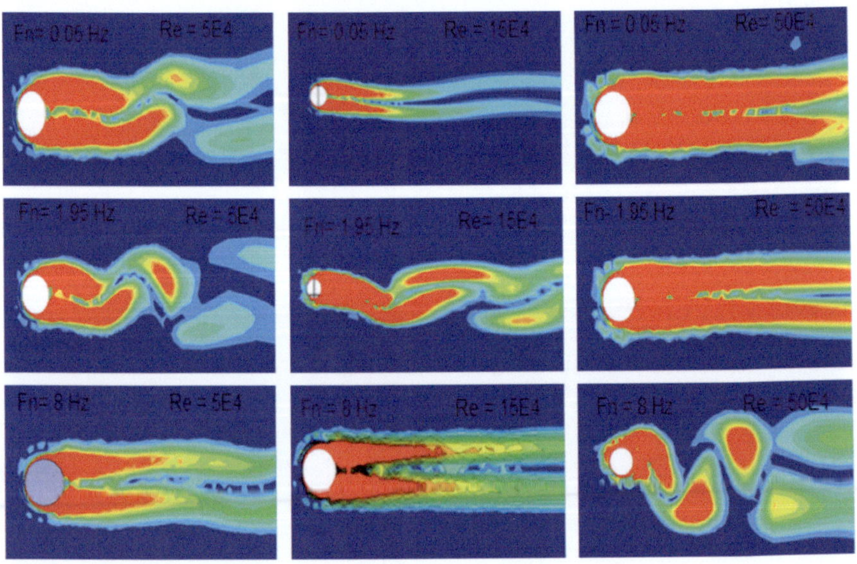

Fig. 8 Vorticity behind cylinder for different natural frequency and Reynolds numbers

5 Conclusion

Vortex-induced vibration phenomenon induces fatigue damage in structures such as underwater pipeline, cables, stacks. Hence, study of VIV for structural safety is important. In present work, effect of natural frequency of structure on its vortex-induced vibration is studied. Two-dimensional RANS equation with k-ε model is used for computation in Fluent. Numerical analysis of cylinder shows that maximum response occurs at frequency ratio ranging from 0.88 to 0.98. This zone is referred as lock-in zone. Numerical analysis shows that frequency ratio is close to unity when reduce velocity (U_r) ranging from 4–10. These results are in good agreement with results obtained from experiment conducted by Feng [3]. Vortex shedding pattern generated behind cylinder varies depending on natural frequency of structure at given Re. At lock-in zone, a clear alternate shedding of vortices is observed.

References

1. Belloli M, Giappino S, Morganti S, Muggiasca S, Zasso A (2015) Vortex induced vibrations at high Reynolds numbers on circular cylinders. J Ocean Eng 94:140–154
2. Cui Z, Zhao M, Teng B (2015) Cheng, L, Two-dimensional numerical study of vortex-induced vibration and galloping of square and rectangular cylinders in steady flow. Ocean Eng 106:89–206
3. Feng CC (1968) The measurements of the vortex induced effects in flow past stationary and oscillating circular and D-section cylinders. Thesis, The University of British Columbia
4. Mittal S, Kumar V (2001) Flow-induced oscillations of two cylinders in tandem and staggered arrangements. J. Fluids Struct.; 15(5):7–17.
5. Han X, Lin W, Tang Y, Zhao C, Sammut K (2015) Effects of natural frequency ratio on vortex-induced vibration of a cylindrical structure. J Comp Fluids 110:62–76
6. Huera-Huarte FJ, Bearman PW (2011) Vortex and wake-induced vibrations of a tandem arrangement of two flexible circular cylinders with near wake interference. J Fluids Struct 27:193–211 (2011)
7. Pan ZY, Cui WC, Miao QM (2007) Numerical simulation of vortex-induced vibration of a circular cylinder at low mass-damping using RANS code. J Fluids Struct 23:23–37
8. Stappenbelt B, Johnstone A (2015) Reynolds number influence on the vortex-induced vibration critical point of a pivoted cylinder; 19th Australasian Fluid Mechanics Conference, Melbourne, Australia 8–11 December 2014

Evaluation of Vibration of a Crankshaft and a Driveshaft Using FEM

Apoorv Rathi, Joy Banerjee, Anurag Dixit, R.K. Misra and H.S. Mali

1 Introduction

All automobiles have a crankshaft for converting the reciprocating motion of the piston into the rotary motion and a driveshaft for power transmission from engine to wheels. Metals have much lower strength, specific stiffness and strength to weight ratio as compared to composites, as a result, metals have much more weight as compared to composites [1]. In recent days, there has been a huge demand for lightweight materials in which composites seems to be a promising solution to this arising demand [2]. Weight reduction of the driveshaft and crankshaft results in the general weight reduction of the vehicle and is a highly desirable goal for better efficiency as there is a direct proportionality between weight of car and its fuel consumption.

A. Rathi (✉) · J. Banerjee · R.K. Misra
Department of Mechanical Engineering, School of Engineering,
Gautam Buddha University, Greater Noida, India
e-mail: apoorv.rathi@gmail.com

J. Banerjee
e-mail: Joyban014@gmail.com

R.K. Misra
e-mail: raghvendrakumarmisra@gmail.com

A. Dixit
Department of Mechanical & Automation Engineering,
G.B. Pant Govt. Engineering College, New Delhi, India
e-mail: anuragdixitiitd@gmail.com

H.S. Mali
Department of Mechanical Engineering, Malaviya National Institute of Technology,
Jaipur, Rajasthan, India
e-mail: harlal.singh@mnit.ac.in

S. Singh et al. (eds.), *Proceedings of the International Conference on Modern Research in Aerospace Engineering*, Lecture Notes in Mechanical Engineering,
https://doi.org/10.1007/978-981-10-5849-3_25

Crankshaft is one of the analytical components of an internal combustion engine, failure of which may turn into disfunction of the engine [3]. It possesses complicated geometry and while being in operation experiences irregular loading pattern. It converts the reciprocating displacement of the piston into a rotary motion of the crank. The most common factors responsible for the failure of the crankshaft are repeated bending (σ) and shear stresses (τ), which it experiences due to its own frequency and resulting deformation, generated during the rotatory motion of the crankshaft [4]. A driveshaft is a mechanical component that is used for transmission of torque and power between engine and rear wheels and is used to connect other components of a drive train that cannot be connected directly because of distance or the need to allow for relative movement between them [5].

Recent developments in the applications of composite materials have shown that a composite material structural member, if used in power transmission, can be of great assistance in overcoming many of the problems faced with conventional driveshafts. The problems mainly are low values of stiffness and strength, heavier weight, low corrosion resistance and less damping capacity as compared to that of composite materials [6]. They are subjected to shear stress (τ) and torsion (**T**) due to rotation, and must therefore be strong enough to bear these stresses. Also in the case of crankshaft, the reduction in weight can be a drastic factor for improvements in efficiency as it eases the load on the piston. Due to less weight of alloys and composites, they can surely be an alternative in place of conventional ongoing materials.

The present study focuses on identifying an alternate material for crankshaft and driveshaft by performing vibrational analysis of the FE models created using dimensions from previous studies [7–10] of crankshaft and driveshaft using ANSYS. Maximum deformation (δ) and natural frequency (ω_n) were determined from the analysis, and alternative material has been proposed from the results.

2 Modelling of Crankshaft and Driveshaft

The FE models of crankshaft and driveshaft were prepared using commercial software CATIA V5, in which dimensions of both the models were obtained from previous literatures. Tables 1 and 2 illustrate the parameters of crankshaft and driveshaft used in the modelling and their respective values.

Table 1 Dimension of the FE model of crankshaft

Parameter	Value (mm)
Crank pin radius (r)	22.6
Crank web thickness (t)	21.336
Shaft diameter (d)	34.925
Main journal diameter (D)	49
Crank pin axial length (l)	43.6
Main journal axial length (L_o)	40

3 Finite Element Analysis

For the complex problems of design and analysis, the most useful tool available to engineers is finite element analysis. Its success rate has always been increasing since its origin. Elements are the several small parts formed by dividing the body for examination. It is an approximate medium but also the successful one. In order to obtain accurate values of δ and ω_n in the FE analysis, automatic meshing technique was used. In order to achieve accurate results from the analysis, the loading and boundary conditions were taken as in the case of real-life situations. The crankshaft was given a cylindrical support on both ends at the journals and was fixed in the radial and axial planes, to restrict its movement in these directions and free in the tangential direction to facilitate its movement while being in rotational motion at 4500 rpm. In case of driveshaft, one end was fixed, and speed (N) of 3000 rpm was given to the model. The schematics of the FE models of crankshaft and driveshaft with loading and boundary conditions are shown in Figs. 1 and 2.

Table 2 Dimension of the FE model of driveshaft

Parameter	Value (mm)
Outer diameter (D_o)	90
Inner diameter (D_i)	83.36
Length (L)	1250

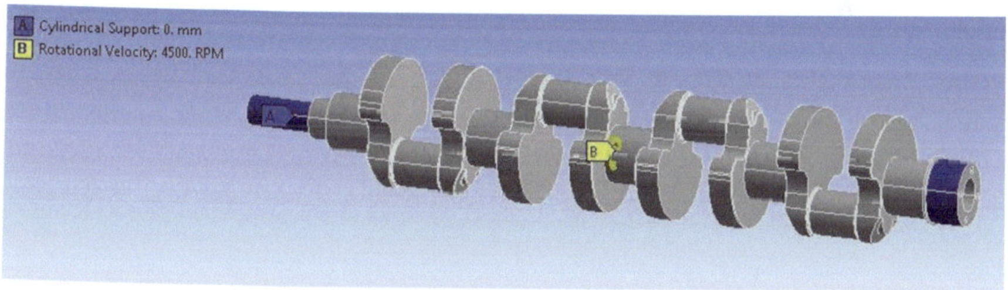

Fig. 1 FE model of crankshaft with loading and boundary conditions

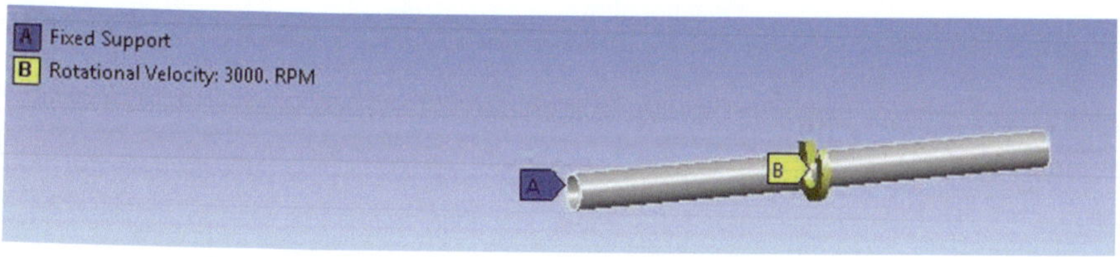

Fig. 2 FE model of driveshaft with loading and boundary conditions

4 Material and Definition

In the present investigation, five different types of materials have been used—two metals, steel and gray cast iron, an alloy of titanium and two composites namely E-glass/epoxy unidirectional composite and carbon/epoxy unidirectional composite. Table 3 shows the mechanical properties of metals and alloys, and Table 4 shows the mechanical properties of composites. Steel is being used for manufacturing of both driveshaft and crankshaft from a long time due to low cost and easy availability in the market and good corrosion resistance. Gray cast iron is not used for manufacturing moving engine components due to its brittle and corrosive nature. It has been included in the present study in order to compare it with the present material being used as it was the manufacturing material for crankshaft in the earlier times of automobile industry. Composites and alloys being the material of the new age have been accommodated in the study to identify an alternative of steel. Table 3 and 4 shows the mechanical and engineering properties of materials being used in the analysis.

5 Results and Discussion

This section focuses on the analysis of crankshaft and driveshaft. The present modelling scheme was executed using solid conventional model. As expected, the maximum deflection for the driveshaft occurred at the free end for every material whereas for the crankshaft the point of maximum deflection remained constant for every single mode of vibration with different material assigned. The natural frequency kept increasing with every passing mode of vibration for all materials. Figures 3 and 4, 5, 6, 7 show the natural frequency (ω_n) and deformation (δ) of crankshaft for different materials obtained at mode 1 after conducting the modal analysis. Figures 8, 9, 10, 11 and 12 show the ω_n and δ of driveshaft for different materials obtained at mode 1 after conducting the modal analysis.

Table 3 Mechanical properties of metals and alloy

Metal/Alloy	E(GPa)	υ
Steel	200	0.3
Gray cast iron	110	0.28
Titanium alloy	96	0.36

Table 4 Engineering properties of carbon and E-glass fiber epoxy

Property	Carbon/epoxy	E-glass/epoxy
E_{11} (GPa)	121	45
$E_{22} = E_{33}$ (GPa)	8.6	10
$G_{12} = G_{23}$ (GPa)	4.7	5
υ_{12}	0.27	0.2

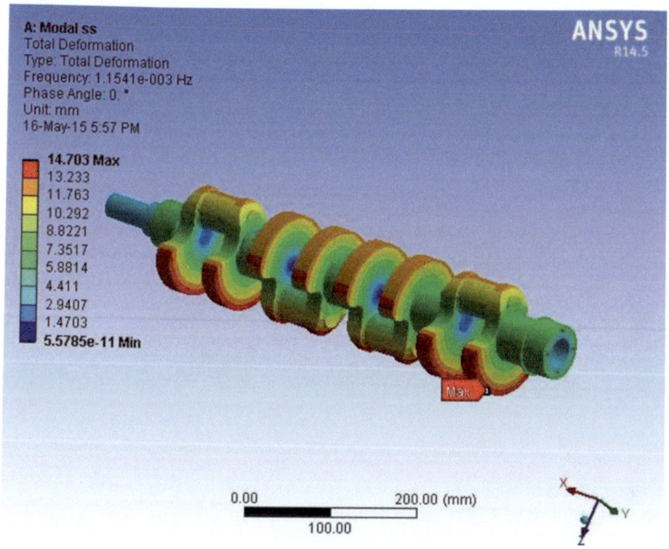

Fig. 3 Natural frequency (ω_n) and total deformation (δ) of steel at mode 1

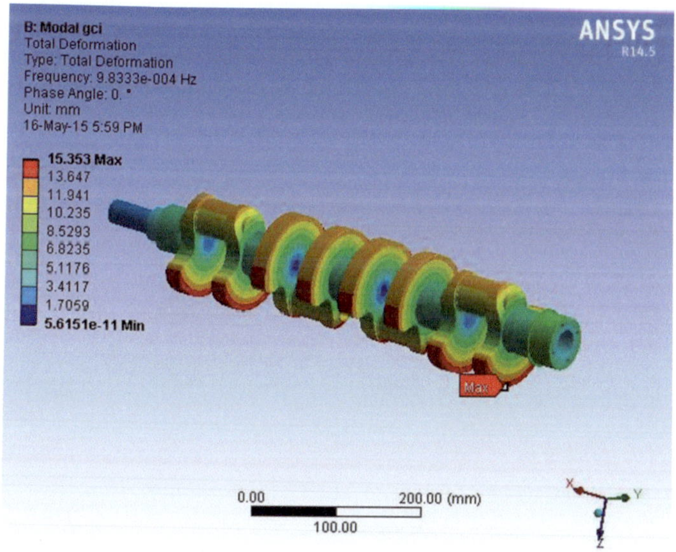

Fig. 4 Natural frequency (ω_n) and total deformation (δ) of gray cast iron at mode 1

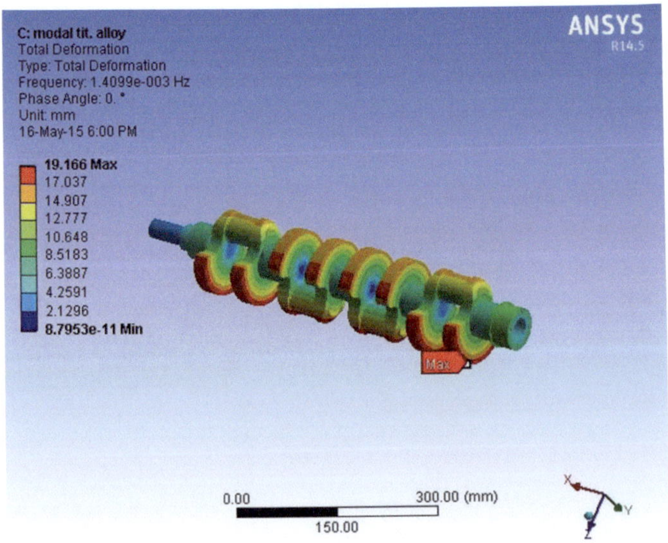

Fig. 5 Natural frequency (ω_n) and total deformation (δ) of titanium alloy at mode 1

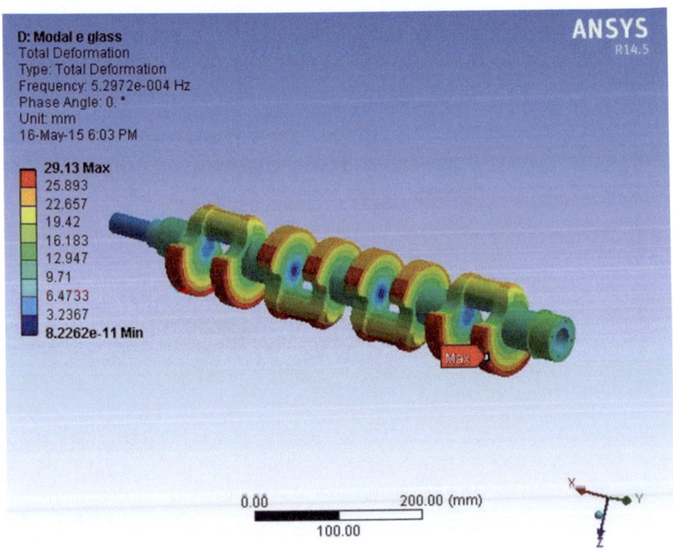

Fig. 6 Natural frequency (ω_n) and total deformation (δ) of E-glass/epoxy at mode 1

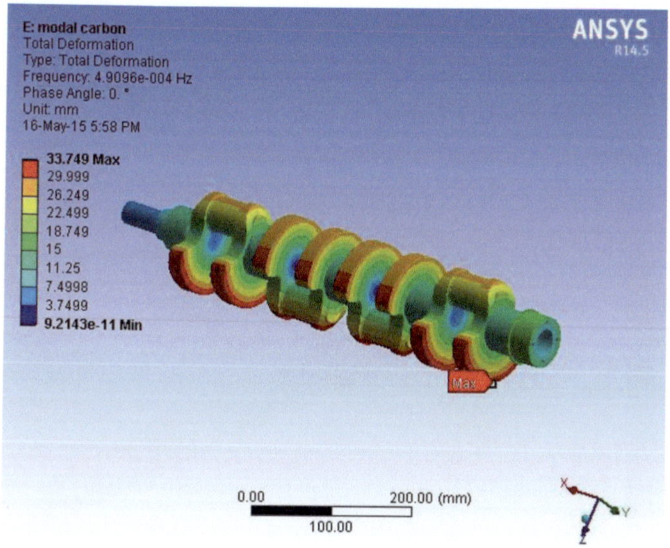

Fig. 7 Natural frequency (ω_n) and deformation (δ) of carbon/epoxy UD 230 GPa at mode 1

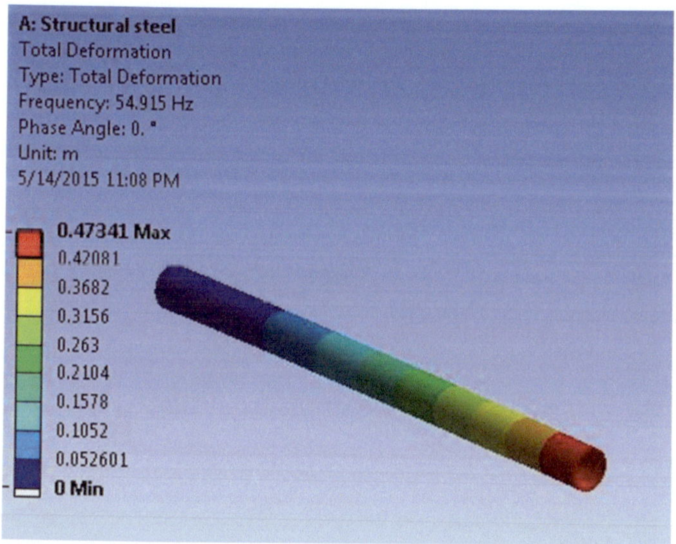

Fig. 8 Natural frequency (ω_n) and total deformation (δ) of steel at mode 1

Fig. 9 Natural frequency (ω_n) and deformation (δ) of gray cast iron at mode 1

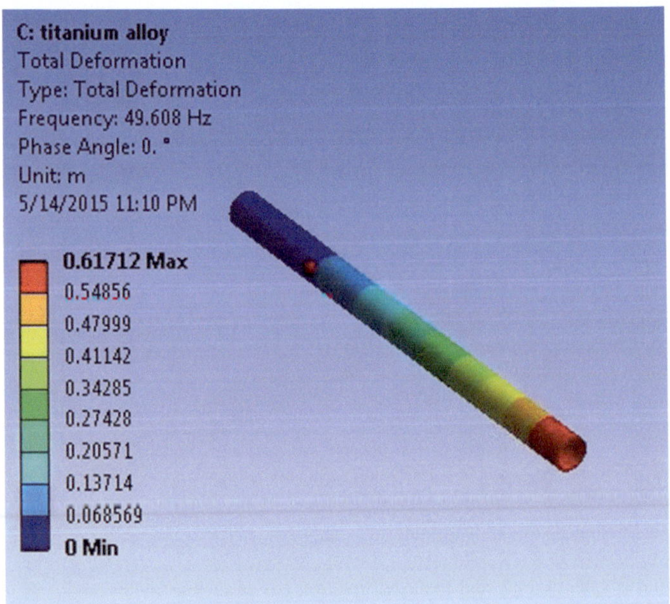

Fig. 10 Natural frequency (ω_n) and deformation (δ) of titanium alloy at mode 1

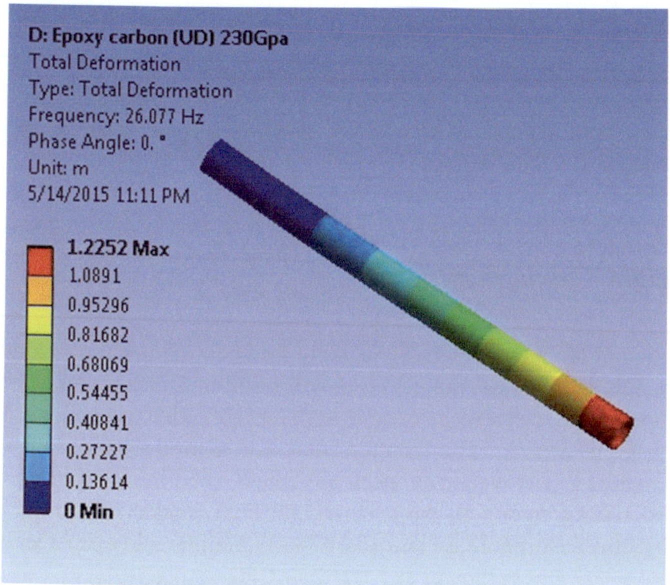

Fig. 11 Natural frequency (ω_n) and deformation (δ) of carbon/epoxy at mode 1

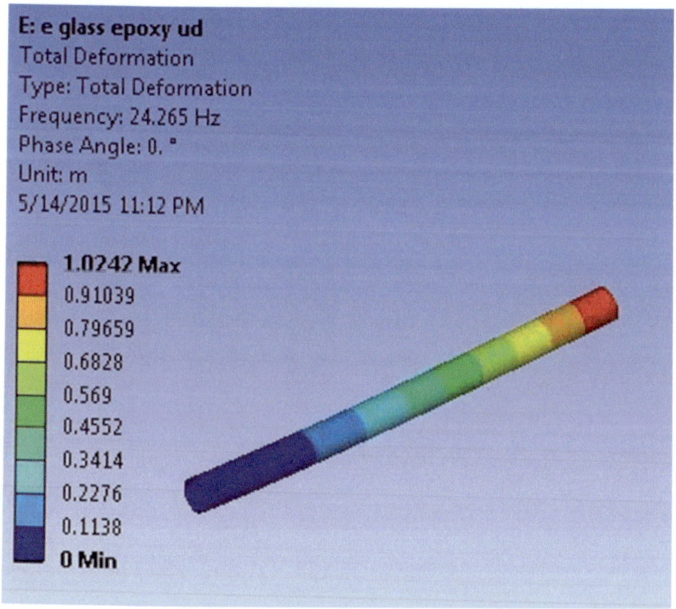

Fig. 12 Natural frequency (ω_n) and deformation (δ) of E-glass/epoxy at mode 1

6 Parametric Studies

A parametric study has been carried out with the aim to compare and find out the most effective and suitable material by evaluating the difference in weight (W), natural frequency ω_n and deformation δ.

6.1 Weight Comparison (W)

Weight is one of the most important deciding factor for a selection of a material for a particular job. In this case, more weight will result in difficult operation, decrease in efficiency due to increase in overall kerb weight. The comparison for all the different materials was made for both the crankshaft as well as driveshaft and shown in Figs. 13 and 14. The weight of steel was found to be the maximum, valuing at 8.87 and 15.108 kg among all the materials for both crankshaft and the driveshaft followed by gray cast iron (8.13 and 13.587 kg), titanium alloy (5.22 and 8.89 kg), E-glass/epoxy UD and carbon/epoxy UD 230 GPa, respectively.

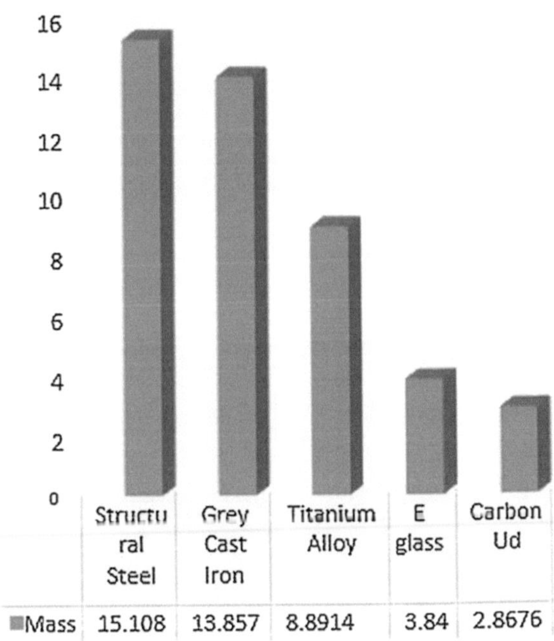

Fig. 13 Weight comparison (W) of different materials for crankshaft

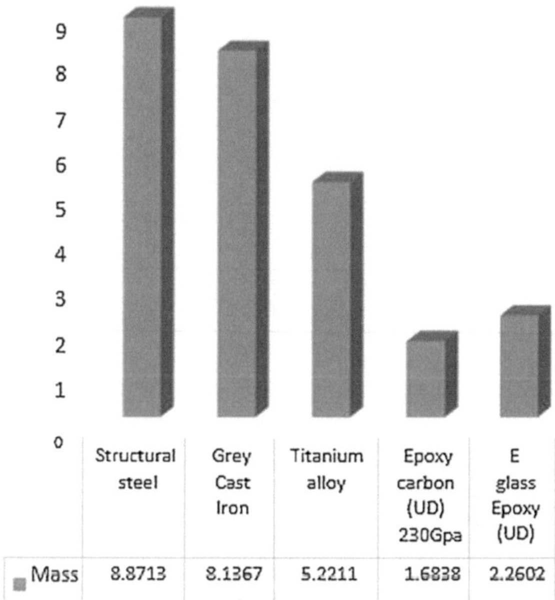

Fig. 14 Weight comparison (W) of different materials for driveshaft

6.2 Natural Frequency (ω_n)

Natural frequency (ω_n) of a component should always be less than its resonant frequency in order to keep the motion in control and prevent any break in operation and catastrophic failures. The frequency of a machine part is also affected by the material it is made up of—geometry of the component, boundary conditions and operating conditions. The imbalance or disproportion present in the driveshaft can be improved by using lighter weight material. If the imbalance still persists, it can have an influence on the driveshaft critical speed (N at which the driveshaft will resonate and begin to vibrate vigorously). Driveshaft and crankshaft of different materials were simulated, and their natural frequency were obtained. The natural frequency at different modes and for different materials was found to be increasing with every passing mode, and ω_n of composite material was found to be lowest followed by alloy and metals, respectively, at each mode as shown in Figs. 15 and 16.

6.3 Deformation (δ)

The deformation (δ) of metals and alloy was found out to be lesser than that of composites, among which carbon/epoxy had the highest deformation followed by E-glass/epoxy, titanium alloy, gray cast iron and steel, respectively, as shown in Figs. 17 and 18. Larger δ in composites can be of advantage in case of driveshaft,

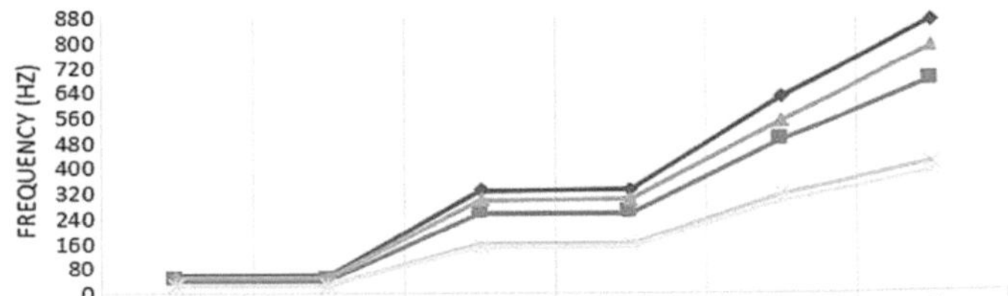

MODES	1	2	3	4	5	6
Structural steel	54.915	55.189	328.63	330.4	626.04	865.99
Grey Cast Iron	42.488	42.762	254.36	256.14	488.56	671.13
Titanium alloy	49.608	49.88	296.48	298.24	552.76	780.38
Epoxy carbon (UD) 230Gpa	26.077	26.362	156.27	158.8	316.77	411.18
E glass Epoxy (UD)	24.265	24.544	145.59	147.72	294.9	384.25

Fig. 15 Natural frequency (ω_n) of materials at different modes of a driveshaft

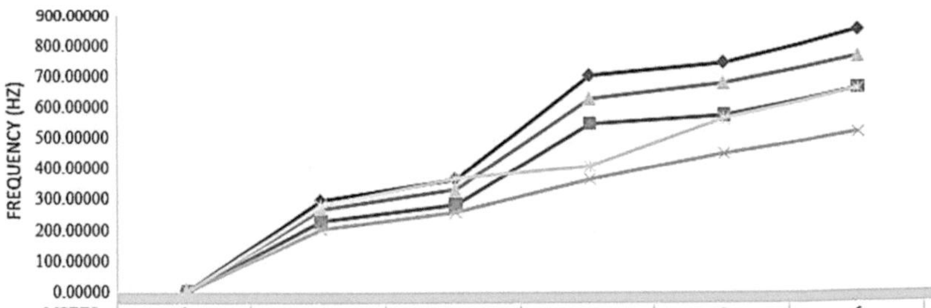

MODES	1	2	3	4	5	6
Structural Steel	0.00115	292.86	361.28	696.64	736.54	841.95
Grey Cast Iron	0.000983	226.96	279.17	541.71	568.88	653.96
Titanium Alloy	0.0014099	264.14	328.97	621.25	671.51	754.64
E glass	0.00052972	200.06	254.41	360.47	442.58	510.24
Carbon Ud	0.00049096	276.15	364.51	402.57	557.3	650.69

Fig. 16 Natural frequency (ω_n) of materials at different modes of a crankshaft

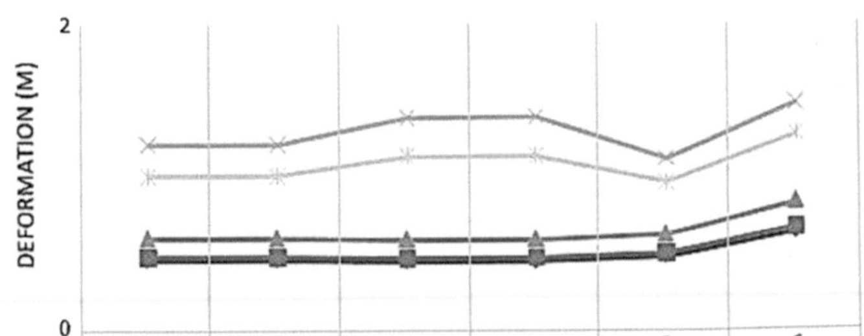

	1	2	3	4	5	6
Structural steel	0.4734	0.4731	0.4657	0.4667	0.4927	0.6479
Grey Cast Iron	0.4943	0.494	0.4862	0.4875	0.5144	0.677
Titanium alloy	0.6171	0.6168	0.6066	0.608	0.6422	0.8437
Epoxy carbon (UD)	1.2252	1.2244	1.398	1.4049	1.1328	1.4855
E glass Epoxy (UD)	1.0242	1.0232	1.1176	1.1521	0.9772	1.2845

Fig. 17 Deflection (δ) at different modes for materials for a driveshaft

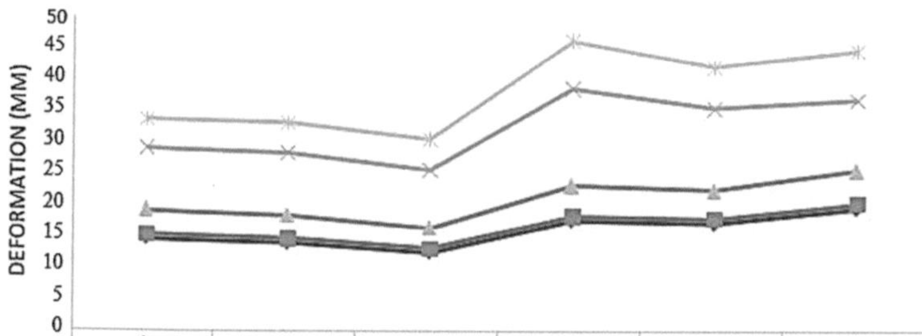

	1	2	3	4	5	6
Structural Steel	14.703	14.105	12.594	17.699	17.258	19.62
Grey Cast Iron	15.353	14.721	13.148	18.396	17.97	20.457
Titanium Alloy	19.166	18.408	16.42	23.205	22.509	25.633
E glass	29.13	28.448	25.773	38.726	35.638	36.899
Carbon Ud	33.749	33.235	30.592	46.32	42.239	44.798

Fig. 18 Deflection (δ) at different modes for materials for a driveshaft

where flexibility is a required characteristic but is a matter of disadvantage in case of crankshaft as large amount of deflection can lead to striking of crankshaft with the crankcase leading to abruption in functioning of the transmission system.

7 Conclusion

In this study, ω_n and δ of a driveshaft as well as of a crankshaft for different metals, alloys and composites have been calculated and simulated using a finite element software ANSYS and designed using CATIA V5. The following significant conclusions from the results obtained were drawn:

(I) δ in the composites was the maximum followed by the alloy and finally the metals which showed the least deflection. (II) ω_n, which needs to be as least as possible so as to obtain maximum possible difference from the resonant frequency, was found out to be least in E-glass/epoxy followed by carbon UD, titanium alloy, gray cast iron and finally structural steel with the maximum. (III) As crankshaft functions in a crankcase with specific dimensions, δ more than desired will alter the functioning and hence composites will not serve the purpose as they have high δ. Hence, titanium alloy is the best option as it easily outperformed structural steel as the base material for the crankshaft. Considering the price factor compared to the steel, it can be used as an alternate material for the manufacturing. (IV) In case of driveshaft, carbon UD composites showed the best results and can be used for current and future manufacturing as it has the best strength to weight ratio and will provide sufficient damping as well.

References

1. Gibson R (2012) Principles of composite material mechanics, 3 edn, CRC Press
2. Shenoy P, Ali F (2005) Connecting rod optimization for weight and cost reduction, SAE-2005-01-0987
3. Farrahi GH, Hemmati F, H-Gangaraj SM, Sakhaei M (2011) Failure analysis of four cylinder diesel crankshaft engine crankshaft made from nodular cast iron, J Eng Res, 22 Spring
4. Bhaumik SK, Rangaragu R, Venkataswamy MA, Bhaskaran TA, Parameswara MA (2001) Fatigue fracture of crankshaft of an aircraft engine. Eng Fail Anal 9:255–263
5. Vogwell J (1998) Analysis of a vehicle wheel shaft failure. Eng Fail Anal, vol 5, 4 edn, pp 271–277
6. Rastogi N (2004) Design of Composite Driveshafts for Automotive Applications. SAE Technical Paper 1:485
7. Novo FMF, Souza MMS, Savoy J, Silva MAC (2012) Analysis of the vibration modes of an automotive propeller shaft using FEM and analytical models. SAE 36:224
8. Jayanaidu P, Hibbatullah M, Baskar P (2013) Analysis of a drive shaft for automobile applications, IOSR J Mech Civil Eng, e-ISSN: 2278-1684, p-ISSN: 2320-334X, 10(2):43–46
9. Yingkui G, Zhibo Z (2011) Strength analysis of diesel engine crankshaft based on PRO/E and ANSYS, third international conference on measuring technology and mechatronics automation, Shangai
10. Montazersadgh FH, Fatemi A (2007) Stress analysis and optimization of crankshafts subjected to dynamic loading, AISI

Magnetic Analysis of an Electromagnetic Band Brake (EMBB)

Anupam Tiwari, Jayanta Sinha, Vishakha Baghel, Gouresh Sood and Aayushi Agrawal

1 Introduction

The basic problem is to analyze the reasons of an ineffective electromagnetic braking system of the legacy Russian weapon system used in the Pechora missile launcher as an electromagnetic band brake. Russian weapon system used in Pechora missile launcher is a support structure for launching a surface to air missile (SAM-3). The electromagnetic band brake is designed to hold the Pechora missile launcher's beam in azimuth as well as elevation angles.

Electromagnetic band brakes follow the working principle of solenoid actuators, used in braking systems to pick up 12 kgf weights and hold 30 kgf weights. They

A. Tiwari (✉)
Institution of Electronics and Telecommunication Engineers (IETE),
New Delhi, India
e-mail: at1972@rediffmail.com

J. Sinha (✉)
Amity Institute of Aerospace Engineering, Amity University,
Noida, Uttar Pradesh, India
e-mail: jsinha1@amity.edu

V. Baghel
Amity Institute of Space Science and Technology, Amity University,
Noida, Uttar Pradesh, India
e-mail: vishakhabaghel05101993@gmail.com

G. Sood
UTC Aerospace Systems, Bengaluru, India
e-mail: gouresh.sood@utas.utc.com

A. Agrawal
Department of Aerospace Engineering, University of Petroleum
& Energy Studies, Dehradun, India
e-mail: aayushi.agrawal13@gmail.com

S. Singh et al. (eds.), *Proceedings of the International Conference on Modern Research in Aerospace Engineering*, Lecture Notes in Mechanical Engineering,
https://doi.org/10.1007/978-981-10-5849-3_26

are being used to manage valves with compressed air. Solenoid is a type of electromagnet where controlled magnetic field is generated when electricity is supplied to the conducting copper coil, and this phenomena helps in moving the load attached to the iron core. Actuator is a mechanical instrument that shifts the inner workings of the air valves. A general solenoid actuator carries the following parts such as coil, armature, spring, and stem [1, 2]. It can be equally run by air and electricity. Electricity is externally provided to the coil. The spring is retained on the armature to retain it down. The armature moves up and down inside the coil and transmits its gesture from the stem to the valve. As the current is being passed in the coil, a magnetic field is generated around it. The so-called magnetic field formed yanks the armature to the center of the coil. The upward movement of the armature causes the spring to compress and the valves to get opened. As soon as the current flow ceases, the very reverse of this happens. As without the magnetic field armature moves down, and so the spring expands thus closing the valve. The most vital benefit of the solenoid actuators or EMBB is their quick processing. Solenoid actuators carry a few demerits. They have either a fully opened or a fully closed position, and solenoid actuators don't produce much force, hence cannot be used in large valves. The usage of solenoid actuator is extensive in all kinds of machinery and even in vehicles in the form of brake actuator [3]. The electric current density is the amount of electric current flowing per unit cross-sectional area of a material. The magnetic flux density of a magnet is also called "B field" or "magnetic induction". It is measured in tesla (SI unit) or gauss (10 000 gauss = 1 T) [4] (Fig. 1).

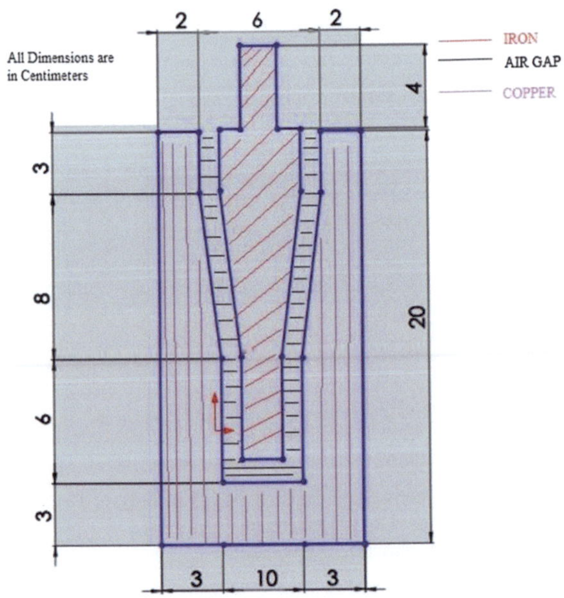

Fig. 1 Schematic solid works model of an EMBB

Table 1 Problem specifications of the model

1	Simulation tool	ANSYS mechanical APDL 14.5
2	Applicable ANSYS products	ANSYS multiphyscis, ANSYS Emag, ANSYS ED
3	Discipline	Electromagnetic
4	Analysis type	Linear static
5	Element type	PLANE13
6	Element behavior	Axisymmetry
7	Loads	Current density
8	Meshing	Fine meshing
9	ANSYS features demonstrated	Axisymmetry, magnetic field intensity vector plots

EMBB is electromagnetic equipment, so its analysis is performed using ANSYS mechanical APDL version 14.5 as explained in methodology. For analysis, the model is considered to be linear (follows the principle of superposition and homogeneity) and static (position of equipment is considered constant throughout the simulation) [5]. The outcome of the simulation elaborates the magnetic flux lines, the magnetic field intensity vector plots, and the magnetic flux density inside the EMBB. First thing is to understand the problem specifications for this project as explained in Table 1.

2 Methodology

2.1 Design and Processing

Methodology adopted in electromagnetic analysis of EMBB can be explained with the help of a process flow diagram as shown in Fig. 2.

Initial step in electromagnetic analysis is the formation/development of geometry of EMBB. At this stage, key points are created which is X, Y, Z coordinates of the point in a three-dimensional plane, and these points are considered according to the measurements of the EMBB as shown in Fig. 1. Key points are joined in specific manner to create areas A1 for air gap, A2 for iron, and A3 for copper coil [6]. By this a close spaces/areas are created in order to perform required mechanical analysis as shown in Fig. 3. The light blue color space depicts air gap, dark blue color region is for iron core, and red color is for copper coil.

In electromagnetic analysis material properties are the relative permeability of the material. So, once basic geometry is ready, thereafter material properties are assigned A1 as M1, A2 as M2, and A3 as M3 material property where A1, A2, and A3 are areas for air, iron core, and copper coil, respectively, and M1, M2, and M3 are the value of relative permeability for the various materials as shown in Table 2 [7].

Second part of the analysis is to finely mesh the developed model and analyze the degree of correctness of the results and post-processing. Prior to meshing element type is defined as PLANE13, considering it to be a lower order element, and

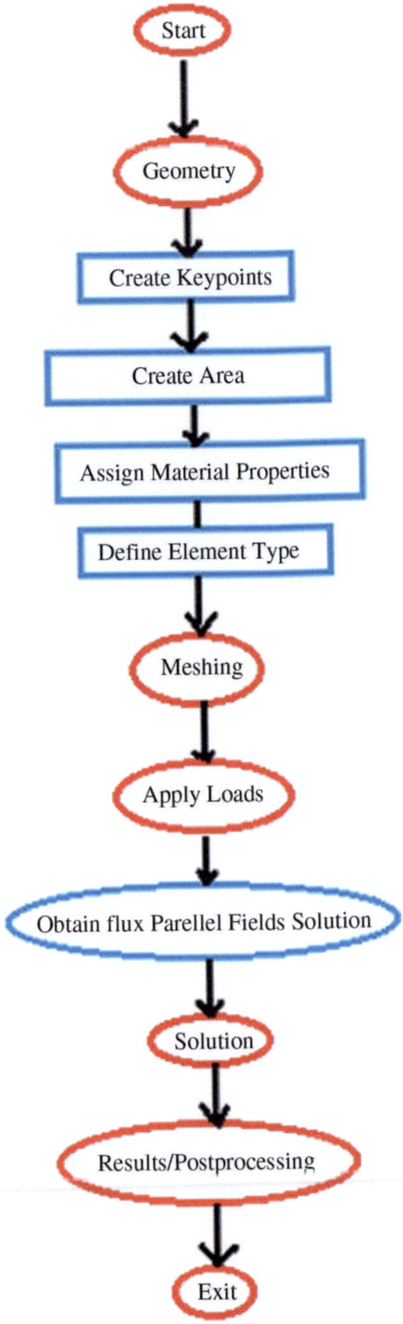

Fig. 2 Process flow diagram for electromagnetic analysis of EMBB using mechanical APDL

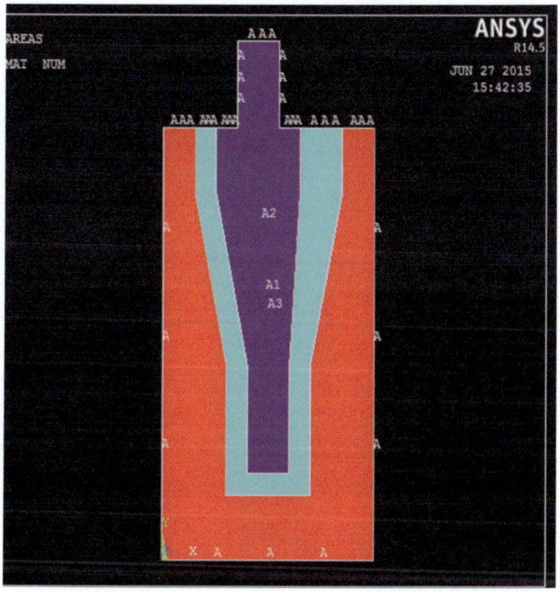

Fig. 3 EMBB model showing areas

Table 2 Material properties

No.	Material (area)	Relative permeability
1.	Air (A1)	1 (M1)
2.	Iron (A2)	1000 (M2)
3.	Copper coil (A3)	1 (M3)

Quad 4-node 13 is selected as magnetic vector and element behavior as axisymmetric. For the purpose of this research, meshing is done considering the element edge length to be 0.25 as shown in Fig. 4.

Third stage of simulation is to apply the necessary loads. Since it is an electromagnetic equipment, so magnetic excitation of copper coil will be considered as load. And as a measure of this, current density is the load applied onto the model for the purpose of simulation, and this is calculated as follows.

Voltage applied to EMBB is 26 V. The resistance of full copper coil is 12.5 Ω.

$$V = I * R \text{ (Ohm's law)} \tag{1}$$

$$I = V/R \tag{2}$$

$$I = 26/12.5 = 2.08\,A$$

Current (I) $= 2.08\,A$ and as number of turns of the coils is 1191, so current per unit turn is 1.7×10^{-3} A/turn. Coil is of circular cross section with a diameter 0.72 mm, i.e., the radius of the coil is 0.36 mm. Therefore, the area of the coil is 0.40715 mm^2. Hence, area of the full copper coil (1191 turns) is $4.849 \times 10^{-4}\ \text{m}^2$.

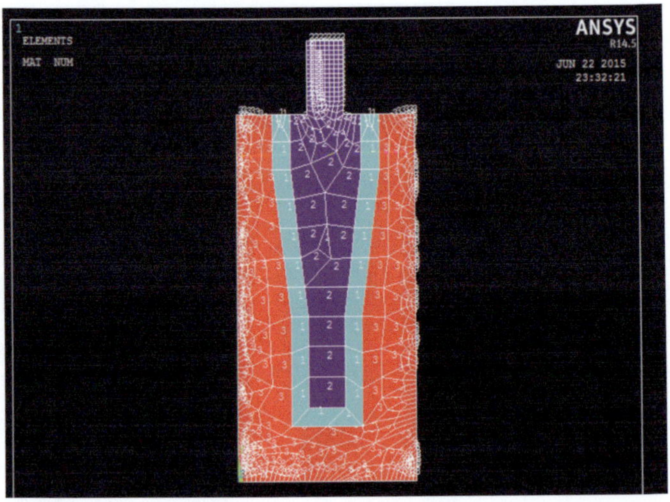

Fig. 4 Model showing meshed areas

Current density is defined as the electric current per unit area of cross section, so it is calculated as 4176.50 A/m^2. Also, a parallel flux field solution is obtained by applying the perimeter boundary conditions. The boundary conditions are assumed so that flux does not leak out of the iron at the perimeter of the model.

Fourth stage is the solution of the simulation and modeling. At this step, solution at each key point is obtained. Finite element method (FEM) is adopted for solving the stiffness matrix. Solution is found based on the material properties and the applied loads.

Based on solution, results and post-processing is performed where magnetic flux lines, magnetic field intensity vector plots, and magnitude of magnetic flux density is plotted and analyzed.

2.2 *Results and Post-Processing*

The results of analysis are carried out for the variation in the magnetic field distribution against voltage. For the given input voltage 26 V and resistance 12.5 Ω, the following results have been obtained.

In Fig. 5, magnetic flux lines show the distribution of magnetic flux inside the EMBB. Non-uniform flux lines are observed in the upper part of the equipment, whereas uniformity can be visualized in the bottom part of the equipment, also density of flux lines varies inside, i.e., at some places large amount of electromagnetic field is present and at other the field strength is visualized as depleting because of the long duration usage of the equipment.

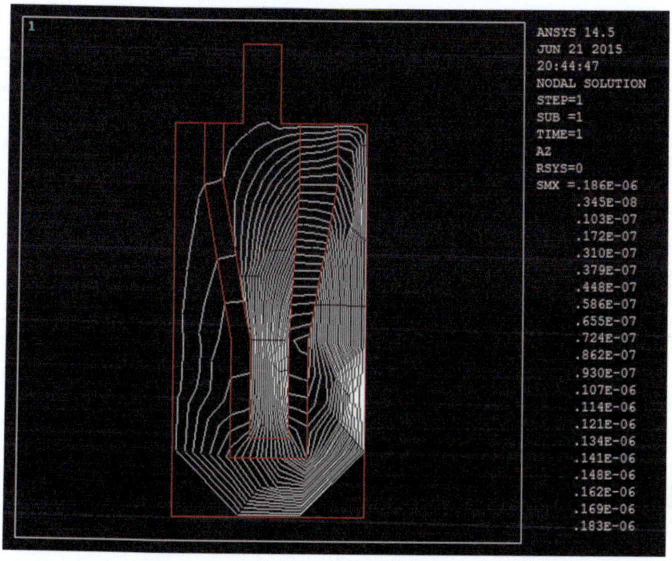

Fig. 5 Magnetic flux lines pattern of EMBB for voltage (V) = 26 V and current (I) = 2.08 A

Fig. 6 Magnetic field intensity vector plot for voltage (V) = 26 V and current (I) = 2.08 A

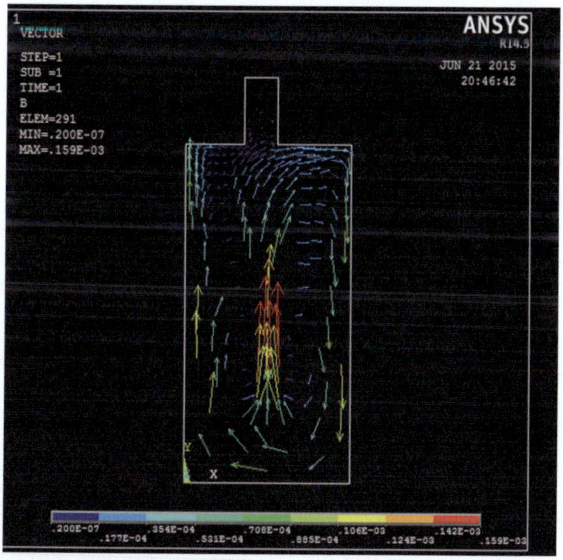

In Fig. 6, magnetic vector plot shows the magnitude, distribution, and direction of the electromagnetic field inside the EMBB. The head of the arrows show the direction of the field, and their respective color and its magnitude shows the magnetic field strength at that point inside the equipment. The direction of field is outward to inward with its maximum strength being depicted by the red color arrows whose numerical value can be referred from the legend as 0.159E-03.

Strength (number of arrows) of the arrows is less in the bottom outer region and is more in the upper region, so non-uniformity is analyzed in electromagnetic field.

Figure 7 depicts the magnitude of the magnetic flux density. This shows the strength of the magnetic field at a given point as shown by red color with maximum value 0.173E-03 and minimum value of flux density is 0.419E-09, i.e., non-uniform magnetic flux density distribution is observed inside the EMBB as the equipment is used for such a long time that the permeability property of the iron core has been depleted due to which the EMBB is unable to perform its required function of holding the launcher's beam.

From the above observation, it is believed that increasing those limits, i.e., current and voltage supply will enable the EMBB works more accurately and as the magnetic flux increases between the iron core and copper coil of the EMBB, more electromagnetic energy can be achieved [8].

Also, it will be able to analyze and predict the magnetic field intensity. Magnetic field of a solenoid can be expressed as given in Eq. (3) [9, 10].

$$B = \frac{\mu I N}{2} \, Tesla \tag{3}$$

So to increase voltage, power electronic devices like voltage doublers or simple choppers circuit can be used. Voltage doublers are electronic circuit which charges the capacitors from input voltage and switches these charges in such a way so that it can produce twice of the exact input voltage at the output [11]. Input given to EMBB is 26 V DC as available in the launcher and doublers produce a twice DC voltage of input AC voltage, so a single phase inverter circuit is used which changes

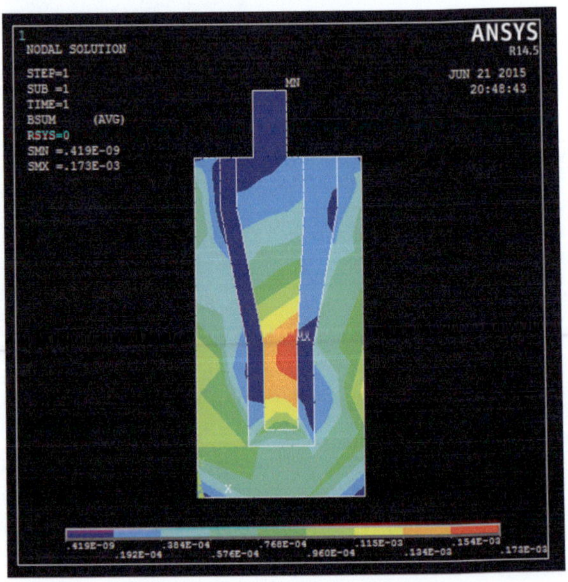

Fig. 7 Magnitude of magnetic flux density for voltage (V) = 26 V and current (I) = 2.08 A

a DC input voltage to a symmetrical AC output voltage of the desired magnitude and frequency [11]. Here the inverter works as a voltage controller which helps in providing an appropriate AC voltage to the voltage doubler. Also a simple chopper circuit is sufficient to serve the purpose. DC–DC converter or chopper is an electronic device that converts fixed DC input to a variable DC output voltage directly. The output voltage of chopper is controlled by varying the conduction time of the transistors. So by finding a suitable conduction time, the chopper's output voltage can be doubled as per requirement. In present analysis as chopper circuit is used to change the voltage, so it is called voltage regulator. As per the requirement, twice the voltage is required at the input of the EMBB, so step-up chopper is placed prior to the EMBB.

Calculating the value of current density when voltage doublers with inverter or simple choppers circuit is implemented keeping the resistance, number of turns and cross-section area of coil constant. So, in this case voltage applied to EMBB is 26*2V = 52V. The resistance of full copper coil is 12.5 Ω. From Eqs. (1) and (2)

$$I = 52/12.5 = 4.16\,A$$

So here, current I = 4.16 A and as number of turns is 1191, so current per turn is 3.49×10^{-3} A/turn. Coil is of circular cross section with a diameter of coil as 0.72 mm, i.e., the radius of coil is 0.36 mm. Therefore, the area of the Coil is 0.40715 mm^2. Hence, area of full copper coil (1191 turns) is 4.849×10^{-4} m^2. Current density for voltage doublers case is calculated as 8572.05 A/m^2.

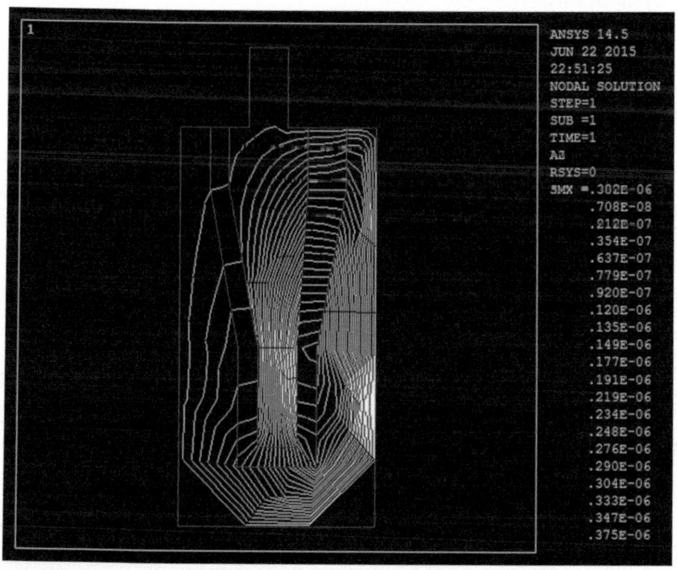

Fig. 8 Magnetic flux lines pattern of EMBB for voltage (V) = 52 V and current (I) = 4.16 A

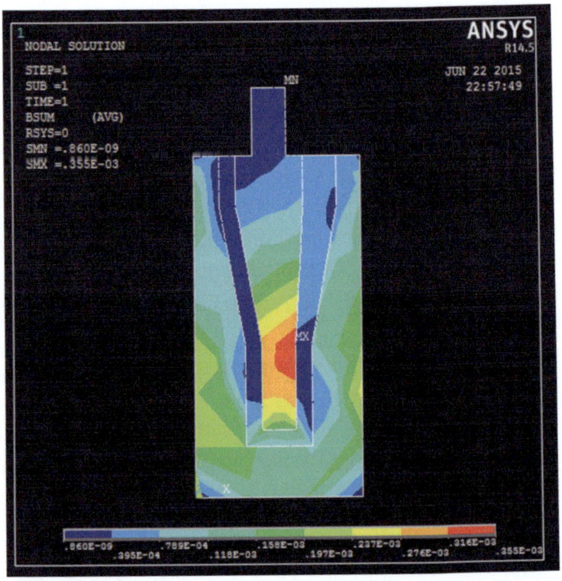

Fig. 9 Magnitude of magnetic flux density for voltage (V) = 52 V and current (I) = 4.16 A

Electromagnetic analysis has been performed for the new value of current density, and the following results as shown in Figs. 8 and 9 are analyzed, and it is evident that all the parameters have been doubled by using a voltage doubler concept which shows that the effectiveness and efficiency of the equipment can be improved by adopting these methods.

3 Conclusions

Magnetic analysis of the model has been developed for predicting the magnetic field distribution in the electromagnetic band brake. From the analysis of the EMBB ANSYS simulation the following observations have been made.

1. Generated magnetic field and magnetic flux lines are not symmetrical.
2. Magnetic field lines are missing in some regions for the given input voltage.
3. Magnitude of magnetic flux density is not proper.

But after increasing the voltage, positive concurrence has been achieved. From Figs. 8 and 9, it can be observed that magnetic field and magnetic flux density varies appreciably with the increase in voltage. It has also been analyzed that magnetic field intensity can be enhanced if the air gap between copper and iron is relatively small. Also, it is evident from the results that the electromagnetic energy has been depleted as the iron core has lost its permeability property, so as a solution to it a new iron core can be placed by knowing its proper metallurgy. The overall conclusion of the research work is that magnetic strength could be increased if voltage supplier is increased which can be done by using step-up choppers circuit or voltage doublers with single phase inverter circuit. Also, magnetic field strength can be increased by increasing number of turns of coil windings. For that either thin copper coils with more number of turns or thick copper coil with less number of turns, by keeping resistance constant, can be used.

References

1. Patel BR (1985) Permanent magnet bistable solenoid actuator. US Patent No 4533890, 6 Aug 1985
2. Banick GS, John JH (1991) Solenoid actuator having a magnetic flux sensor. US Patent No 5032812, 16 Jul 1991
3. Solenoid Actuator http://www.solenoidactuator.org/
4. Magnetic Flux https://www.supermagnete.de/eng/faq/How-do-you-calculate-the-magnetic-flux-densitys
5. ANSYS Mechanical APDL Structural Analysis Guide. Release 14.5
6. Seshu P (2003) Textbook of finite element analysis. Prentice-Hall of India
7. Chandrupatla T, Belegundu R, Ashok D (2011) Introduction to finite elements in engineering. Upper Saddle River, Prentice Hall
8. Kushida T (1985) High speed powerful and simple solenoid actuator DISOLE and its dynamic analysis results. No 850373. Technical Paper, SAE
9. Jorden EC, Balmain KG (1968) Electromagnetic wave and radiating systems. Prentice- Hall of India
10. Prasad KD (2003) Antenna and wave propagation. Satya Prakashan, New Delhi
11. Rashid MH (2007) Power electronics: circuits, devices, and applications. Eastern Economy Edition, Prentice-Hall of India

Experimental Evaluation of Circular Cylindrical Shells with Cutouts Under the Combined Effect of Structural and Thermal Loads

S. Narendar, R. Gopikrishna, T. Srinivasa Kumar and M. Varadanam

1 Introduction

Circular cylindrical shells have vast applications in the construction of missile, rocket, and aircraft structures [1–3]. Apart from aerospace applications, examples of shells include pressure vessels, pipes, automobile tires, incandescent lamps, caps, roof domes, factory or car sheds, and a variety of containers. Each of these has walls that are curved. Such airframe sections need to be qualified at ground level to simulate the design conditions for checking and qualifying its structural integrity.

An effort is made to design, develop, and test a cylindrical shell under a combined structural and thermal load applications [4, 5].

2 Test Article Preparation

The following activities were carried out as a part of test setup and control airframe assembly test article preparation:-

(1) Protective paints from all external and internal surfaces of the test article being heated were removed for thermocouple bonding
(2) Temperature sensors were bonded at the designated locations on the test article
(3) All surfaces to be exposed to heat were coated with high-emissivity black paint (emissivity = 0.94)

S. Narendar (✉) · R. Gopikrishna · T. Srinivasa Kumar · M. Varadanam
Defence Research and Development Laboratory, Dr. APJ Abdul Kalam Missile Complex, Hyderabad 500058, Telangana, India
e-mail: nanduslns07@gmail.com; narendar@drdl.drdo.in

S. Singh et al. (eds.), *Proceedings of the International Conference on Modern Research in Aerospace Engineering*, Lecture Notes in Mechanical Engineering,
https://doi.org/10.1007/978-981-10-5849-3_27

(4) Bottom bulkhead and the test rig were separated by the reflective sheet adhered onto the test rig and were not heated during the test
(5) The LVDTs were protected from heat using silica fabric.

3 Test Setup

This section describes the components of the thermo-structural test setup of circular cylindrical airframe assembly.

3.1 *Test Fixture*

A special setup was designed for the thermo-structural test on cylindrical airframe assembly. This setup was made to make sure that the bottom surface of the airframe assembly to mate with the fixture base plate. An interface plate has been designed to use in between shell airframe and the test rig. The heating system segments are mounted concentrically with the airframe assembly. The radial distance between the outer surface of airframe assembly and the inner surface of the heating system was kept at 75 mm distance for temperature profile simulations. A schematic test setup was shown in Fig. 1.

One 10-ton capacity single-acting actuator and one 5-ton capacity double-acting actuators are used to simulate the desired structural loading scheme on to the airframe assembly. These two actuators apply differential load on to the dummy section, which will be transferred to the shell airframe assembly, which simulates the equivalent bending moment coming on to the airframe section. These two actuators are hydraulic type. These two loading directions are shown in Fig. 1.

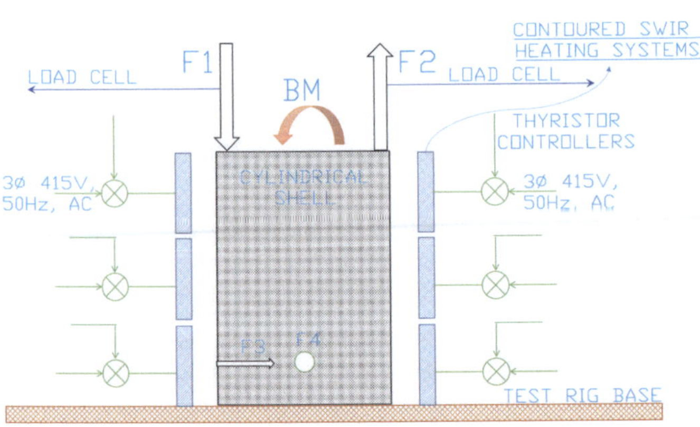

Fig. 1 A Schematic test setup for thermos-structural test on a circular cylindrical shell

Two perpendicular loads (F3 and F4) due to the thrusters are applied on of cutouts 90 deg. apart. For this load application, an in-house fixture is designed and fabricated.

3.2 Power Controllers

For application of the controlled voltage to the heater cassettes, thyristor-based 3-phase AC power controllers are used. These controllers have a fixed input of 3-phase 415 V AC 50 Hz supply which is converted into variable 3-phase 0–415 V AC 50 Hz supply. The power controllers are capable of giving power outputs in response to a gate voltage of 0–10 V, where 0 V corresponds to no output and 10 V corresponds to full output. The power output varies linearly with the applied gate voltage.

3.3 Control System

To apply a variable heat input to the test article, a control system was used. Typically for temperature control, closed-loop control systems are used. In the present case, software-based controller is used to achieve the closed-loop control of temperature. The schematic of the closed-loop temperature control is shown in Fig. 2.

3.4 Data Acquisition and Control System

To control and acquire the heat flux data National Instruments (NI) make data acquisition and control system (DACS) was used. The DACS was configured with a voltage channel to read the sensor output from the heat flux gauge during the heat flux simulations. A low-pass filter of cutoff 2 Hz was used in each channel to

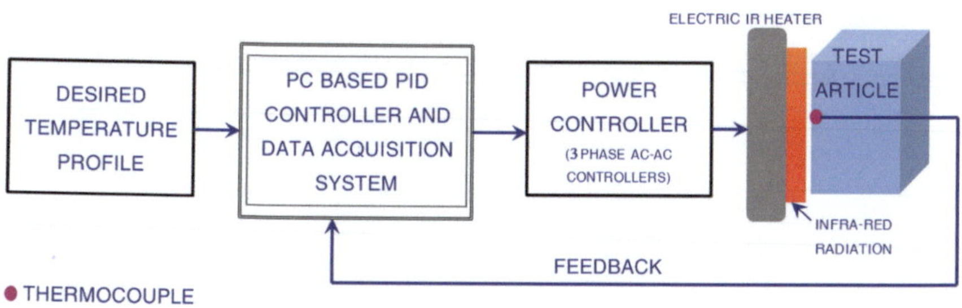

Fig. 2 Schematic of the control and data acquisition system for closed-loop temperature control

remove high frequency noise. The data was recorded at a sampling rate of 5 samples per second.

3.5 LVDTs

Four numbers of LVDTs are mounted on the cylindrical airframe assembly. Two LVDTs (D3 & D4) are mounted on the inner side of the airframe section, where the side loads F3 and F4 are applied, to measure the radial deformations of the section. LVDTs (D1 & D2) are mounted axially on the dummy section of the airframe to measure the deformation due to bending.

4 Test Procedure

After assuring the readiness of the test setup including hose connections from power pack to the actuators, power connections to power pack and the DAQ system, and the health of the LVDTs, the test was started after considering all the safety requirements. Initially, the airframe was loaded to 20% and 40% of the design load and released back to 0%, to check the linearity in LVDT measurements and the pressure buildup in the power pack to actuator connecting hoses. After this activity, the actual test has been started, the section is loaded up to 100% of the design load in steps of 20%, and this load data is plotted in Fig. 3. The deflection data recording during this process was in continuous mode.

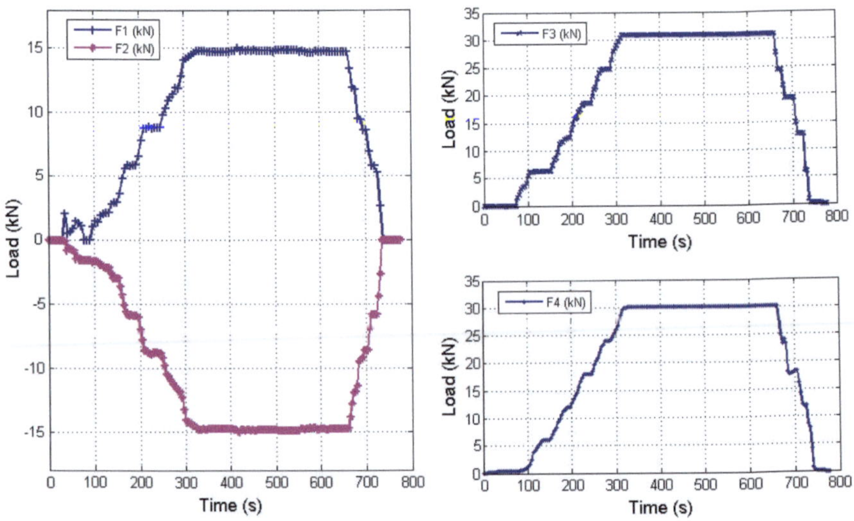

Fig. 3 Structural loads applied during the thermo-structural test

After assuring the readiness of the test setup including heater power supply connections, water supply for heater cassette cooling, and the health of the thermocouples, the test was started after considering all the safety requirements.

Once the airframe section is loaded up to 100% of the design load, then the thermal load has been applied on to the section. Control thermocouple data for the heating system was monitored online during the application of the thermal profile. The temperature data at all the other location on the airframe section is recorded in continuous mode at a sampling rate of 100 s/s.

5 Results and Discussion

The results of the structural loading on the cylindrical shell airframe assembly are plotted. Four LVDTs are used for measuring the deformation of the airframe section, and the locations of these sensors are shown schematically in Fig. 4. The deformations of the airframe assembly at locations D1, D2, D3, and D4 are shown in Fig. 5, during both structural loading and thermal loading. The plot represents the deformation of the section with respect to the percentage of structural load application. A maximum of 0.78 mm deflection is observed at D1 location and

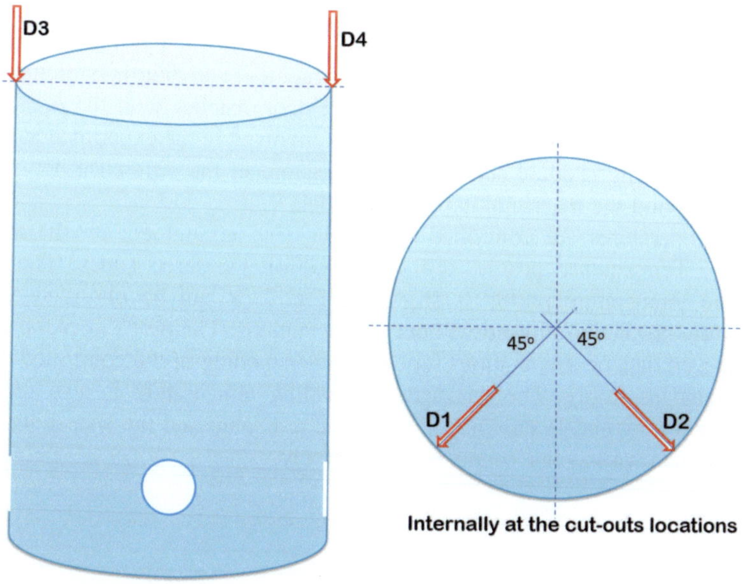

Fig. 4 Schematic diagram of the location of LVDTs

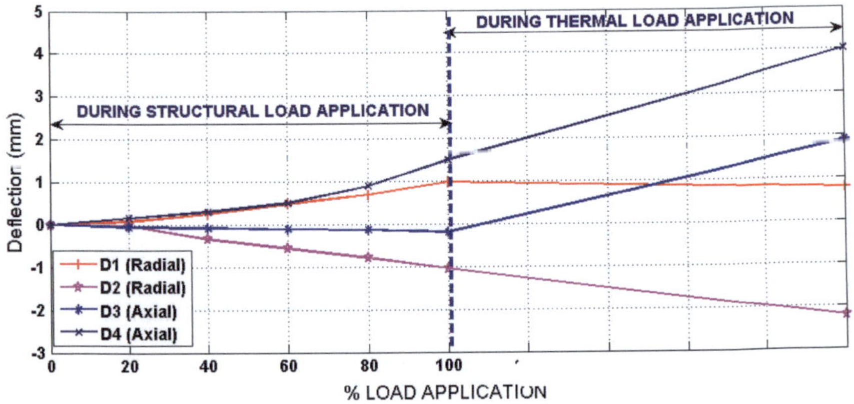

Fig. 5 Deflections measured at radial and axial directions of the cylindrical airframe assembly

−2.23 mm at D2 location. Both D1 and D2 represent the radial deformation of the airframe assembly at the thrust loading points. Deformations of 1.89 and 4.01 mm are observed at the axial loading points and are designated as D3 and D4, respectively. These two values are observed when the section is loaded at 100% structural load along with the 100% thermal load.

The linear and cubic curve fits for the load versus deflection plot for LVDT D4 is shown in Fig. 6. It shows that, for both structural and thermo-structural load application cases, the experimental deflection data matches with the cubic curve fitting. This is mainly due to the nonlinearities involved in the experimental testing. The corresponding residual shows the clear picture of the difference between the curve fitting and the measured experimental data.

The temperatures are controlled as per the test requirement on the airframe assembly. Temperature profiles are controlled on the three zones. The zone-I controlled temperature profile is shown in Fig. 7. It can be observed that the temperature profile is followed without any deviations. The plot shows the desired and achieved data of temperatures for clear understanding of the controlled data on the airframe assembly. The package surrounding temperature is also measured during the testing and is shown in Fig. 8. The test setup and the snapshots during the test are shown in Figs. 9 and 10, respectively.

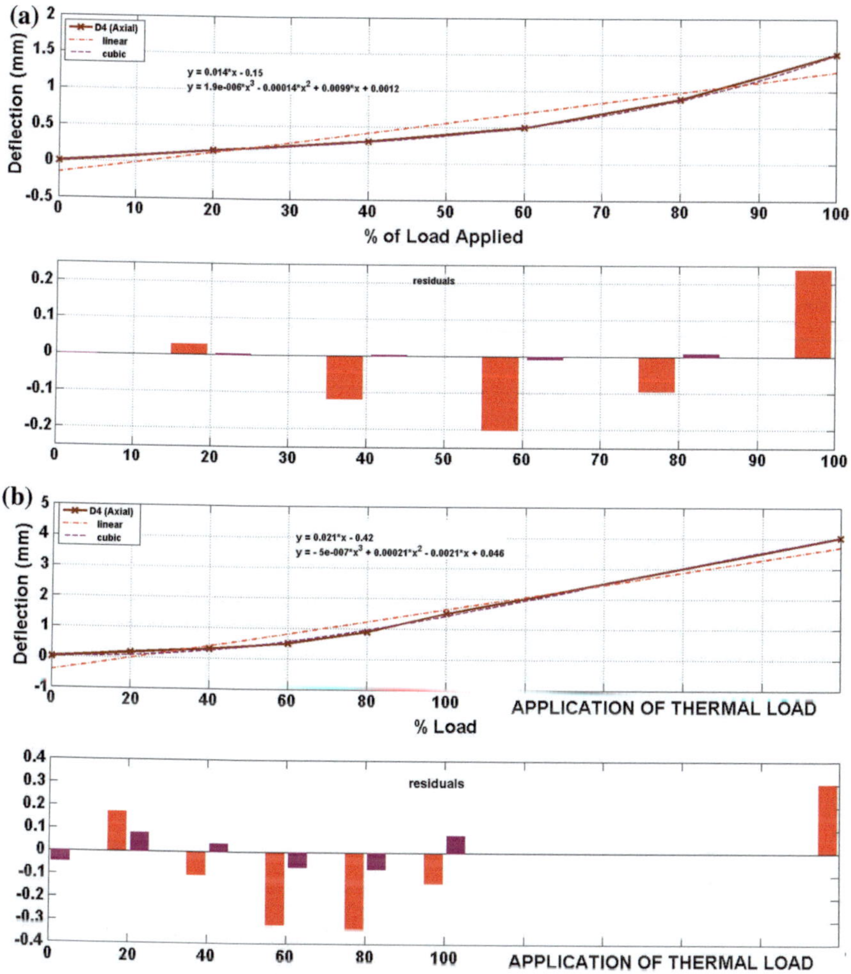

Fig. 6 Basic curve fitting of experimental deflection data along with the residuals for **a** structural load test case and **b** thermo-structural load test case

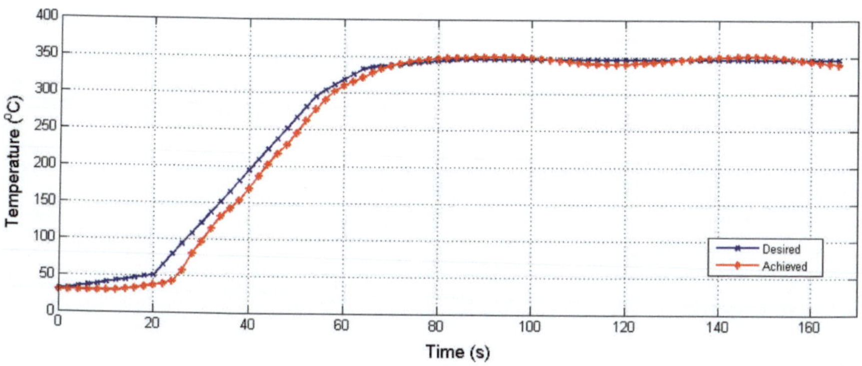

Fig. 7 Controlled temperature at the circular zone-I

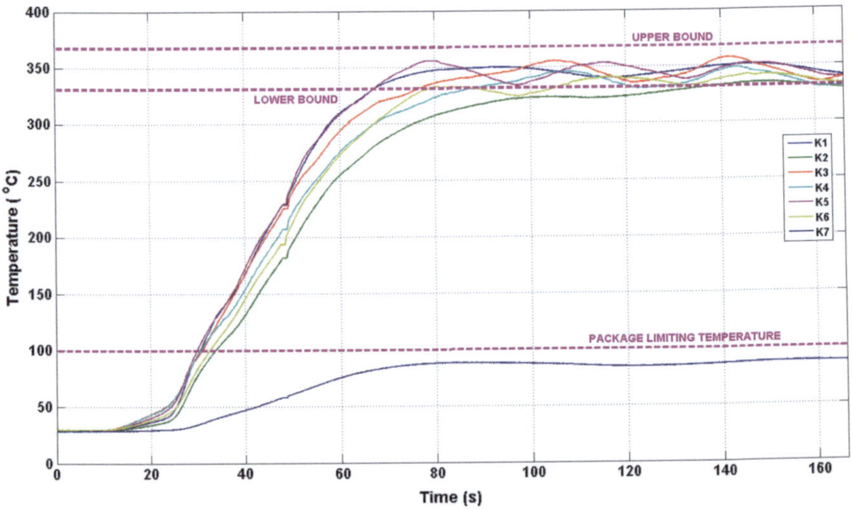

Fig. 8 Measured temperatures at various locations on the airframe

Fig. 9 Internal 90° loading arrangement and the complete test setup

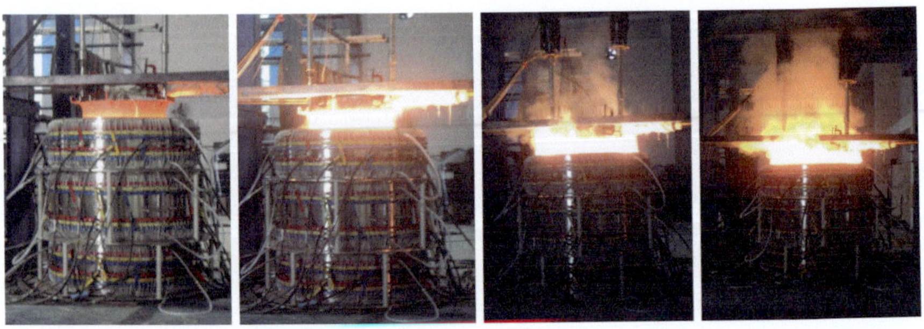

Fig. 10 Snapshots during thermo-structural testing

6 Conclusions

Combined structural and thermal loads are applied on a circular cylindrical airframe shell for qualifying the structural integrity. The structural load, namely combination of axial force and bending moment, was applied onto the section under test using a dummy section attached to the front bulkhead of the airframe sections. The loads were generated in accordance with the desired values within the tolerance mentioned. The thermal load, specified in the form of temperature distribution, is produced on the test article faithfully using the control of incident radiative heat flux with respect to closed-loop temperature control. The airframe section withstood the thermo-structural loads during the tests. The permanent deflections are within allowable limits. No part of the test section exhibited visual failures. It will be further processed with NDT tests for final decision.

References

1. Vinson Jack R (1989) The behavior of thin walled structures: beams, plates, and shells. Kluwer Academic Publishers, Boston
2. Yoshimura Y (1955) On the mechanism of buckling of a circular cylindrical shell under axial compression. NACA-TM-1390
3. Simitses GJ (1991) Analysis of shell-type structures subjected to time-dependent mechanical and thermal loading. NASA-CR-189051, NAS 1.26: 189051
4. A technical report on Establishment of Thermo-Structural Test Facility at DRDL, India (2010) DRDL, Hyderabad
5. Thermo-Structural test on Circular Cylindrical Airframe Shell. Test Report No. DRDL/DOFS/STF/TSTF/2014/01, Rev. 0.0, Date 21/01/2015

Weight Reduction of Structural Members for Ground Vehicles by the Introduction of FRP Composite and Its Implications

Medhavi Sinha, R.K. Tyagi and P.K. Bajpai

1 Introduction

Petroleum is the most common fossil fuel for automobiles all over the world. Its consumption is increasing day by day, and the fossil fuel is depleting at a very fast rate. The middle of twentieth century was the period of sudden growth, industrialization and development. It was also the phase of immense growth in the field of automotive industry. According to a recent survey, the current number of automobiles all over the world has risen to about 3.6 millions [1]. Such a large number requires substantially large amount of fuel. With industrialization and development, the automobile has enabled phenomenal transformation with the ease in transportation in the lives of human beings. But this transmutation is completely dependent upon fuel for endurance. Therefore, large number of automobiles for the facilitation of human beings presents a challenging energy and environmental problem for the world. It is consequently required that some measures must be taken in the field of automotive industry for decreasing the quantity of fuel consumption for the mobility of the automobiles.

The present paper discusses about the various strategies that should be adopted for decreasing the fuel consumption of automobiles. The main focus of the paper is the utilization of lightweighting techniques to improve fuel economy in vehicles. The current paper presents an opportunity to reduce fuel consumption from the transportation sector by reducing the weight of structural members of the vehicle. The structural members are imperative to any ground vehicle. The contemporary

M. Sinha (✉) · R.K. Tyagi
Amity School of Engineering and Technology, Amity University,
Noida, Uttar Pradesh, India
e-mail: medhavisinha.02@gmail.com

P.K. Bajpai
NSIT, Delhi, India

S. Singh et al. (eds.), *Proceedings of the International Conference on Modern Research in Aerospace Engineering*, Lecture Notes in Mechanical Engineering,
https://doi.org/10.1007/978-981-10-5849-3_28

engineering technology requires that the structural members must be designed considering the safety requirement and at the same time should serve their intended function without any interruptions during their designed service life [2]. Therefore, the response of the structural members to the applied loading under working condition must be completely analyzed. The present paper discusses about the fuel pipe as a structural member of automobiles. The main focus is vehicular lightweighting and consequently the decreasing fuel consumption. Various analyses on ANSYS have also been performed on the fuel pipe taking FRP as its constituting material for safety examination.

Many rules and standards have been imposed to limit the fuel consumption and emissions from the ground vehicles. CAFÉ 2016, Environmental Protection Agency (EPA), National Highway Traffic Safety Administration (NHTSA), US Federal Standards, and the Clean air Act are among a few agencies which have imposed very strict restriction over automotive industries to limit the fuel consumption and emissions of automobiles. The lighter-weight and other more fuel-efficient vehicles have many beneficial implications on the energy saving and the environment [3–7].

The ground vehicles manufactured in the late 70s were very bulkier, but the introduction of newer materials and the importance of fuel economy by the imposition of various standards in the 80s and 90s enforced the automotive industries to reduce the weight of the automobile. In the 1980s, the manufactures of the automotive industries focused on finding new materials for weight reduction in the ground vehicles, and a favorable climate was created for the development of materials and fabrication techniques for composite body panels and structures for ground vehicles [8]. This has continued for the current trend prevailing these days to opt for heavier vehicles such as SUVs which are like lightweight trucks in comparison with small cars. Also, the inclusion of many extra feature safety and comfort has increased the weight of the vehicle. Figure 1 shows the trend of weight of the automobiles with respect to the time in years. The current demand these days is the vehicles must not only become lighter, but should have power improvements. They should revolutionarily use advanced, more fuel-efficient powertrains.

The lightweighting of ground vehicles is a challenging and intricate issue. Various materials are suggested during the study, and the characteristics of each

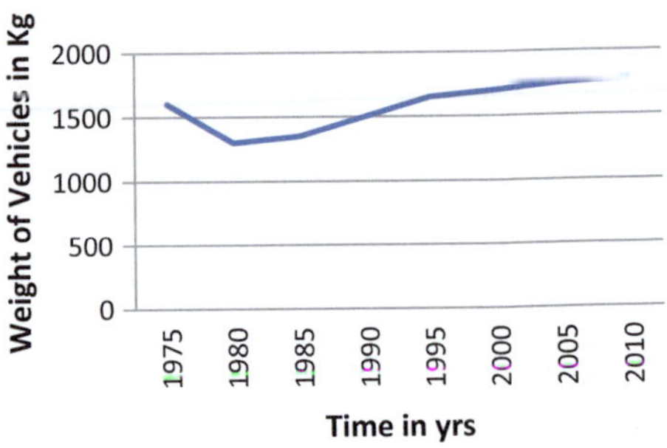

Fig. 1 Weight trend of automobiles with time. *Data source* [14]

material are also examined. The present paper emphasized on the use of composite materials for the weight curtailment of the automobiles. The composites have special characteristics of high-strength-to-low-weight ratio which makes them ideal for the application. The fiber-reinforced composites are among the most demanding composites in various modern technological applications. The present paper discusses about the carbon fiber-reinforced polymer (CFRP) composite where the reinforcement is carbon fiber (T300) and the thermosetting plastic is epoxy (Araldite LY5052). The FRP has their unique properties such as lightweight, high strength, high stiffness, high biological and chemical resistance, high surface finish as well as low cost and maintenance at the same time. The properties of the FRP have been utilized and applied in the present paper in the field of automobile engineering [9, 10]. The current analysis reveals the process of achievement of energy savings via vehicle lightweighting under the new fuel economy mandate.

2 Consequences of Vehicular Weight Reduction

The weight reduction of vehicles can have various desirous consequences. The present section discusses about a few of them.

2.1 *Impact on Fuel Consumption*

One of the major acquisitive consequences is its implications on fuel consumption of the vehicle. The vehicles are designed to reduce the drag on the vehicle. It can be understood from Fig. 2 that the drag in any vehicle will reduce as the weight of the vehicle will reduce. And the weight of the automobile has a direct relationship with its fuel consumption. The weight reduction of the vehicle will directly reduce the fuel consumption of the vehicle. When the vehicle starts moving, it has to overcome various inertial and frictional forces. The power to overcome these forces is extracted from the combustion of the fuel. Therefore, it is a prime concern that the vehicle weight should be reduced as much as possible for low inertial force and frictional force on the vehicle so that the fuel consumption at the starting of the vehicle is as low as possible.

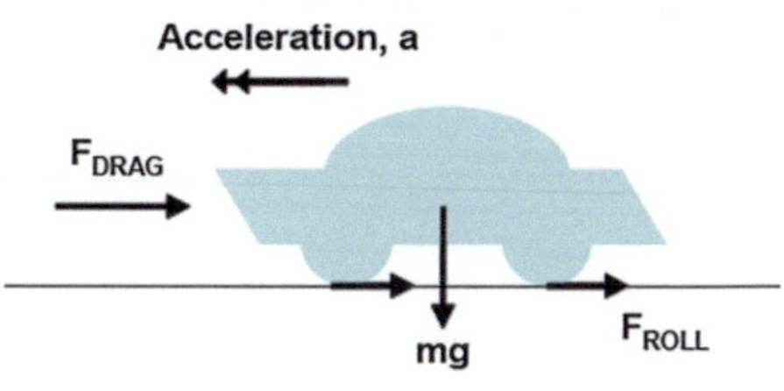

Fig. 2 Role of drag in automobiles

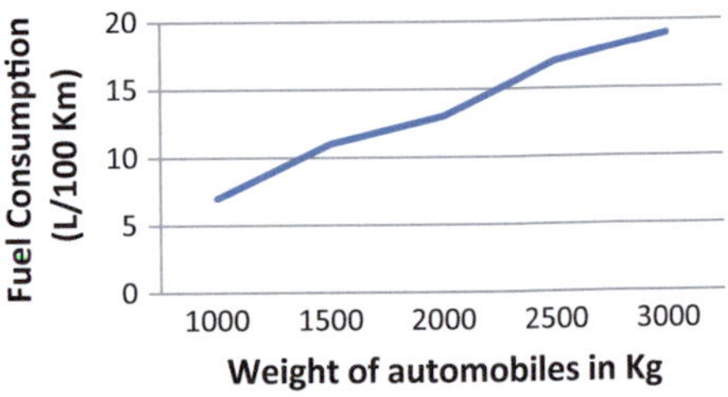

Fig. 3 Trend of fuel consumption of automobiles with weight. *Data source* [15]

Similarly, when the vehicle is moving, a force is again required for its acceleration. This force for its acceleration will significantly be reduced if the weight of the vehicle is reduced, and consequently, the fuel consumption of the vehicle will also reduce. Figure 3 shows the effect on fuel consumption of automobiles with respect to their weight. The reduction in the weight of the automobiles will reduce the inertial forces that the engine has to overcome for its acceleration, and thus, the energy required for the vehicle by combustion of fuel will get lowered. For every 10% reduction in vehicle weight, the fuel consumption of vehicles is reduced by 4–7%. It is observed from the literature that a 100 kg weight reduction on a medium-sized passenger car reduces the fuel consumption of the automobile by 0.35 L of fuel per 100 km.

3 Measures for Vehicular Weight Reduction

A common medium-weighted vehicle has its major parts composed of heavy iron and steel. Figure 4 shows the mass distribution of various subsystems of an automobile. It can be observed from the figure that the most bulky subsystem of an automobile is its body following which is the subsystem chassis. These are the subsystems where iron and steel predominate as a major constituent. The other subsystems such as powertrain, interior, and power systems also constitute these bulkier materials, but their overall contribution in the weight of the vehicle is comparatively less than the above-mentioned subsystems. The prime concern therefore should be to reduce the weight first in these two subsystems.

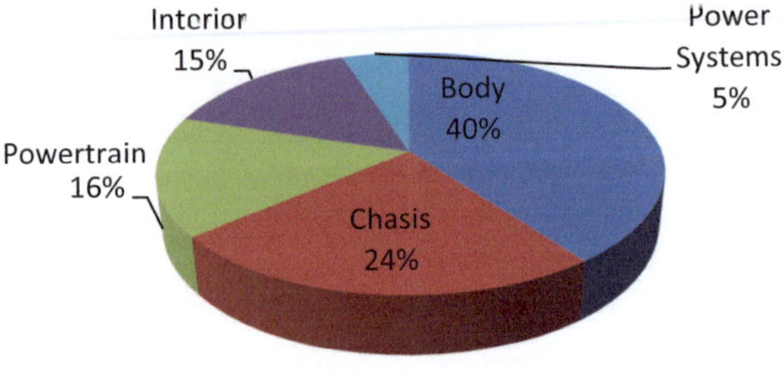

Fig. 4 Mass distribution of different subsystems of an automobile. *Data source* [16]

As vehicle weight reduction has become a serious concern, various researches and studies are being performed on the matter because of its proficient implications. Various measures are been adopted for the same. It can be observed from the literature that the following methods if applied properly will lead to a significant weight reduction in the vehicle.

3.1 Introduction of Alternative Materials

Various alternative lightweighted materials such as aluminum, high-strength steel, polymer composites, and plastics are identified as adroit substitution materials for heavier and bulkier iron and steel components of automobiles.

3.1.1 Properties of Alternative Substitution Materials

Any material cannot be selected as a perfect substitution for the traditionally used iron and steel for the manufacturing of majority of parts of an automobile. Every material must be contemplated and scrutinized for its proficiency as an alternative substitution material. The alternative substitution material must constitute in itself all the properties of iron and steel subject to the function they are performing in the automobile, and besides that, they should also have the property of low weight for performing the extra function of lightweighting an automobile. The various alternative substitution materials mentioned above are thus selected by executing several tests for their appropriation. The most remunerative performance characteristics of the above selected alternative substitution materials is low weight-to-strength ratio. It is a perquisite for these materials to have a lightweight and a high strength simultaneously to fulfill the purpose of the application without any compromise on the performance behavior of the traditionally used materials. Table 1 shows the properties of the materials for the purpose of the study.

Table 1 suggests that among the various tested the best alternative substitution materials are fiber composites. They have highest tensile strength and at the same time have the lowest weight, thus fulfilling each perquisite perfectly.

3.2 Fiber Composites as Alternative Substitution Materials for Automobiles

A composite material is a material that is composed of two or more different materials or constituents with distinct interface between its constituents to give a unique combination of properties of the constituents [11]. Mostly, composites consist of two constituents, a 'matrix' and a 'reinforcement.' The reinforcement

Table 1 Properties of materials for the purpose of the study (*data source* 20)

Material	Specific gravity	Tensile strength (MPa)	Ratio of tensile strength to weight (10^6 m)
SAE 1010 steel	7.87	365	4.72
AISI 4340 steel	7.87	1722	22.3
AL 6061-T6 aluminum alloy	2.70	310	11.7
AL 7178-T6 aluminum alloy	2.70	606	22.9
17-7 stainless steel	7.87	1619	21.0
INCO 718 alloy	8.2	1399	17.4
High-strength carbon fiber–epoxy (unidirectional)	1.55	1550	101.9
E-glass fiber–epoxy (unidirectional)	1.85	965	53.2
Kevlar 49 fiber–epoxy (unidirectional)	1.38	1378	101.8
Boron fiber-6061 Al alloy (annealed)	2.35	1109	48.1
Carbon fiber–epoxy (quasi-isotropic)	1.55	579	38

usually have properties superior to the matrix, and the matrix is used to hold the reinforcement together. The composite material has its characteristics derived from the constituents involved in its composition.

It can be understood by the literature that composite materials have been utilized to solve various engineering and technological problems from a very long time, but, in the middle of twentieth century, with the introduction of polymeric-based composites, these materials started capturing the attention of various industries. It was then, in about 1960s, that composites became common engineering materials and various devices, objects, and components were all manufactured by one or the other kind of composites. This growth in the usage of composites emerged because the awareness regarding the performance of the product increased suddenly during the era.

The fiber-reinforced composites are the most common types of composites. The fibers are the primary load carrying constituent and therefore are stronger and stiffer than the matrix material. The matrix holds the fibers together and redistributes the loads from one fiber to another. The fibers can be continuous, long, or short. The most commonly used form of fiber-reinforced composites consists of polymer matrix and fiber reinforcement. Thus, this kind of fiber-reinforced composites is also called as fiber-reinforced polymer (FRP) materials, as they consist of matrix made up of polymer-based resins. The material properties of fiber-reinforced composite are determined by properties of matrix and fibers as well as the volumetric ratio and the angle of orientation of the composite. If the volumetric ratio is high, the composite will have its properties closer to the fiber, and if the volumetric ratio is low, then the composite will have its properties closer to the matrix. The

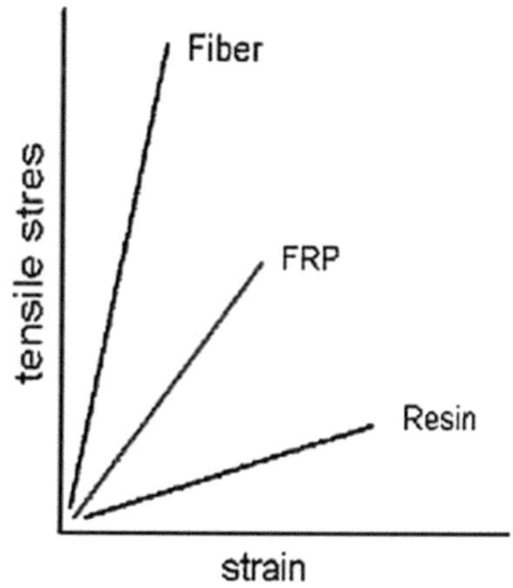

Fig. 5 Comparison of mechanical properties of fiber, FRP, and resin [17]

angles of orientation of the fibers also play an important role in determining the properties of the composite, since the fibers have superior mechanical properties along its length. Figure 5 suggests how the properties of FRP lie in between the properties of the fiber and the matrix polymer.

The matrix materials of FRP can be thermoset or thermoplastic resins. The fiber which is reinforced into the matrix provides strength and stiffness to the composite, and the matrix provides rigidity and environmental resistance. There can be many different forms of an individual laminated composite layer depending upon the arrangement of the fiber constituent. Figure 6 gives the different fiber arrangement patterns in a layer of a fiber-reinforced composite.

The glass fibers, carbon fibers, aramid fibers, and the boron fibers are the most common types and typical fibers used in industries and various other applications nowadays. Each type of fiber has different mechanical properties as well as

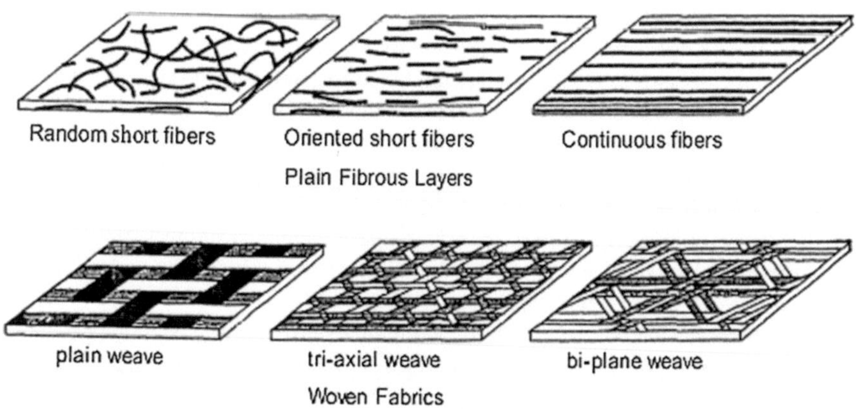

Fig. 6 Fiber arrangement patterns in a layer of fiber-reinforced composite [11]

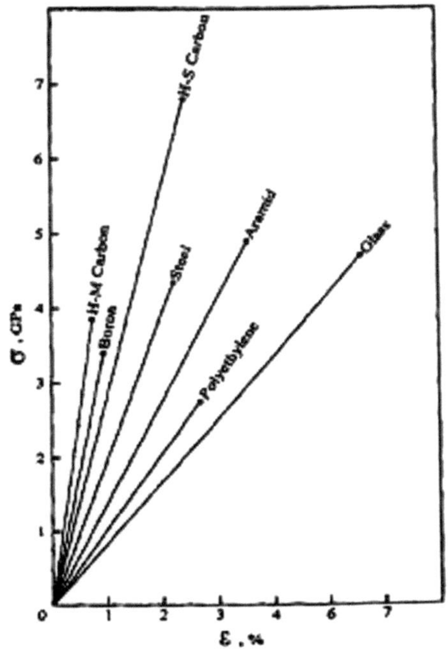

Fig. 7 Stress–strain diagram for typical fibers of composite materials. *Data source* [18]

characteristics. The stress–strain diagram for the typical fibers of composite materials is shown in Fig. 7.

Glass was the first type of fiber for advanced composites. Nowadays, glass in the form of continuous fiber is used most commonly. The typical mechanical characteristics as well as the density of glass fibers are depicted in Table 2. The glass fibers have high strength, and the strength is maintained at even humid conditions, but the strength decreases rapidly as the temperature increase. This property of the glass fiber is shown very clearly in Fig. 7. Besides this, the glass fibers have low cost and high chemical and biological resistance. Also, the glass fibers are brittle and have poor impact resistance. The E-Glass and the S-Glass are the two major types of glass fibers being used commercially in the industry these days.

The carbon fibers find their applications in modern composite products used very often these days. The carbon fibers demonstrate modulus which is about four times higher than the modulus of steel with the same density. The drawback associated with the carbon fibers is its strength. The strength of the carbon fibers is lower than the glass fibers, but the modern carbon fibers demonstrate strength which is about 40% higher than the glass fibers. At the same time, the density of the carbon fiber is about 30% less than the glass fibers.

These fibers have a porous structure, and their specific gravity is about 1.8. They have characteristic properties such as high tensile and compressive modulus, high tensile and compressive strength, and high flexural modulus. The stress–strain diagram for high-modulus (HM) and high-strength (HS) carbon fiber is shown in

Table 2 Mechanical properties of fiber for advanced composites [18]

Material	Diameter (μm)	Ultimate tensile stress (MPa)	Modulus (GPa)	Specific gravity	Maximum specific strength	Maximum specific modulus
Glass	3–19	3100–5000	72–95	2.4–2.6	200	3960
Quartz	10	6000	74	2.2	2700	3360
Basalt	9–13	3000–3500	90	2.7–3.0	130	3300
Aramid	12–15	3500–5500	140–180	1.4–1.47	390	128000
High-strength carbon	5–11	7000	300	1.75	400	17100
High-modulus carbon	5–11	2700	850	1.78	150	47700
Polyethylene	20–40	2600–3300	120–170	0.97	310	1750
Boron	100–200	2500–3700	390–420	2.5–2.6	150	16800
Alumina	20–500	2400–4100	470–530	3.96	100	13300
Silicon carbide	10–15	2700	185	2.4–2.7	110	7700
Titanium carbide	280	1500	450	4.9	30	9100
Boron carbide	50	2100–2500	480	2.5	100	100000
Boron nitride	7	1400	90	1.9	70	4700

Fig. 7. The properties of carbon fibers changes with the changes in its crystallite size, orientation, porosity, and purity of carbon structure. Also, these fibers possess high fire and impact as well as electrical resistance. Besides this, thermal expansion coefficient of these fibers is very low. Because of this, these fibers change there dimensions negligibly under high heat and temperatures. These fibers are one of the most expensive types of fibers used these days.

The aromatic polyamide fibers also known as the aramid fibers are an organic fiber, which is used quite commonly in textile applications as reinforcing elements of advanced composites. The aramid fibers are also known by a commercial name or trademark name of 'Kevlar.' The aramid fibers possess high stiffness and high strength at a very low density.

The properties of aramid fibers are listed in Table 2, and the stress–strain for the aramid fibers is shown in Fig. 7. The aramid fibers have typical characteristics such as low density with high specific strength and high stiffness. The strength of these fibers depends directly on temperature. They have low thermal conductivity which provides them with high thermal resistance. Besides this, these fibers possess negative thermal expansion coefficient which gives them very low thermal expansion and the dimensions of these fibers remain approximately the same under heating. Also, the aramid fibers have very high resistance to damage, that is, why it has wide scope in ballistic applications. The only drawbacks of these fibers are there low compressive strength and low flexural strength.

Many other types of fibers such as polyester, polyethylene, quartz, and boron are also used as fibers in various composite products. These fibers have limited applications. Boron is manufactured by deposition of boron over the tungsten or carbon fiber of about 12 μm. Because of this, the diameter of the boron fibers is quite higher (about 100–200 μm) in comparison with other fibers. These fibers are very brittle and have very low resistance to surface damage. The mechanical properties of these fibers are listed in Table 2. The stress–strain diagram for the boron fiber is depicted in Fig. 7.

The increased competition in the global market for lightweight materials is the prime motivation for the adoption of composite materials as alternative materials for the substitution of various parts composed of heavier and bulkier materials in automobiles. Composite materials can replace many engineering materials such as iron and steel, as well as aluminum, and that too with many more times better performance.

4 Methodology

In the present study, the fuel pipe in automobiles which is composed of steel 4340 and aluminum 6061 alloys is substituted by T300/LY5052 composite [12]. It is a carbon fiber composite, also referred as carbon fiber-reinforced polymer composite (CFRP). In this type of CFRP composite, the class or grade of carbon fibers is T300 and the polymer is composed of epoxy Araldite LY5052. The T300 carbon fiber is a baseline carbon fiber which has its widest application in aerospace industry. It has a fiber diameter of about 7 μm [13]. It is a PAN-based carbon fiber and has high strength and high modulus. The Araldite LY5052 is a low viscosity epoxy, and it has high-temperature resistance, excellent mechanical, and dynamic properties after ambient cure. The basic properties of T300 carbon fibers and LY5052 epoxy Araldite are given in Table 3.

The dimension of the fuel pipe is given in Table 4. After the substitution, the weight of steel 4340 alloy, aluminum 6061 alloy, and CFRP pipe is compared.

Further, two types of analysis are performed for the working internal pressure of 35 MPa on the fuel pipe. The composite pressure vessels are designed for various fiber orientations in ANSYS 11. In the first part and second part of the analysis, the fibers are oriented helically at different fiber orientations, such as ±5°, ±10°, ±15°, ±20°, ±25°, ±30°, ±35°, ±40°, ±45°, ±50°, ±55°, ±60°, ±65°, ±70°, ±75°, ±80°,

Table 3 Basic properties of T300/LY5052

S. no.	Properties	Carbon fiber T300	Epoxy araldite LY5052
1.	Elastic modulus	230 GPa [19]	3 GPa [13]
2.	Tensile strength	3.5 GPa [19]	71 MPa [13]
3.	Density	1760 kg/m^3 [19]	1.14 g/cm^3 [20]

Table 4 Real constants and dimensions of multilayered CFRP pressure vessel [12]

S. no.	Real constants and dimensions	Values (mm)
1.	Outer radius	52
2.	Inner radius	50
3.	Length	400

±85°, and ±90° at symmetrical and anti-symmetrical fiber orientations. The stress distributions for all these fiber orientations and stacking sequences are then recorded.

5 Results and Discussions

The results of the analysis are depicted in Fig. 8. It can be observed from the graph of Fig. 8 that the weight of the fuel pipe decreases by 65% when a fuel pipe of steel 4340 alloy is substituted by fuel pipe of aluminum 6061 alloy. It can be further perceived that a weight reduction of about 83% can be obtained when a fuel pipe of steel 4340 alloy is substituted by CFRP fuel pipe.

The use of CFRP pipes will not only reduce the weight of the automobile but also add on to other benefits to the automobile. As it is non-permeable, it will serve as an excellent fuel barrier and help in ceasing the vaporization fuel into the atmosphere. It will be free from corrosion and chemical reactions with the fuel.

The results for the first phase of analysis are shown in Fig. 9 which is the comparisons of the maximum stress obtained under working pressure for symmetrical and anti-symmetrical stacking sequence and shows that maximum stress for various fiber orientations of both the stacking sequence is minimum at ±45° fiber orientation

The results for the second phase of the analysis are shown in Fig. 10. It shows the comparisons of burst pressure for symmetrical and anti-symmetrical stacking sequence. The burst pressure of composite pressure vessel for various fiber orientations of both the stacking sequences is maximum at ±45° fiber orientation.

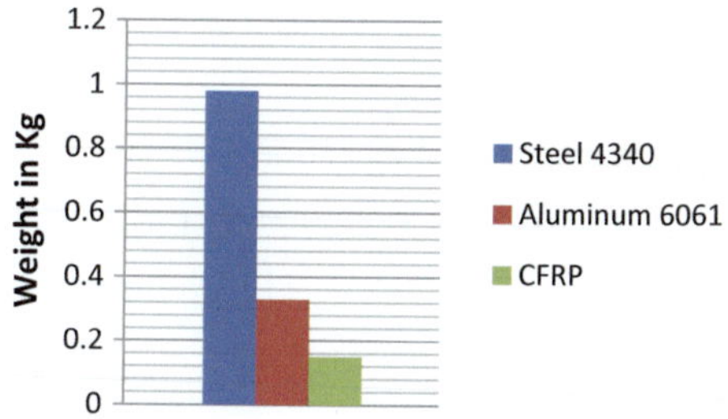

Fig. 8 Comparative analysis of fuel pipe weight of Steel 4340 alloy, aluminum 6061 alloy, and CFRP fuel pipe

Fig. 9 Comparison of maximum stress at working pressure variation for symmetrical and anti-symmetrical stacking sequence

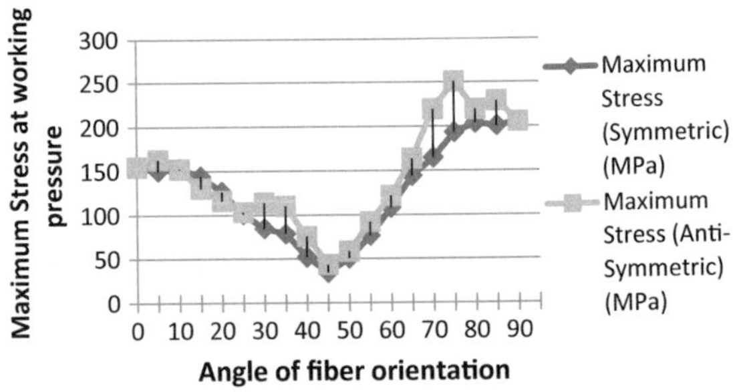

Fig. 10 Comparison of burst pressure for symmetrical and anti-symmetrical stacking sequence

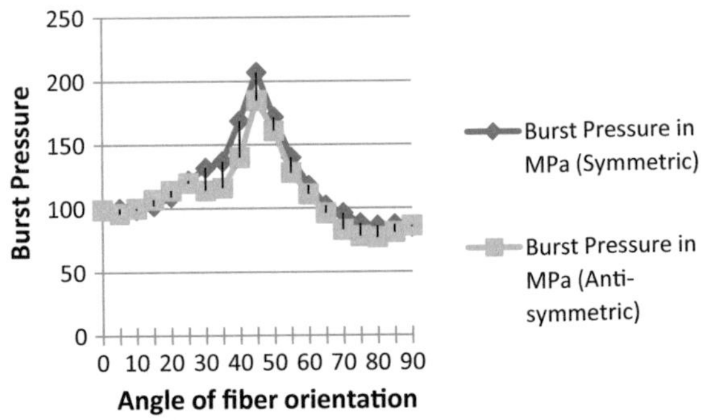

6 Economics of Vehicular Weight Reduction

The most profitable vehicular weight reduction by substitution requires the cost of substitution to be as low as possible. Table 3 shows the cost of materials used for the purpose of the study.

It can be observed from Table 5 that the replacement of heavier iron will require least cost with mild steel and maximum cost with fiber composites but as depicted by Table 1. The ratio of tensile strength to weight of SAE 1010 steel is minimum and that of fiber composites is maximum. The substitution of bulkier iron and steel with fiber composites will serve best purpose in terms of performance of the automobile part but will be most expensive.

Table 5 Cost of materials used for the purpose of study

Material	Relative cost per part [21]
Mild steel	1.0
High-strength steel	1.0–1.5
Iron	–
Aluminum	1.5–2.5
Fiber composites	2.0–10.0

7 Conclusion

The prime motivation behind the study was to analyze the effect of replacement of structural members of the ground vehicles to reduce fuel consumption by materials having low weight particularly composites, and to scrutinize the various implications associated with it. The following inferences are obtained from the investigation:

- Iron and steel being very heavy by weight should be replaced to increase the fuel economy of the ground vehicles.
- Various lightweight materials such as aluminum alloys and carbon fiber composites have been identified as ideal for the substitution fuel pipe of automobile.
- The fuel pipe of Aluminum 6061 alloy will lead to the weight reduction of the pipe by 65%.
- The fuel pipe of CFRP will lead to the weight reduction of the pipe by 83%.
- Among all the identified materials, the fiber composites have been identified as the best suited substitution material for the best low-weight to high-strength property.
- After studying the results of different analyses, it can be considered that the composite pressure vessel can work most effectively at $\pm 45°$ fiber orientation in anti-symmetrical stacking sequence. Therefore, $\pm 45°$ fiber orientation is considered as the optimum fiber orientation for the multilayered cylindrical CFRP composite pipe in this study.
- During the study for the economics of the substitution, it was further identified that the cost of substitution of fiber composites is maximum among the identified materials for substitution.
- The most effective weight reduction by composite materials in conjunction with vehicle redesigning and downsizing will lead to maximum saving in fuel consumption and thus will result in high amount of energy saving.

References

1. Tyagi RK, Sharma SK, Chandra A, Maheshawari S, Goyal P (2015) Improved intake manifold design for I.C. engine emission control. J Eng Sci Technol 10:1188–1202
2. Thandavamoorthy TS (2009) Analysis of structures. Oxford University Press
3. EPA (2010) Joint technical support document: rulemaking to establish light-duty vehicle greenhouse gas emission standards and corporate average fuel economy standards. US Environmental Protection Agency and National Highway Traffic Safety Administration
4. Knittel CR (2009) Automobiles on steroids: product attribute trade-offs and technological progress in the automobile sector. Institute of Transportation Studies, University of California, Davis
5. Feychting M, Svensson D, Ahlbom A (1998) Exposure to motor vehicle exhaust and childhood cancer. Scand J Work Environ Health 8–11

6. Guibet JC, Faure-Birchem E (1999) Fuels and engines: technology, energy, environment, vol 1. Paris: Editions Technip, pp 1–385
7. Ganesan V (1996) Internal combustion engines (4th edn). McGraw-Hill
8. Rudd CD (2001) Composites for automotive applications. iSmithers Rapra Publishing
9. Xu Z, Huang Y, Song Y, Zhang C, Liu L (2007) Surface characteristics of rare earth treated carbon fibers and interfacial properties of composites. J Rare Earth 25(4):462–468
10. Stickel MJ, Nagarajan M (2012) Glass fiber-reinforced composites: from formulation to application. Int J Appl Glass Sci 3(2):122–136
11. Zafer G, Haftka RT, Hajela P (1999) Design and optimization of laminated composite materials. Wiley
12. Bakaiyan H, Hosseini H, Ameri E (2009) Analysis of multi-layered filament-wound composite pipes under combined internal pressure and thermomechanical loading with thermal variations. Compos Struct 88:532–541
13. Wu W, Jacobs E, Verpoest I (2006) Experimental evaluation of interphase region of carbon fiber composites with plasma polymer polymerized coating. In: ECCM-7 seventh european conference on composite materials, realizing their commercial potential, composites: testing & standardization, vol 2, pp 153–158
14. EPA (2009) Light-duty automotive technology, carbon dioxide emissions, and fuel economy trends: 1975 through 2009. US Environmental Protection Agency
15. Ward's Communications (2006–2008) Model Light vehicle US specifications and prices. Prism Business Media Inc
16. Volkswagen AG (2009) In: Proceedings on the international conference on innovative developments for lightweight vehicle structures, Wolfsburg, Germany
17. Introduction to composites (1990) Reference handbook (4th edn). Composites Institute, NY
18. Vasiliev V, Morozov EV (2001) Mechanics and analysis of composite materials (1st edn). Elsevier
19. Fritz Larsson, Lars S (2001) Carbon polyethylene and PBO hybrid fiber composites for structural light weight armour. Compos A 33:221–231
20. Marianna TR (2008) Evaluation of effect of thermal stresses on results obtained from single fiber fragmentation test. Master's Thesis, Lulea University of Technology, Department of Applied Physics and Mechanical Engineering
21. Powers WF (2000) Automotive materials in the 21st century. In: Advanced materials and processes. ASM International, Materials Park, Ohio, pp 38–41

Optimization of Surface Grinding Process Parameters Through RSM

Harshita Khangarot, Shubham Sharma, Umesh Kumar Vates, Gyanendra Kumar Singh and Vivek Kumar

1 Introduction

Surface grinding process is frequently used in various industries from last fifteen decades (1880). Lot of surface finishing processes is being used to fulfill the industries demand and customer satisfaction of perfect surface finish. Various unconventional and hybrid surface finishing approaches are being adopted by the modern industries as per the customer demand, if surface finish improves diminish the productivity. In present scenario, every manufacturing industry demands for a higher precision and productivity at the same time. Present research aimed to elaborate the further enhancements of the flat surface finishing process optimization of the critical influencing parameters with influence of response surface methodology (RSM). The compromise between surface roughness (SR) and material removal rate (MRR) is needed in conventional surface grinding to produce the satisfactory surface quality of the product with adequate productivity per unit. It is verdict that combined optimization of material removal rate (MRR) and surface

H. Khangarot · S. Sharma · U.K. Vates (✉) · G.K. Singh · V. Kumar
Department of Mechanical Engineering, ASET, Amity University, Noida, Uttar Pradesh, India
e-mail: ukvates@amity.edu

H. Khangarot
e-mail: harshitakhangarot@gmail.com

S. Sharma
e-mail: ssharma32@amity.edu

G.K. Singh
e-mail: gksingh@amity.edu

V. Kumar
e-mail: vkumar7@amity.edu

S. Singh et al. (eds.), *Proceedings of the International Conference on Modern Research in Aerospace Engineering*, Lecture Notes in Mechanical Engineering,
https://doi.org/10.1007/978-981-10-5849-3_29

roughness (SR) of the hard tool and die steel (AISI D3) under grinding processes is need to be done to fulfill the customer demand and economic production.

For the above purpose, many attempts have been made by researchers for modeling and optimization of grinding process to enhance the surface quality and MRR. Combined responses, SR and MRR, are still challenging problems in surface grinding. Kumar et al. [3] studied the effect of abrasive tools on EN 24 steel surface using three parameters viz. grinding wheel speed of 850 RPM, table speed of 15 m/min, and depth of cut of 11.94 μm. The work was done on a surface grinding machine and optimizes the surface roughness and material removal rate using RSM technique. The predicted values were almost similar to the experimental values with R^2 of 0.9164 for surface roughness and R^2 of 0.99 for material removal rate. The error among predicted and experimental values at the combination of all the input parameters for MRR and SR lie within 4.96% and 4.30%, respectively. Dasthagiri and Goud [2] investigate the parameters such as cutting speed, feed rate, and depth of cut influence the MRR and SR of mild steel, and this was optimized using response surface methodology model and tested through F-test and analysis of variance (ANOVA). Lijohn et al. [4] present the influence of input machining parameters as work speed, depth of cut, and hardness of material on the surface roughness of various alloy steel materials such as EN 24, EN 31, and EN 353 using Taguchi parametric optimization as the optimization technique. The experiments were done on cylindrical grinding machine with L_9 orthogonal array. SJ-400 surface roughness tester was used to measure the surface roughness of the various alloy steel materials. Melwin et al. [7] report that with optimum grinding parameters such as wheel speed of 150 RPM, depth of cut of 0.02 mm, and 1 number of pass, OHNS steel produces good surface finish when it is machined on cylindrical grinding machine. Besides this, precise tolerance can be gained on OHNS steel during cylindrical grinding. In this process of cylindrical grinding, to achieve larger metal removal rate, an important role is played by the number of passes. Mohanet et al. (2008) selected best feasible network of interactive parameters with learning rate, number of neurons in hidden layer, and error goals were decided automatically to investigate their effect on surface roughness by developing the neural network and fuzzy-based methodology. The network model was trained using back propagation algorithms. The experiment was carried out for grinding of alloy steel using a black carbide silicon grinding wheel on traverse cylindrical grinding machine. He has considered only five controllable parameters in the grinding process to develop neural network model. The accuracy can be increased by indulging some other controllable parameters like dressing depth and dressing lead. Alagumurthi [1] investigates that the input parameters like depth of cut, wheel speed, and work speed play a major role in the quality of surface during grinding. It may be opted for the most suitable design of experiment to perform the work. As it verdict that the work speed has more influence on surface roughness and suggested that the factorial method is flexible, so it is more advantageous over the Taguchi method as the development of mathematical models and regression models is not feasible with the Taguchi method. When there are only a few numbers of tests required then Taguchi method is efficient, but when the interactions between the process variables are present then the factorial method is more efficient and systematic. The results obtained from the grinding cycle time indicate that we can get more productivity if the optimal condition obtained by Taguchi technique is applied. Good quality products are obtained if the use of optimal parameters is obtained through the factorial method. *Suresh* Thakor and Patel [6] came out with the result of the cylindrical grinding process using the signal-to-noise ratio approach, regression analysis approach for EN8 steel, the value of surface roughness, and MRR water soluble oil as the cutting fluid is the most influencing parameter followed by work piece speed as 120 RPM and high depth of cut of 500 μm. Successful verification of optimum cutting parameters, full factorial method was used for the conformation test for surface roughness and material removal rate. When water soluble oil was used as the cutting fluid with the higher work piece speed and higher depth of cut, an improved surface finish was obtained after grinding. Vinay and Rao [7] did experiment to get the better set of cutting forces and surface finish by doing the experiment under two different conditions, dry as well as pool cooling conditions, and it was found that the AISI D3 yields good surface finish of 0.14 microns when dry grinding is done keeping low feed, low depth of cut, and low dressing depth. Mahajan et al. [4] propose that the table speed has the major effect on the surface roughness, whereas depth of cut has the major effect on MRR of the surface grinding of AISI D2 steel. It was found that for minimum surface roughness the optimized parameters were wheel grit size (46), grinding wheel speed (2300), table speed (0.834), depth of cut (0.05). On the other hand, for maximum MRR, the optimized parameters were wheel grit size (36), grinding wheel speed (1650), table speed (0.834), and depth of cut (0.075). The optimized minimum surface roughness is 0.438 μm and maximum MRR is 60.231. Manimaran and Kumar [5] conducted experiments on AISI 316 stainless steel using hydraulic surface grinder under three different grinding environments. It was found that the grey relational analysis and the Taguchi method are useful techniques for the optimization of multiresponse problems. Grinding under cryogenic cooling has more effect than conventional cooling. It was proved from the ANOVA testing those environments, DOC and work speed are influencing factors which have an impact on grinding of AISI 316 stainless steel. The effect of environment was 45.38% followed by work speed 29.41% and DOC 15.52%. It was suggested that the Taguchi–Grey technique can be applied to improve the grinding performance and the application of cryogenic cooling in the manufacturing industries. It is verdict that the process variables like grit size of the grinding wheel depending on its hardness, feed rate, and depth of cut are the most viable influencing parameters on SR & MRR.

Present research aimed to elaborate the further enhancements of the flat surface finishing process to optimization of the critical influencing parameters using response surface methodology (RSM). The compromise between SR and MRR is needed in conventional surface grinding to produce the satisfactory quality of the product with adequate productivity per unit. It is verdict that combined optimization of MRR and SR of the AISI D3 under grinding processes is needed to be done to fulfill the customer demand and economic production. In this work, horizontal spindle and reciprocating table type surface grinder is used on AISI D3 steel work piece with an Al_2O_3 abrasive with different grit size grinding wheels. The most influencing parameters grit size, depth of cut, and feed rate are selected based on the literature survey, which directly influences on SR and MRR.

2 Experimentations and Methodology

AISI D3 steel was chosen for machining due to its capability and application. Work piece faces (dimensions of 800 mm x 150 mm x 25 mm) were used to surface finish with application of Al_2O_3 grinding wheels on surface grinder. The horizontal spindle and reciprocating table surface grinding machine was applied to do the tests. The compositions of work material AISI D3 and properties of Al_2O_3 grinding wheels are illustrated in Tables 1 and 2 and Figs. 1 and 2.

Table 1 Composition of AISI D3 die steel

C	Si	Mn	P	S	Cr	Ni	Mo	Al	Cu	Zn	Fe
2.06	0.55	0.449	0.036	0.056	11.09	0.277	0.207	0.0034	0.13	0.27	85

Table 2 Properties of Al_2O_3 grinding wheels

Manufacturer identification numbers	Type of abrasive	Grit numbers	Hardness	Porosity	Bond materials (resinoid reinforced)	Manufacturer identifier
51	A	140	K	8	BF	5
51	A	100	K	8	BF	5
51	A	60	K	8	BF	5

Fig. 1 AISI D3 material work piece

Fig. 2 Al_2O_3 grinding wheels

2.1 *Surface Roughness Measurements and MRR Calculation*

Surface Roughness measurements were carried out using a portable stylus-type profile meter (Mitutoyo Surface Tester SJ-210). The profile meter was set to a cut-off length of 0.8 mm, filter 2CR, traverse speed 1 mm/s, and 4 mm evaluation length. Center line average (CLA) values of surface roughness (SR) were measured in the transverse direction of work piece. The values of SR measurement were repeated three times for each reading and average value was recorded. The parameters that affect MRR and surface roughness are spindle speed, feed, coolant concentration, and flow rate. Optimal parametric combination may deal the technological quality of a product, which mostly influences the manufacturing cost of the product. SR is defined as the arithmetic value of the profile from the centerline along the length. This can be express as

$$R_a = 1/L \int y(x)|dx|| \tag{1}$$

where L is the sampling length, y is the profile curve, and x is the profile direction. The average 'R_a' is measured within stylus traveling length of 0.8 mm. Center line average 'CLA' value of SR measurements was taken to provide quantitative evaluation of the effect of drilling parameters on surface finish. The average value of three reading up to two decimal places of microns will be obtained the least count 1 nm (nanometer) as Fig. 3.

The amount of material removal was obtained by finding the weight difference before and after machining using a precision electronic digital weight balance with 0.1 mg resolution. The MRR is calculated using the following formula in Eq. 2:

$$MRR\,(g/s) = \left(W_i - W_f\right)/t \tag{2}$$

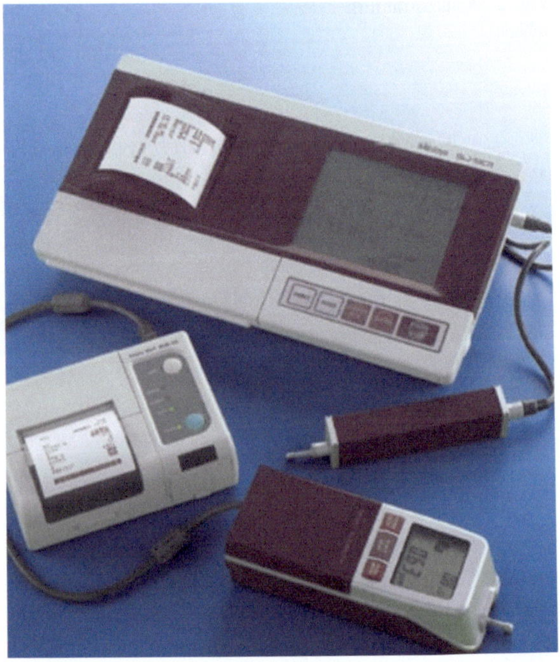

Fig. 3 Surface roughness measuring instrument (SJ 210)

where W_i is initial weight of workpiece in gram (before machining); W_f is final weight of workpiece in gram (after machining); t is machining time in seconds.

2.2 *Response Surface Methodology*

RSM is a collection of mathematical and statistical techniques that are useful for modeling and analysis of problems in which output or response is influenced by several input variables, and the objective is to find the correlation between the response and the variables investigated. It is one of the design of experiments (DOE) methods used to approximate an unknown function for which only a few values are computed. These relations are then modeled by using least square error fitting of the response surface. A central composite design (CCD) is used since it gives a comparatively accurate prediction of all response variable averages related to quantities measured during experimentation. CCD offers the advantage that certain level adjustments are acceptable and can be applied in the two-step chronological RSM. In these methods, there is a possibility that the experiments will stop with few runs and decide whether the prediction model is satisfactory or not.

Table 3 Factor and their levels for surface grinding of AISI D3

Factor	Name	Units	Levels		
			−1	0	+1
A	Grit size	microns	46	60	80
B	Depth of cut	mm	0.2	0.3	0.4
C	Feed rate	mm/min.	0.5	1.0	1.5

Table 4 Experimental observation

S. No.	Grit size (μm)	Depth of cut (mm)	Feed rate (mm/s)	M.R.R. (g/s)	Avg. R_a (μm)
1.	60	0.3	1.5	0.0470	0.3666
2.	80	0.4	0.5	0.0148	0.2733
3.	80	0.4	1.5	0.0184	0.4033
4.	60	0.4	1	0.0291	0.0733
5.	80	0.2	1.5	0.0222	0.1566
6.	60	0.3	1	0.0322	0.1000
7.	60	0.3	1	0.0394	0.1300
8.	80	0.2	0.5	0.0259	0.0866
9.	46	0.4	1.5	0.0307	0.1533
10.	46	0.3	1	0.0273	0.2300
11.	60	0.2	1	0.0666	1.0066
12.	46	0.4	0.5	0.0206	0.6866
13.	46	0.2	1.5	0.0625	0.0533
14.	60	0.3	1	0.0352	0.0466
15.	60	0.3	1	0.0303	0.0500
16.	60	0.3	1	0.0540	0.4933
17.	46	0.2	0.5	0.0092	0.0333
18.	60	0.3	1	0.0327	0.3533
19.	80	0.3	1	0.0327	1.0300
20.	60	0.3	0.5	0.0392	0.4566

In CCD, the limits of the experimental domain to be explored are defined and are made as wide as possible to obtain a clear response from the model. The spindle speed, feed, coolant concentration, and flow rate are the machining variables selected for this investigation. The different levels have been taken for this study which depicted in Table 3. An experiment in series of test called runs. L_{20} runs DOE at three levels were selected critically as per the feasibility and scope of setup. Material removal rate and surface roughness values are given in Table 4 for 20 tests according to different control levels.

The second-order model is normally used when the response function is not known or nonlinear. In the present study, a second-order model has been utilized. The experimental values are analyzed, and the mathematical model is then developed that illustrates the relationship between the process variable and response. The second-order model in Eq. 2 explains the behavior of the system.

$$y = a_0 + \sum_{i=1}^{n} a_i x_i + \sum_{i=1}^{n} a_{ii} x_i^2 + \sum_{i=1}^{n} \sum_{i=1}^{n} a_{ii} x_i x_j \quad i<j \tag{3}$$

where Y is the corresponding response, X_i is the input variables, X_i^2 and $X_{i,}$ X_j are the squares and interaction terms, respectively, of these input variables.

The input parameters applied in the experiment are presented in Table 3.

2.3 Regression Models

Based on the experimental data, statistical regression analysis enabled to study the correlation of process parameters with the MRR and SR. Both linear and nonlinear regression models were examined and high to very high coefficients of correlation (R^2). In this study, three variables are under consideration to obtain the polynomial regression modeling. For simplicity, a quadratic model of MRR and SR are proposed. The coefficients of regression model can be estimated from the experimental

Table 5 Estimated regression coefficients for R_a (micron)

Term	Coef.	SE Coef.	T	P
Constant	4.24,072	1.25,559	3.377	0.007
Grit size (microns)	−0.09,173	0.04,079	−2.249	0.048*
Depth of cut (mm)	0.09,242	5.61,289	0.016	0.987**
Feed rate (mm/min.)	−2.76,224	0.84,555	−3.267	0.008*
F Grit size (microns) * Grit size (microns)	0.00,087	0.00,031	2.804	0.019
Depth of cut (mm) * Depth of cut (mm)	−8.31,364	8.63,494	−0.963	0.358**
Feed rate (mm/min.) * Feed rate (mm/min.)	1.16,745	0.34,540	3.380	0.007*
Grit size (microns) * Depth of cut (mm)	0.06,178	0.02,969	2.081	0.064*
Grit size (microns) * Feed rate (mm/min.)	0.01,569	0.00,594	2.642	0.025
Depth of cut (mm) * Feed rate (mm/min.)	0.72,250	1.01,254	0.714	0.492**
S = 0.143,194	PRESS = 2.97,923			
R-Sq = 98.19%	R-Sq (adj) = 96.56%			

Table 6 Estimated regression coefficients for MRR

Term	Coef.	SE Coef.	T	P
Constant	0.020,449	0.048,239	0.424	0.681
Grit size (microns)	0.003,392	0.001,567	2.164	0.056*
Depth of cut (mm)	−0.556,204	0.215,645	−2.579	0.027*
Feed rate (mm/min.)	−0.102,502	0.032,485	−3.155	0.010*
Grit size (microns) * Grit size (microns)	−0.000,037	0.000,012	−3.119	0.011*
Depth of cut (mm) * Depth of cut (mm)	0.445,909	0.331,750	1.344	0.209**
Feed rate (mm/min.) * Feed rate (mm/min.)	0.018,836	0.013,270	1.419	0.186**
Grit size (microns) * Depth of cut (mm)	0.004,016	0.001,141	3.521	0.006*
Grit size (microns) * Feed rate (mm/min.)	0.000,681	0.000,228	2.984	0.014
Depth of cut (mm) * Feed rate (mm/min.)	0.135,250	0.038,901	3.477	0.006*
S = 0.00,550,145	PRESS = 0.00,525,351			
R-Sq = 93.06%	R-Sq (adj) = 86.82%			

results. The effects of these variables and the interaction between them were included in this analysis. The unknown coefficients are determined from the experimental data as presented in Tables 5 and 6. The standard errors on estimation of the coefficients are tabulated in the column 'MRR coefficient' and 'SR coefficient.' The P & T values are calculated for 95% level of confidence, and the factors having p-value more than 0.05 are considered insignificant (shown with ** in p-column). The model made to represent MRR and SR depicts that grit size, feed, feed*feed, and interaction of grit size and depth of cut are the most influencing parameters in order of significance. The final response equation for MRR and SR are nonlinear in nature; a linear polynomial will not be able to predict the response accurately. Therefore, the second-order model (quadratic model) is found to be adequate for the grinding process.

The ANOVA for the curtailed quadratic model depicts the value of coefficient of determination of MRR and SR are R^2 as 93.06% and 98.19%, which signifies that how much variation in the response is explained by the model. The higher of R^2 indicates the better fitting of the model with the data. However, R^2adj. is 86.82 and 96.56% which accounts for the number of predictors in the model describes the significant coefficient relationship. It is important to check the adequacy of the fitted model, because an incorrect or under-specified model can lead to misleading conclusions. By checking the fit of the model, one can check whether the model is

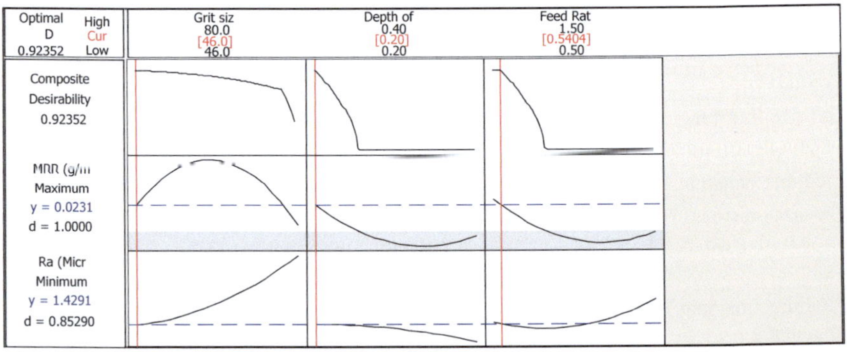

Fig. 4 Response optimization plot (R_a & MRR)

under specified. The model adequacy checking includes the test for significance of the regression model, model coefficients, and lack-of-fit, which is carried out subsequently using ANOVA on the curtailed model. The total error on regression is sum of errors on linear, square, and interactions terms. The residual error is the sum of pure and lack-of-fit errors. The fit summary recommended that the quadratic model is statistically significant for analysis of SR. In the table, p-value for the lack-of-fit is 0.186, which is insignificant, so the model is certainly adequate. Moreover, the mean square error of pure error is less than that of lack-of-fit.

3 Result and Discussion

It is very clear that surface roughness is minimum at certain input parametric combinations, and on the other hand, MRR is also maximum at same combinations. It is very difficult to obtain such influencing parametric combinations which are applicable to achieve the optimal responses as surface roughness and material removal rate. Lot of modeling and optimization techniques are frequently used in the grinding processes for different materials, but the RSM is the most important for modeling and optimization tool which is applicable for the multi-objective response optimization. The responses SR & MRR have successfully predicted up to the adequate level 1.4291 microns and 0.0231 g/s at combination of variable as grit size (46), depth of cut (0.20), and feed rate (0.5404) using multi-objective response optimization technique as Fig. 4.

The effect of the machining parameters (grit size, feed, and depth of cut) on the response variables MRR & SR have been predicted. It can be seen from Fig. 5, the MRR tends to increase significantly with increase in feed rate and depth of cut for any value of grit size. However, the R_a tends to decrease significantly with decrease in feed rate and grit size for any value of depth of cut as Fig. 6.

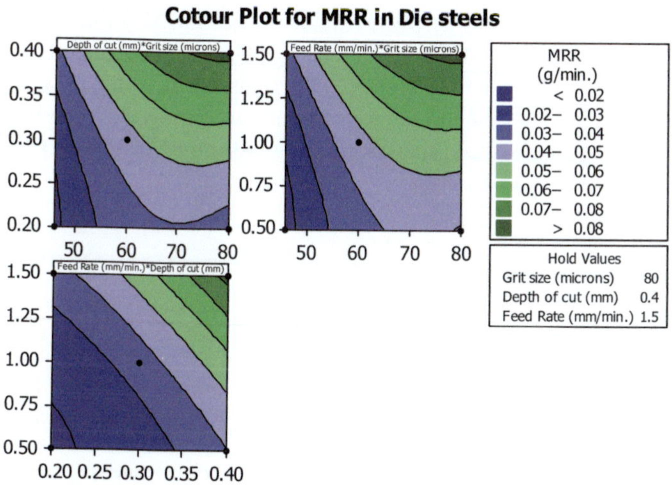

Fig. 5 Contour plot for MRR

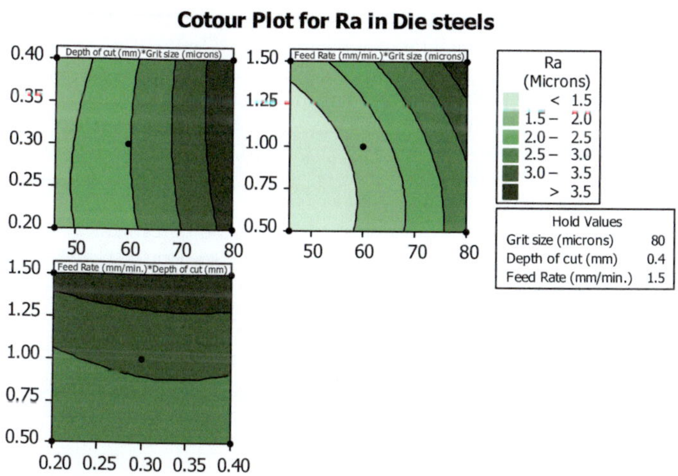

Fig. 6 Contour plot for R_a

4 Conclusions

In the present study, the process parameters are significantly influencing on SR and MRR. A second-order response model of these parameters is developed and found that grit size, feed, speed*speed, and interaction of speed*feed and grit size*depth of cut with other parameters significantly affect the SR & MRR. RSM techniques were also implemented to predict the SR and MRR. Correlation coefficient (R^2) values were observed 93.06% and 98.19% for MRR and SR, respectively.

The responses have predicted as 1.4291 microns and 0.0231 g/min at combination of variable as grit size (46), depth of cut (0.20), and feed rate (0.5404). This research can also help for the researchers' and industries personal to predict the SR & MRR within the experimental range, without performing the experiments using RSM model. Response surface methodology is the robust modeling technique for optimization of grinding of steels.

References

1. Alagumurthi N (2006) Optimization of grinding process through design of experiment (DOE) —a comparative study. Mater Manuf Process 19–21
2. Dasthagiri B, Goud EVG (2015) Optimization studies on surface grinding process parameters. Int J Innov Res Sci Eng Techno 4(7):6148–6156
3. Kumar P, Kumar A, Singh B (2013) Optimization of process parameters in surface grinding using response surface methodology. Int J Res Mech Eng Technol 3(2):245–252
4. Lijohn PG, Job KV, Chandran IM (2013) Study on surface roughness and its prediction in cylindrical grinding process based on Taguchi method of optimization. Int J Mach Tools Manuf, 3(5):ISSN 2250–3153
5. Mahajan TV, Nikalje AM, Supale JP (2015) Optimization of surface grinding process parameters for AISI D2 steel. Int J Eng Sci Res Technol 4(7):944–949
6. Manimaran G, Kumar MP (2013) Multi response optimization of grinding AISI 316 stainless steel using grey relational analysis. Mater Manuf Process 28(4):418–423
7. Melwin Jagadeesh MS, Manickam M, Kalaiyarasan V, Abdul Ghani Khan M, Kannan TTM (2014) Optimization of cylindrical grinding process parameters of OHNS steel (AISI 0-1) rounds using design of experiments concept. Int J Eng Trends Tech (IJETT), 17(3): ISSN:2231-5381. http://www.ijettjournal.org
8. Thakor SP, Patel DM (2014) An experimental investigation on cylindrical grinding process parameters for EN 8 using regression analysis. Int J Eng Dev Res 2(2):2486–2491
9. Vinay PV, Rao CH (2014) Experimental analysis and modeling of grinding AISI D3 steel. Int J Recent Adv Mech Eng 4(1):47–60

Evaluating Significance of Green Manufacturing Enablers Using MOORA Method for Indian Manufacturing Sector

Varinder Kumar Mittal, Rahul Sindhwani, Punj Lata Singh, Vivek Kalsariya and Faizan Salroo

1 Introduction

Green manufacturing (GM) is generically defined as 'the process of redefining the processes in order to reduce or minimize the waste' [1]. GM is a relatively new concept that can be viewed as a product of the 1990s. It is defined as an economically driven, system-wide, and integrated approach to the reduction and elimination of all waste streams associated with the design, manufacture, use and/or disposal of products and materials [2]. Several other jargons such as end-of-line management, sustainable manufacturing, and environmentally conscious manufacturing mean the same thing [3]. The common observation of recent trends in manufacturing technology reveals that every organization is taking their problem solving method to an advanced level by adopting different techniques and solutions. GM is one such solution which could enhance the scope of problem solving by taking part in each and every activity untaken by the organization. Such solutions lead us to reduced cost by reducing material handling and reducing effluents and

V.K. Mittal (✉) · R. Sindhwani · V. Kalsariya · F. Salroo
Department of Mechanical Engineering, Amity School of Engineering and Technology, Amity University Uttar Pradesh, Noida 201313, Uttar Pradesh, India
e-mail: vkmittal@amity.edu

R. Sindhwani
e-mail: rsindhwani@amity.edu

V. Kalsariya
e-mail: vivekkalsariya@yahoo.in

F. Salroo
e-mail: faizansalroo1@gmail.com

P. Lata Singh
Civil Engineering Department, Amity School of Engineering and Technology, Amity University Uttar Pradesh, Noida 201313, Uttar Pradesh, India
e-mail: plsingh@amity.edu

S. Singh et al. (eds.), *Proceedings of the International Conference on Modern Research in Aerospace Engineering*, Lecture Notes in Mechanical Engineering,
https://doi.org/10.1007/978-981-10-5849-3_30

other wastes. All such efforts are applications of GM. This manufacturing concept not only addresses the environmental issues, but also reduces redundancy and improves ergonomics and cost too. Nowadays, cheaper production at faster pace is not the only criterion which is followed by manufacturers, but the type of material used, the processes involved, and the futuristic fate of the product is taken into consideration. So, in order to make our system fully eligible for such criteria, green manufacturing systems need to be adopted.

The GM paradigm is an outcome of technological and market motivations. People are getting more and more aware about the environmental risks, and thus, they support the shaping of a new system which meets their demands without affecting other aspects as well. This is achieved through new eco-friendly product designs which will actually realize the objectives of green manufacturing systems [4]. In recent years, the GM proved to be a very important edge for its implementation in corporate sector to become competitive. However, there were certain complications which arose because of the initial investments which are huge and irreversible. Thus, the managers had to take into account all the factors before the implementation of GM. The accuracy of the decision to adopt GM is very crucial as it should fulfill the objectives of the company and the world. In reality, manufacturing organizations directly impact the environment, and it is sometimes overshadowed by the economic aspects [5]. Manufacturing firms not only generate harmful wastes, but also consume a huge portion of available energy. Thus, we need to foresee the effects related to the environment as well as on the economic aspects. As the adoption of GM requires many major decisions such as disruption of routine production, huge initial investment, and fear of success [6], so it is required to make this change very smooth and gradual. It is realized that the analysis of the factors which enables this change needs to be investigated so that government can act as strong catalyst to facilitate the change without much negative impact.

Section 2 presents literature survey followed by enablers of green manufacturing in Sect. 3. Section 4 provides methodology, results, and discussions. The study is concluded in Sect. 5 followed by acknowledgements and references.

2 Literature Review

The ‘Industrial Revolution’ came into picture in late 1700s and has been through several phases [7]. The real challenge begun when different new manufacturing systems were adopted and the sustainability issues came later on. The major issues such as cost and the environmental aspects such as reduction of greenhouse gases forced developed economies such as USA, Canada, Great Britain, and European countries to shift their manufacturing base to China and India. The problem needs a solution based on the facts related to the cost optimization as well as environmental safety. One possible solution to this problem is the adoption of GM [3]. GM is defined as total elimination of wastage and redefining existing process to minimize the carbon emissions during each process without increasing cost and affecting production targets [1, 8]. Nowadays, it is unescapable situation for the companies to think of a system which do not respond to environmental consequences. So, it is necessity for the companies to consider GM system as a solution for environmental safety and for optimize the cost of the products [9, 10].

GM is not only limited to production stages of the manufacturing system, but also includes the design development and management aspects as well [11]. It also emphasizes the reverse engineering in the manufacturing processes for implementation of closed-loop system like recycling and reconsidered the waste product with an aim to reduce the wastage produced by the industries [12, 13]. GM is not only limited to environmental aspects, but also includes the best use of available resources in the industry. It helps the company in long run to compete in global market scenario and able to increase the production quality, produce the product at optimum cost, and increase the efficiency of the company [14]. Adoption of GM gives various benefits to the company in term of brand enhancement with customers, long-term cost savings, etc.

Studies on GM can be divided into two groups, first, the studies that dealt with the overall theoretical concept of GM and second, the studies that provided various analytical tools and models to realize GM at different levels [4]. Various researchers have done lot of work on both types of studies separately. In today's competitive scenario, there is need to combine these two types of research together so that a comprehensive GM model can be framed which includes the total concepts of GM. There exists a lot of literature on drivers for GM in different countries including India, European, and American. Similarly, many studies were conducted on barriers to GM. These studies discussed a set of influencing factors affecting the implementation in either way and suggested various solutions relevant to different countries, different industry sizes, different industry sectors, etc. The research studies are very limited that discussed the GM enablers which is very important for GM implementation in the industry. This is clearly evident from the literature that GM adoption in the industry is not possible without mitigating the barriers to GM which is not possible without investigating the GM enablers. This study aims to identify the GM enablers through literature survey and brain storming with various researchers in the field of industry and academia. At the end, most effective and high-impact enablers are proposed, so that GM systems can easily implement in the industry more effectively and efficiently.

3 Enablers of Green Manufacturing

Any new change in the existing manufacturing systems requires some motivations and facilitation for implementation. Similarly, the adoption of GM in the industry needs enablement in addition to motivations for implementation. Hence, fifteen enablers which can facilitate the implementation of GM in the industry are identified through a review of literature and opinion of experts from industry and academia (see Table 1). Additionally, 6 assessment criteria were identified which

Table 1 Enablers of green manufacturing system

Sr.	Enabler	Description
1	Waste management system [22–27]	Waste is disposal of resources at a rate beyond the environment can absorb. Waste includes the resources in the form of material, energy, etc.
2	General awareness [24, 25, 28–30]	Awareness among the customers and common masses about GM—its need, benefits, etc. It can be achieved by special training, awareness programs, promotions, and advertisements
3	Competitiveness [24, 25, 28, 31, 32]	The environmentally conscious quality culture, minimal waste, innovative green products and processes and flexibility will be able to trim the product costs and hence becomes more competitive
4	Strategic green management [25, 33, 34]	Incorporation of green initiative in the top management agenda will be able to motivate the middle and lower management for whole-hearted participation
5	Green supplier management [25, 35–39]	The green and clean supplier organizations will enable the parent organization to adopt similar green initiatives
6	Strategic management commitment [25, 30, 36, 40]	The top management beliefs, voluntary initiatives, and commitment are very crucial for enabling the implementation of green manufacturing
7	Environmental performance goals [36, 41, 42]	The setting of environmental performance goals in terms of reduction of carbon footprints, use of less virgin material, proper disposal of effluents, efficient energy use, etc.
8	Customer feedback system [36, 43, 44]	Customers are getting more and more educated about the environmental degradation, and they want their product to be environment friendly. This feedback from the customers enables the change
9	Public feedback system [36, 45–48]	External forces such as stakeholders and NGOs motivate firms to undertake similar strategies as they want social legitimacy besides the profit of the company
10	Energy and resource crisis [22, 37]	Rising cost of fuels and virgin materials for manufacturing enables the companies to switch to more efficient ways of manufacturing
11	Government support and legislation [25, 32, 49–51]	Government subsidies, government projects, government rewards and recognitions for green practices enable the implementation in addition to compliance with environmental legislation and regulation set up by the government
12	Financial and human resources [25, 39, 52]	The availability of capital for investment in newer green technologies and skilled and trained manpower to install, operate, maintain and handle the newer systems
13	Information technology management [25, 30, 38, 53]	Effective, efficient, and sufficient organizational communication through modern information technology tools may help in sharing of innovative ideas, schedule of training, and other awareness activities about green concept. It is required to handle flow of information associated with flows of materials and other resources to manage green supply chain efficiently

(continued)

Table 1 (continued)

Sr.	Enabler	Description
14	Reverse logistics [54]	The existence of reverse logistics for the purpose of monetary benefits and/or compliance to government regulations leads to adoption of green initiatives of the company
15	Green supply chain management [23, 25, 28]	Use of optimized, clean fueled, and efficient transportation for supply, eco-packing, reusable packing materials, etc. act as enabler for GM implementation

can be used to find the most important, high-impact and influential enablers for faster implementation of GM.

Assessment Criteria's for GM Enablers

1. **Green Reputation**: Green reputation describes the image of the company in the market. High reputation leads to the alignment of customers. Here, we will see the effect of each enabler on these criteria and how much is the effect of this factor in deciding the application of GM in that particular firm.
2. **Profit**: Each and every firm enters into the market with an aim to earn profit. The technology application and each criteria is oriented at increasing the profit. So, these criteria hold the upper hand over all the other criteria's. A firm will prefer only that very technological perspective in which they need not to compromise with their profit.
3. **Productivity**: A firm needs to see that if it is meeting the customer demands at the right time. So, productivity plays an important role. High productivity leads to high sustainability and prevents the stock outs. We will assess as to what effect these enablers bring on the productivity of the firm.
4. **Recurring Cost**: It defines the overall cost of production including the investment cost. Efforts are being made to minimize this cost as it provides the edge of competition in the market. Lower recurring cost leads to lower unit price, which in turn attracts the customers.
5. **Time Period**: A firm need to spend some time to implement GM. That consumes valuable time of the company which can further influence the routine production and finance. So, the time required for the change is important criteria for assessment of enablers.
6. **Human Resources**: The implementation of any new system requires skilled human resources which can install, operate, use, and maintain the new technology. The availability and cost of such human resources is an important criterion in the present case.

4 Methodology, Results, and Discussion

Fifteen enablers are identified through the review of literature and perception of researchers, academicians, and industry executives. The application of MOORA method as explained in the methodology section is done using six assessment criteria i.e., green reputation, profits, productivity, recurring cost, time period, and human resources. The first three criteria evaluates/assesses the green manufacturing on the basis of output after implementation of GM in industry i.e., green reputation, profits, and productivity. The second three criteria evaluates/assesses the green manufacturing on the basis of input required (say consumables) for implementation of GM in industry i.e., recurring cost, time period, and human resources. These criteria are decided while a rigorous reference to the existing literature, in-depth discussion among 10 researchers of our research group, and consultation with 3 industry executives working at the top management level of manufacturing companies. As the impact of all six criteria is different so the weightage of each criteria is decided in such a way in consultation with all involved in this research that the total weightage is 100% (or 1). The rating of each enabler on a Likert's scale of 1–5 is done by the same group of people who decided the criteria weights.

Multi-objective optimization on the basis of Ratio Analysis (MOORA) is also known as multi-criteria or multi-attribute optimization. It is the process of simultaneous optimization of two or more conflicting attributes (objectives) subject to certain constraints. The MOORA method, first introduced by Brauers and Zavadskas [15] is a multi-objective optimization technique that can be successfully applied to solve various types of complex decision making problems in the manufacturing environment by providing weightage to each and every analysis. This method is observed to be simple and computationally easy which helps the decision makers to eliminate the unsuitable alternatives while selecting the most appropriate alternative to strengthen the existing selection procedures [16].

Advantages of the MOORA method in comparison with other MCDM method are: easy to understand as comparison to other MCDM techniques; less time require for mathematical calculation work; and no extra assumed parameters required such as 'v' in VIKOR method and 'n' in GTA method [17].

This methodology has been successfully used by various researchers for different production life cycle such as stages selection, flexible manufacturing system selection, welding process selection, supplier selection [17], material selection [18], optimizing milling process [19, 20], and privatization in a transition economy [21]. The various steps used by Brauers and Zavadskas [15] in MOORA method as follows:

Step 1: Identification of enablers, criteria, and criteria weightage

Table 1 presents the identified 15 enablers of GM and then finalized 6 assessment criteria for evaluating each enabler i.e., green reputation, profits, productivity, recurring cost, time period, and human resources.

Table 2 Ranking to enablers

Criteria weight	Outputs of GM			Inputs to GM		
	Green reputation [C1]	Profit [C2]	Productivity [C3]	Recurring cost [C4]	Time period [C5]	Human resources [C6]
	0.09	0.28	0.13	0.2	0.15	0.15
Enabler 1	3	2	3	3	2	2
Enabler 2	3	1	1	2	1	1
Enabler 3	1	3	2	1	1	1
Enabler 4	2	1	2	2	1	1
Enabler 5	3	2	2	2	1	1
Enabler 6	3	2	3	2	1	1
Enabler 7	3	2	2	3	2	1
Enabler 8	3	1	1	2	1	1
Enabler 9	3	1	1	1	1	2
Enabler 10	2	3	3	1	1	1
Enabler 11	3	2	2	2	2	1
Enabler 12	3	2	2	1	1	1
Enabler 13	2	2	2	2	1	1
Enabler 14	3	2	2	2	1	2
Enabler 15	3	2	2	2	1	2

Likert's scale rating on 1–5 scale

Step 2: Allocation of ranking to enablers with respect to all six criteria

Ranking to each enabler on Likert's scale of 1–5 with respect to outcome and input of GM as shown in Table 2. Criteria weightage and ranking to each enabler decides after the discussion with experts from industry and academia.

Step 3: Normalization of matrix

Normalization of enablers can be done by different methods such as total ratio, Schärlig ratio, Weitendorf ratio, Jüttler ratio, Stop ratio, and Körth ratio suggested by Brauers and Zavadskas [15] in MOORA method. However, it is recommended that for this case, total ratio method is suggested. This ratio can be expressed as follows:

$$Xij = \frac{\mathrm{Xij}}{\sum_{\mathrm{i}=1}^{\mathrm{m}} \mathrm{Xij}} \quad (j = 1, 2, 3 \ldots n) \tag{1}$$

Normalized matrix is generated with the help of Eq. 1 from columns 2–7 as shown in Table 3.

Step 4: Weightage Calculation (Y_i)

In case of multiple goals, normalized value added for all beneficial goals (case of maximization) and subtract for all non-beneficial goals (case of minimization). Then, final equation is for Y_i is

Table 3 Normalizing, weightage, and ranking of enablers

Normalizing matrix							Weightage (Y_i)	Rank
	C1	C2	C3	C4	C5	C6		
Enabler 1	0.075	0.071	0.100	0.107	0.111	0.105	−0.014	15
Enabler 2	0.075	0.036	0.033	0.071	0.056	0.053	−0.009	13
Enabler 3	0.025	0.107	0.067	0.036	0.056	0.053	0.018	2
Enabler 4	0.050	0.036	0.067	0.071	0.056	0.053	−0.007	10
Enabler 5	0.075	0.071	0.067	0.071	0.056	0.053	0.005	5
Enabler 6	0.075	0.071	0.100	0.071	0.056	0.053	0.009	4
Enabler 7	0.075	0.071	0.067	0.107	0.111	0.053	−0.011	14
Enabler 8	0.075	0.036	0.033	0.071	0.056	0.053	−0.009	12
Enabler 9	0.075	0.036	0.033	0.036	0.056	0.105	−0.010	11
Enabler 10	0.050	0.107	0.100	0.036	0.056	0.053	0.024	1
Enabler 11	0.075	0.071	0.067	0.071	0.111	0.053	−0.003	9
Enabler 12	0.075	0.071	0.067	0.036	0.056	0.053	0.012	3
Enabler 13	0.050	0.071	0.067	0.071	0.056	0.053	0.003	6
Enabler 14	0.075	0.071	0.067	0.071	0.056	0.105	−0.003	7
Enabler 15	0.075	0.071	0.067	0.071	0.056	0.105	−0.003	8

$$Y_i = \sum_{j=1}^{g} Xij - \sum_{j=g+1}^{n} Xij \quad (2)$$

where g is the number of beneficial goals, (n − g) is the number of non-beneficial goals, and Y_i is the normalized assessment value for *i*th enabler. In some cases, it is often observed that some goals are more important than the others. In order to give more importance to a goal, it could be multiplied with its corresponding weight (significance coefficient) (Brauers et al. 2009). When these goal weights are taken into consideration then new equation for Yi is

$$Y_i = \sum_{j=1}^{g} Wj * Xij - \sum_{j=g+1}^{n} Wj * Xij \quad (j = 1, 2, 3 \ldots n) \quad (3)$$

where W_j is the weight of *j*th goal. Values of Y_i are calculated for each enabler with the help of Eqs. 2 and 3 in column 8 as shown in Table 3.

Step 5: Final Ranking of Enablers

The Y_i value can be positive or negative depending of the totals of its beneficial goals and non-beneficial goals in the Table 3. An ordinal ranking of Y_i shows the final preference. Thus, the best enabler has the highest Y_i value, while the least effective enabler has the lowest Y_i value. Values of Y_i and ranking for each enabler are calculated in column 9 as shown in Table 3.

Graph between the values of Y_i and enablers are represented as shown in Fig. 1. The middle line (where $Y_i = 0$) represents that the criteria input of the enabler is

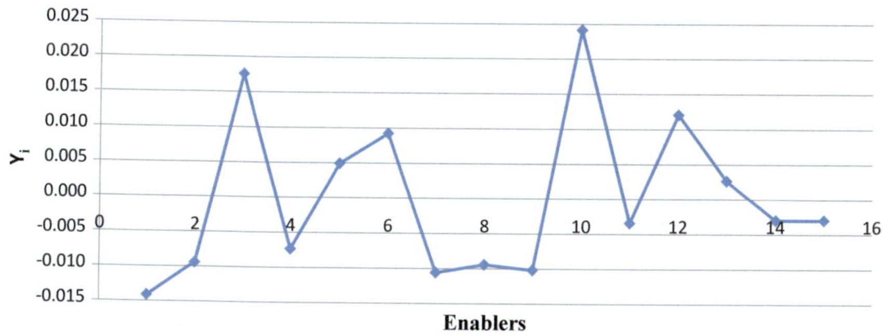

Fig. 1 Graph between the Y_i and enablers

equal to criteria output of the same enabler. This graph represents enablers 3, 5, 6, 10, 12, and 13 lie above the middle line means ($Y_i \geq 0$). It means all these enablers have higher output comparison to input given. Other enablers 1, 2, 4, 7, 8, 9, 11, 14, and 15 ($Y_i \leq 0$) mean that they require more input value for getting output value. The enabler 13 ($Y_i = 0.003$) and enablers 11, 14, and 15 ($Y_i = -0.003$) lie very near to the middle line, and little variation in the criteria values may change the enabler location from positive value of Yi to negative value of Yi and vice versa.

5 Conclusions

The application of 5 step methodology yielded the ranking of the GM enablers. The objective of the research is to identify the most important and high-impact enablers which can greatly enable or facilitate the GM implementation in industry. The research came with six most important enablers, viz. energy and resources crisis; competitiveness; financial and human resources; strategic management commitment; green supplier management; and information technology management.

The enabler 10 (energy and resource crisis) has got highest Yi value (i.e., 0.024) which means it enables the GM implementation more as comparison to other enablers. Waste management system (enabler 1) has got the minimum Yi (i.e., −0.014) which means that it enables the system by consuming more inputs in term of recurring cost, time period, and human resources and achieves less outputs. So, management needs to have primary focus on the three top enablers i.e., energy and resource crisis, competitiveness, and financial and human resources, because it will enable the GM implementation in industry more as comparison to other enablers.

The natural resources are limited in nature, so their cost is increasing with time so it puts extra financial burden on the manufacturing systems which motivate the companies to use these resources in efficient manner while adopting various material recovery strategies like recycling to reduce the use of virgin material for production. It also includes various energy resources which are backbone of any manufacturing system. The cost of energy is increasing worldwide and significantly contributes to the cost of production. The industry works for profits, so energy and resources impact the manufacturing and hence the most important enabler for GM implementation.

The world market is open for all companies, so competitiveness is the primary factor which governs the survival of manufacturing units in the market. The companies have limited scope of reducing their profit margins, so the only way to be competitiveness is by reducing the cost of production which can be achieved by implementation of GM. The adoption of newer technology is not possible without capital investment and human resources which can implement, maintain, and operate the newer technology. The newer initiatives require the commitment among the top management which will allocate funds and direct the organization in the direction of implementation of GM.

The companies are dependent to many other vendors and suppliers for supply of various sub-assemblies, parts and consumable and if all such suppliers and vendors are having active green initiatives that would enable the companies to implement GM. Lastly, the information technology plays a vital role in adoption of newer technology like GM. It helps in operations of newer systems by providing support in data storage, analysis, simulations, interpretations, etc.

This research utilized MOORA method which is one of the multi-criteria decision making (MCDM) method to evaluate the importance of enablers of GM. Various other MCDM methods can be applied such as TOPSIS and AHP for assessment of the enablers by taking inputs from different group of stakeholders for evaluation of enablers from different perspectives. A structural model of enablers using interpretive structural modeling or structural equation modeling like techniques will put more light on the impact, relationship, and hierarchy of the enablers to understand the GM implementation process better.

References

1. Balan K (2008) Introduction to green manufacturing. Shot Peener 22(3):4–6
2. Handfield RB, Walton SV, Seegers LK, Melnyk SA (1997) Green' value chain practices in the furniture industry. J Oper Manage 15(4):293–315
3. Sangwan KS, Mittal VK (2015) A bibliometric analysis of green manufacturing and similar frameworks. Manage Environ Qual Int J 26(4):566–587
4. Deif AM (2011) A system model for green manufacturing. J Clean Prod 19(14):1553–1559
5. Vachon S, Klassen RD (2008) Environmental management and manufacturing performance: the role of collaboration in the supply chain. Int J Prod Econ 111(2):299–315
6. Mittal VK, Sangwan KS (2014) Prioritizing drivers for green manufacturing: environmental, social and economic perspectives. Proc CIRP 15:135–140
7. Lofrano G, Brown J (2010) Wastewater management through the ages: a history of mankind. Sci Total Environ 408(22):5254–5264
8. Azzone G, Noci G (1998) Identifying effective PMSs for the deployment of "green" manufacturing strategies. Int J Oper Prod Manage 18(4):308–335

9. Atasu A, Sarvary M, Van Wassenhove LN (2008) Remanufacturing as a marketing strategy. Manage Sci 54(10):1731–1746
10. Kleindorfer PR, Singhal K, Wassenhove LN (2005) Sustainable operations management. Prod Oper Manage 14(4):482–492
11. Lai KH, Wong CW (2012) Green logistics management and performance: some empirical evidence from Chinese manufacturing exporters. Omega 40(3):267–282
12. Ferguson ME, Toktay LB (2006) The effect of competition on recovery strategies. Prod Oper Manage 15(3):351–368
13. Rogers DS, Tibben-Lembke R (2001) An examination of reverse logistics practices. J Bus Logist 22(2):129–148
14. Fullerton D, Leicester A, Smith S (2008) Environmental taxes, NBER working paper no 14197, issued in July 2008
15. Brauers WKM, Zavadskas EK (2006) The MOORA method and its application to privatization in a transition economy. Control Cybern 35(2):445–469
16. Brauers WKM, Zavadskas EK, Peldschus F, Turskis Z (2008) Multi-objective decision making for road design. Transport 23(3):183–193
17. Attri R, Grover S (2013) Decision making over the production system life cycle: MOORA method. Int J Syst Assur Eng Manage 5(3):320–328
18. Karande P, Chakraborty S (2012) Application of multi-objective optimization on the basis of ratio analysis (MOORA) method for materials selection. Mater Des 37:317–324
19. Gadakh VS (2011) Application of MOORA method for parametric optimization of milling process. Int J Appl Eng Res 1(4):743–758
20. Shihab SK, Chanda AK (2015) Multi response optimization of milling process parameters using Moora method. Int J Mech Prod Eng 3(4):67–71
21. Brauers WK (2013) Optimization methods for a stakeholder society: a revolution in economic thinking by multi-objective optimization, vol 73. Springer
22. Bhattacharya A, Jain R, Choudhary A (2011) Green manufacturing: energy, products and processes. The Green manufacturing report by The Boston Consultancy Group for Confederation of Indian Industry
23. Bhateja AK, Babbar R, Singh S, Sachdeva A (2011) Study of green supply chain management in indian manufacturing industries: a literature review cum an analytical approach for the measurement of performance. Int J Comput Eng Manage 13:84–99
24. Jhadav JR, Mantha SS, Rane SB (2013) Practice bundles for integrated green-lean manufacturing systems. In: IJCA proceedings on international conference on green computing and technology ICGCT, October 2013
25. Kumar S, Kumar S, Gahlot P, Haleem A (2013) Analysis of interdependence among the enablers of green concept implementation In Indian automobile supply chain. J Eng Res Stud 4(2):05–11
26. Sushil, Vrat P (1989) Waste management policy analysis and growth monitoring: an integrated approach to perspective planning. Int J Syst Sci 20(6):907–926
27. Yamakawa H, Ueta K (2002) Waste reduction through variable charging programs: its sustainability and contributing factors. J Mater Cycles Waste Manage 4(2):77–86
28. Luthra S, Kumar V, Kumar S, Haleem A (2011) Barriers to implement green supply chain management in automobile industry using interpretive structural modelling technique: an Indian perspective. J Ind Eng Manag 4(2):231–257
29. Mittal VK, Sindhwani R, Kapur P. K (2016) Two-way assessment of barriers to lean–green manufacturing system: insights from India. Int J Syst Assur Eng Manage 1–8. https://doi.org/10.1007/s13198-016-0461-z
30. Zhu Q, Sarkis J, Geng Y (2005) Green supply chain management in China: pressures, practices and performance. Int J Oper Prod Manage 25(5):449–468
31. Khan RWA, Khan N, Chaudhary MA (2011) Green supply chain management—global opportunities and challenges: a case study. In: Business innovation and technology management (APBITM), IEEE international summer conference of Asia Pacific, pp 5–9

32. Srivastava SK (2007) Green supply chain management: a state-of-the-art literature review. Int J Manage Rev 9(1):53–80
33. Tsoulfas GT, Pappis CP (2006) Environmental principles applicable to supply chains design and operation. J Clean Prod 14(18):1593–1602
34. Zhu Q, Sarkis J, Cordeiro JJ, Lai K (2008) Firm-level correlates of emergent green supply chain management practices in the Chinese context. Omega 36(4):577–591
35. Delmas M, Montiel I (2009) Greening the supply chain: when is customer pressure effective. J Econ Manage Strategy 18(1):171–201
36. Dubey R, Bag S, Ali SS (2014) Green supply chain practices and its impact on organizational performance: an insight from Indian rubber industry. Int J Logist Syst Manage 19(1):20–42
37. Field JM, Sroufe RP (2007) The use of recycled materials in manufacturing: Implications for supply chain management and operations strategy. Int J Prod Res 45(18–19):4439–4463
38. Singh LP, Singh S, Bhardwaj A (2011) Role of logistics and transportation in green supply chain management: an exploratory study of courier service industry in India. Int J Adv Eng Technol 2(1):260–269
39. Vachon S (2007) green supply chain practices and the selection of environmental technologies. Int J Prod Res 45(18–19):4357–4379
40. Min H, Galle WP (2001) Green purchasing purchases of US firms. Int J Oper Prod Manage 21(9):1222–1238
41. Simpson D, Power D, Samson D (2007) Greening the automotive supply chain: a relationship perspective. Int J Oper Prod Manage 27(1):28–48
42. Sundarakani B, Souza R, Goh M, Wagner SM, Manikandan S (2010) Modeling carbon footprint across supply chains. Int J Prod Econ 128:43–50
43. Lee C (2010) The effect of environmental regulation with green technology innovation using supply chain integration. Int J Sustain Econ 2(1):92–112
44. Soler C, Bergstrom K, Shanahan H (2010) Green supply chains and the missing link between environmental information and practice. Bus Strategy Environ 19(1):14–25
45. Berns M, Townend A, Khayat Z, Balagopal B, Reeves M, Hopkins MS, Kruschwitz N (2009) The business of sustainability: what it means to managers now. MITSloan Manage Rev 51(1): 20–26
46. Ninlawan C, Seksan P, Tossapol K, Pilada W (2010) The implementation of green supply chain management practices in electronics industry. In: Proceeding of international multi conference of engineers and computer scientists, IMECS 2010, vol 3, March 17–19, 2010, pp 17–19
47. Simpson D, Samson D (2008) Developing strategies for green supply chain management. Decis Line 39(4):12–15
48. Zhu Q, Sarkis J (2007) The moderating effects of institutional pressures on emergent green supply chain practices and performance. Int J Prod Res 45(18–19):4333–4355
49. Hosseini A (2007) Identification of green management of system's factors—a conceptualized model. Int J Manage Sci Eng Manage 2(3):221–228
50. Hsu CW, Hu AH (2008) Green supply chain management in the electronic industry. Int J Sci Technol 5(2):205–216
51. Zhu Q, Sarkis J (2006) An inter-sectoral comparison of green supply chain management in China: drivers and practices. J Clean Prod 14(5):472–486
52. Nidumolu R, Prahalad CK, Rangaswami MR (2009) Why sustainability is now the key driver of innovation. Harvard Bus Rev 87(9):56–64
53. Diabat A, Govindan K (2011) An analysis of the drivers affecting the implementation of green supply chain management. Resour Conserv Recycl 55(6):659–667
54. Golinska P, Romano CA (2012) Environmental issues in supply chain management: new trends and applications, Eco production: environmental issues in logistics and manufacturing. Springer, Heidelberg

Military Airworthiness and Certification Procedures: Global Scenario

Kanchan Biswas

1 Airworthiness

Airworthiness can be defined as '*Demonstrated capability of an airborne store to perform satisfactorily and fulfil the mission requirements, throughout the specified life in the prevailing environments with acceptable level of safety and reliability*'. Federal Aviation Administration (FAA) order 8130.2H stipulates two conditions for an aircraft to be airworthy. *An aircraft is airworthy if it conforms to its Type Design and if it is in a condition for safe flight.* [1]. The type design is the design against which the aircraft is certified and therefore describes the initial airworthiness requirement towards 'Certification' and the second part of the definition refers to the aircraft status on date and thus is related to 'continued airworthiness' which regulates the repair, maintenance and operation of the aircraft throughout its lifespan.

To make the aircraft design safe, airworthiness authorities specifies certain design requirements called '***Regulations***'. The regulations are intended to promote safety by eliminating or mitigating unsafe conditions that can cause death, injury or damage of properties. ***Airworthiness Authorities establishes these Regulations***. These regulations differ depending on whether the aircraft is for military or civil applications and also on the country of origin/regulatory bodies. A fundamental difference between civil and military aircraft design objectives are that a civil application will look for *passenger comfort, safety and economy of operation* while

K. Biswas (✉)
CSIR-NAL, Bangalore, India
e-mail: kanchan.biswas@rediffmail.com

K. Biswas
Jain University, Bangalore, India

S. Singh et al. (eds.), *Proceedings of the International Conference on Modern Research in Aerospace Engineering*, Lecture Notes in Mechanical Engineering, https://doi.org/10.1007/978-981-10-5849-3_31

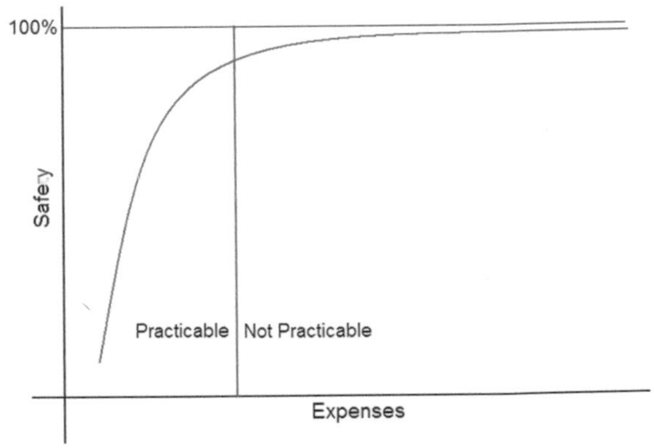

Fig. 1 Acceptable level of safety [3]

its military counterpart will look for *completion of mission tasks* as primary design driver with the safety levels set *as agreed between manufacturer and users* (defence services). Thus to get a *slight edge over the contemporaries*, military aircraft designs often use technologies, which are yet to be proven to their maturity [2–4].

It must be borne in mind that '*No machine can be built which would never fail*' or '*No human could be found who never commits error*'. Consequently, all modes of transportation have to tolerate certain amount of risk or lack of safety. While determining the '***acceptable level of safety***' is must be appreciated that *absolute safety is hypothetical* and can be achieved only *at infinite cost*. Akin to the economics law of diminishing returns, the airworthiness standards have to be a balance between safety concerns and the cost and practicability from design and manufacture point of view. This is explained in Fig. 1 [3].

2 System Safety

'*Safety means freedom from death, injury or damage to people on board as well as collateral damage to human life and property in the ground (accident)*'. Every activity has a *risk threshold beyond which accident will be inevitable.* Airworthiness control is to minimise the risk and maximise the effectiveness. All the airworthiness standards, military or civil, whether that of USA, Europe or Russia, have a common point of reference which is that *an inverse relation should exist between probability of occurrence of an event and the degree of hazard inherent in its effect* (see Fig. 2).

System safety requirements are defined for civil aircraft in SAE ARP 4761 [5] and for military aircraft in US DoD (Department of Defence) document Mil–STD 882D [6]. ARP 4761 considers failure severity as ***catastrophic, severe major, major and minor.*** A 'catastrophic damage' means loss life, aircraft and property, while a minor failure indicates certain inconvenience. The permissible failure probability depends on the severity of the failure. The failure condition severity and

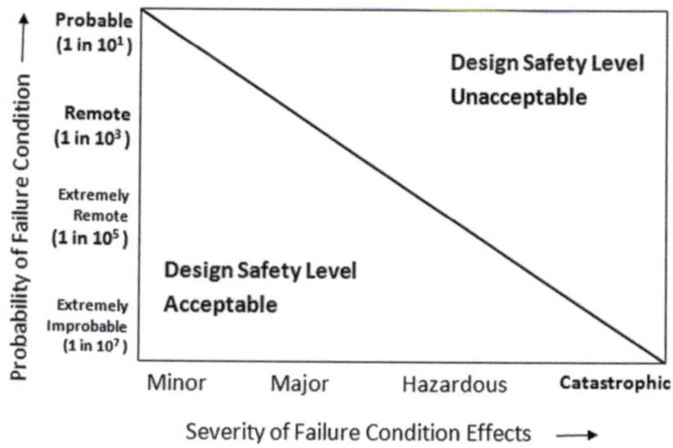

Fig. 2 Inverse relation between damage occurrences and hazard content [3, 25]

Criticality ↑					
Catastrophic I	HAWR	HUD	HUA	HUA	HUA
Critical II	HAWR	HUD	HUD	HUA	HUA
Marginal III	HAWR	HAWR	HUD	HUD	HUA
Negligible IV	HA	HA	HA	HAWR	HAWR
Frequency →	Level E Improbable	Level D Remote	Level C Occasional	Level B Probable	Level A Frequent

Hazard Un Acceptable (HUA) — Hazard Un Desirable (HUD)

Hazard Acceptable With Review (HAWR) — Hazard Acceptable (HA)

Fig. 3 Damage acceptance decision matrix: military aircraft [5, 25]

permissible frequency of occurrence and the corresponding design assurance level (DAL) is shown in Table 1. Mil STD 882 considers four failure severities and five level of frequency of occurrence. The damage acceptance criterion is based on '***Hazard Index***' which is defined as the product of 'damage severity' and the 'frequency of occurrence' of the event. The decision matrix for hazard acceptance as per Mil Standard is shown in Fig. 3.

2.1 *Flight Safety and Airworthiness*

For completing any task certain risk may be required to be taken. Risk level increases with the complexities of the task. Safety decreases with increase in risk taken at some level it may be unsafe to operate. A '***Risk Threshold***' may be defined as the level of risk beyond which accidents are inevitable. This 'Risk Threshold' is,

Table 1 Falure condition severity and permissible probaility of occurrence (SAE ARP 4761)

Falure severity classification	Minor	Major	Severe major (Hazardous)	Catastrophic
Failure condition effects	Slight reduction in safety margin	Significant reduction in safety margin	Large reduction in safety margin	Codition which prevent continued safe flight and landing
Permissible probability of failure	10^{-3} (frequent)	10^{-5} (remote)	10^{-7} (extremely remote)	10^{-9} (extremely improbable)
Development of assurance level (ARP 4754)	Level D	Level C	Level B	Level A

however, not a stationary one and it keeps varying based on the role, function and a host of other associated factors. It needs therefore to be reassessed under each changing scenario. The threshold would vary depending upon the accuracy and resolution levels of the various instruments and other associated systems used for the purpose of flying. For example, if an aircraft depends only on barometric altimeters, the accuracy of altitude prediction may be in the range of ±500 feet. This has an effect on the vertical separation requirement of two aircraft flying in the same flight corridor. With INSGPS, the altitude could be measured at much more accurately and this has facilitated International Civil Aviation Organisation (ICAO) to reduce the requirement of Relative Vertical Separation Minima (RVSM) from 2000 to 1000 feet and ease the traffic congestion.

Thus, it could be appreciated that the safety level would mainly depend on the:

(a) Airworthiness status of the aircraft,
(b) Cost and time of development and the implementation of the maintenance procedures,
(c) Operating crew and their skill,
(d) Maintainability and the maintenance crew skills,
(e) Air traffic control system and its effectiveness,
(f) Effectiveness of the navigational aids,
(g) Effectiveness of weather forecast.

The 'flight safety directorate' is primarily responsible to estimate the risk threshold under all dynamic conditions and take appropriate measures. The basic purpose of flight safety studies is to ensure that the chances of achieving the tasks should be optimal while risks are minimal. The flight safety directorate has a very complex duty to perform. On one hand, the military training must give a high level of exposures to possible war scenario and threats. The directorate has also to ensure that in peace time high level of risks are to be avoided within the stated training or operational tasks. This is because the accidents have very deleterious effects on the morale of the flier. During the war time, however, task achievement is paramount and hence risks are to be taken even at high degrees if the operational requirements dictate.

The main purpose of the flight safety studies can therefore be summarised as follows:

(a) Identify and minimise those risks which may contribute to accidents,
(b) Avoid very high cost of losses and damages,
(c) Identify all risk hazards real or potential at all levels and all phases of flying,
(d) Risk thresholds are dynamic and they need to be reassessed,
(e) Continuously reassess the risks and make necessary readjustments.

3 Certification

Certification is a process of evaluation and documentation of *Compliance of a Product* to its specified requirements and declares '*safe to fly*'. It is a third-party assurance to the user that the product has been designed, developed, evaluated and produced in such a manner that its quality, reliability and integrity meets the requirements. While, certification is a declaration that a product design conforms to the stipulations as defined in specification, airworthiness, on the other hand, is an explicit declaration that the product item built is certified to be airworthy, with limitation as applicable. A product that is certified need not be airworthy, unless each product item built is also individually qualified as airworthy. It is thus seen that airworthiness and certification are complimentary. An aircraft can really be considered airworthy when [7]:

(a) Its type has been designed to and approved as meeting the applicable certification standards,
(b) It has been manufactured by an approved organisation, in accordance with the conditions of its certification of approval to conform with the approved type design,
(c) It has been maintained by qualified people in accordance with an approved system and inspected in accordance with all applicable directives issued by the airworthiness authority,
(d) Further, on compliance of (c) above, no significant defects have been found and not rectified.

ICAO recognises only two types of aircraft, viz. civil and state aircraft. According to ICAO, aircraft used for military, customs, police and firefighter are considered state aircraft and is not within airworthiness control of ICAO. The operational procedure for state aircraft need not conform to ICAO requirements. Governments can therefore 'Self-Certify' their state aircraft as airworthy and compliant to controlled airspace performance requirements. However, if a state aircraft has to fly in civil airspace, it is expected to comply with ICAO norms for airspace access and utilisation. Otherwise, 'State' aircraft can be denied access to civil controlled airspace [8].

In most countries including US, UK, Australia, military certification is carried out by the military airworthiness directorate of the armed forces. It is understood that similar procedures were also followed in the erstwhile Soviet Union. Besides military standards, military policy directive and detailed staff instructions are released for certification. No third-party certification exists and the military air worthiness authorities carry out qualification and performance verifications with a purpose of inducting the aircraft into state service. In India, however, there are two third-party authorities in the Ministry of Defence, namely ***Centre for Military Airworthiness and Certification (CEMILAC)*** for design certification and ***Director General Aeronautical Quality Assurance (DGAQA)*** to provide quality assurance coverage and accord inspection approval of the product and services.

4 Military Certification

Military certification is not equivalent to civil certification (e.g. FAA certification); it concerns the general concept of airworthiness as well as rules for design and performance evaluations. Each government has its own rules to 'Self-Certify' their state aircraft as airworthy and compliant to some specified and controlled airspace performance requirements. In the rules for military aircraft, the operational risk and time of use are defined for each type of aircraft. While there is no conceptual difference on airworthiness, the civil and military certification differs on the fact that:

(a) Governments can 'Self-Certify' their state aircraft,
(b) Operational risk is defined and accepted by the service,
(c) Military certifications differ on the degree and coverage of the evidences needed. This is in general limited by contract, budget, lack of past legal liability and aircraft type and legacy,
(d) Acceptance of specific tasks and the risk levels can vary with aircraft purpose and type.

Military certification differs from civil procedure broadly due to the fact that the military certification takes the new aircraft up to the level of qualification for induction into service by issuing an operational clearance. The certification procedure followed is as follows:

(a) Approval builds up in a building block method,
(b) First it's foundation is built from each components' compliance evidences,
(c) Then systems' performance compliance evidences are added,
(d) Then aircraft performance compliance evidence meeting compliance to staff requirement are recorded,
(e) A compliance document is prepared capturing the evidences generated through analysis, aircraft inspection, system integration test and flight tests.

Table 2 Comparison of civil and military certification

Objectives	Civil aircraft certification	Military aircraft certification
Business outlook	International business through common rules and collaboration	Military aviation is under each individual nation's responsibility/sovereignty
Design standard followed	Internationally accepted airworthiness regulatory standards (FAR/CS)	Different standards followed by different nations e.g. Mil (USA), Def. Stan (UK)
Compliance	Compliance to airworthiness certification criteria (ACC) based on aircraft category	Tailored airworthiness certification criteria. Demonstration restricted to budget provisions and various other criteria
Continued airworthiness	Part of airworthiness certification and documentation	Continued airworthiness is defined by OEM and executed by military
International relation	Mutual acceptance amongst nations through bilateral aviation safety agreement	Each nation operates its own airworthiness certification system
Certification objectives	Verification and demonstration of compliance to regulatory requirement	Qualification and issue operational clearance for induction into state service through issue of Release to Service Documents (RSD)

A comparison of the certification requirements of civil and military aircraft is shown in Table 2 [9, 10].

5 Military Certification Procedures in Different Countries

International regulations and standards for military aviation safety are varied, which leads to diversity in structures of military aviation safety organisations in different countries. Only a few military authorities like those of UK, USA and USSR have published military design standards. However, USSR documents are not readily available in the open literatures. While all civil airplane designs have to meet civil regulations, most military aircraft designs world over follow either civil regulatory standards like FARs, CARs, JARs or military standards US DoD and MoD UK documents. In general for the military airplanes, the military specification and documents nominate some specified elements like:

(a) Handling qualities;
(b) Weapons, ammunition stores and self-defence suites;
(c) Specific operations in wartime;
(d) Military Role, mission and Tasks.

6 Military Airworthiness in UK

With effect from 1 April 2010, the Secretary of State for Defence (SofS) established by Charter the Military Aviation Authority (MAA) as the single independent regulatory body for all Defence Aviation activities in the UK [11]. As the 'Regulator', Director General MAA (DG MAA) is accountable to SofS, through the second Permanent Under-Secretary of State, for:

(a) providing a regulatory framework,
(b) given effect by a certification,
(c) approvals and inspection process for the acquisition,
(d) operation and airworthiness of air systems within the Defence aviation environment.

DG MAA is the owner of the MAA Regulatory Publications (MRP) and has the authority to issue them on behalf of the SofS. MAA carries out Military Air Systems Certification Process (MACP) in six stages as per MAA document 'RA 1500–Certification of UK Military Air System' prior to their Release to Service (RTS) as shown in Fig. 4.

The details of full and tailored MACP are discussed below [12, 13].

(a) **Full MACP**—The MACP should be applied in full for the certification considerations of:
    (1) **Full MACP for New Design** leading to issuing ***Military Type Certificate (MTC)***.
    (2) **Full MACP for major change**—leading to ***Approved Design Change Certificate*** *(ADCC)*.
(b) **Tailored MACP**—For Air systems that were procured prior to Sep 2011.
(c) **Release to Service Recommendations (RTSR)**.
(d) **Urgent Operational Requirements (UORs)**.

Design standard followed is normally Def-Stan-00-970 [14].

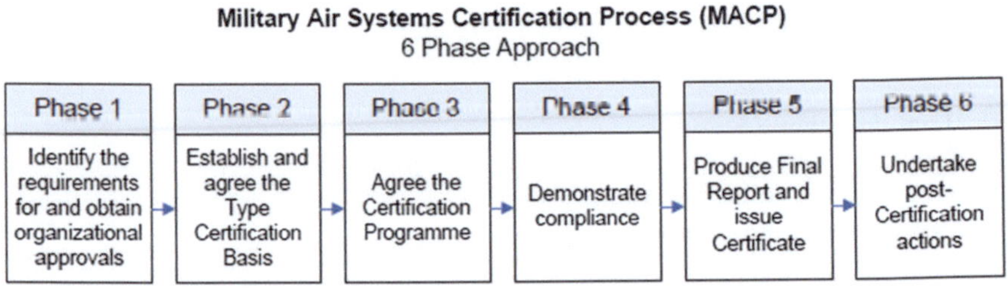

Fig. 4 Six stages military air systems certification process (MACP); MoD, UK [12]

7 Military Airworthiness in USA

The Secretary of the Air Force (AF), USA issued ***AF Policy Directive no. AFPD 62-6 on 11 Jun 2010*** [15]. According to this directive, USAF is responsible for assuring the airworthiness of the aircraft which it operates. The directive establishes policies for formal airworthiness assessments to ensure that AF operated aircraft are airworthy over their entire life cycle and maintain high levels of safety. This policy is implemented by detailed procedures indicated in ***AF Instruction no. AFI 62-601, dated 15 Jun 2010 and supplemented on 12 May 2011*** [16]. According to the supplement, a ***Technical Airworthiness Authority (TAA)*** and an ***AF Airworthiness Board (AB)*** chaired by the TAA have been created in the Air Force Material Command (AFMC) to provide independence in airworthiness determinations. It establishes the requirement for *design-based airworthiness certification* and also to provide *non-design-based special flight release* of aircraft. It enables AF acceptance of FAA certifications, evaluations and inspections and disestablishes the Airworthiness Certification Criteria Control Board (AC^3B).

7.1 USAF Military Aircraft Certification Procedure

According to the USAF policy directive [16], the types of approach will be followed:

(a) A design-based airworthiness Assessment,
(b) A Non-design-based airworthiness Assessment,
(c) Certification of,
(d) Commercial Derivative Aircraft (for transport fleet only).

For non transport aircraft project, Project Managers (PM) shall request the TAA to make a determination to proceed with one of the two above possible alternative assessment processes. Design-based assessments are the preferred approach, while non-design-based assessment is adopted 'by-exception' basis for unique aircraft or situations.

7.1.1 Design-Based Airworthiness Assessment

The design-based certification is carried out when adequate design information is available. The certification is carried out in accordance with the USAF policy directives [15] and USAF Instruction [16] in stages of evaluations. This will lead to issuance of military experimental flight release and finally issuance of Military Type Certificates (MTC). As approved by TAA, the Project Manager shall obtain a MTC before entry into service and maintain Military Certificate of Airworthiness (MCA) for each tail number. Issuance of an MTC indicates that:

(i) Aircraft design documentation accurately defines the configuration which meets the certification basis and the aircraft design is in compliance with requirement.
(ii) The design package includes all necessary technical information required to construct, maintain and operate the aircraft system throughout its approved service life.
(iii) Mission usage has been defined and approved, and the flight manual accurately describes the permissible flight envelope.
(iv) A service life limit has been established and approved for the type design.

7.1.2 Non-design-based Airworthiness Assessment

A non-design-based (NDB) airworthiness assessment is conducted in accordance with USAF policy directive [15] and USAF instruction [16], when it has been determined by the TAA that a design-based airworthiness certification cannot reasonably be accomplished, but there is a compelling military need to operate the air system. When followed to a successful conclusion, the result of this process is TAA issuance of a special flight release. The non-design-based special flight releases process is used to identify and assess the inherent risks of operating these aircraft and to formally acknowledge acceptance of these risks during their flight operations.

7.2 Commercial Derivative Aircraft (CDA)

USAF utilises FAA type certified aircraft for their military transport air applications. USAF accepts these aircraft as Commercial Derivative Aircraft (CDA). In fact FAA type certification is a preferred route by USAF. However, there may be applications which are more severe than civil roles or exceed FAA cleared flight or usage environment. Hence, CDA requires additional Military Type Certification (MTC) to be issued by TAA. The basis for issuance of MTC will be baseline FAA certification and additional certification requirement as assessed by TAA or any item which was not included by FAA in their TC process. ***MIL-HDBK-516*** [17] ***shall be used to define applicable military airworthiness certification criteria, standards and methods of compliance*** and FAA order no 8100.101 [18] procedure will be followed for processing certification. Any item not listed on an FAA form 8130-2 or 8130-31 (i.e. items which are not included in the FAA TC) are to be included in MTC. FAA special airworthiness certificates in the experimental category are not an acceptable certification basis for AF CDA airworthiness certification. Maintenance of CDA is to be dealt with as per AFI 21-107:*Maintaining Commercial Derivative Aircraft.*

8 Military Airworthiness of Australia

Chief of Defence, Australian Defence Forces (ADF) and Secretary of Defence directed that ADF implements self-regulation of its Aviation Practices to meet Defence Safety Objectives. Accordingly, Department of Defence, Australia, issued **Australian Air Publication AAP 7001.048, 'Defence Aviation Safety Programme (DASP) Manual'** on 30 June 2014 [19]. As per the AAP, Chief of Air Force will be 'Defence Aviation Authority' accountable for establishing and managing DASP. He will be supported by Deputy Chief of Air Force (DCAF) as Operational Airworthiness Regulator (OAR) and Director General Technical Airworthiness (DGTA) as Technical Airworthiness Authority. These directorates are responsible for establishing the regulatory framework which includes formulations of Regulations, Acceptable Means of Compliance and Guidance Materials. The functional diagram of the regulatory process is shown in Fig. 5.

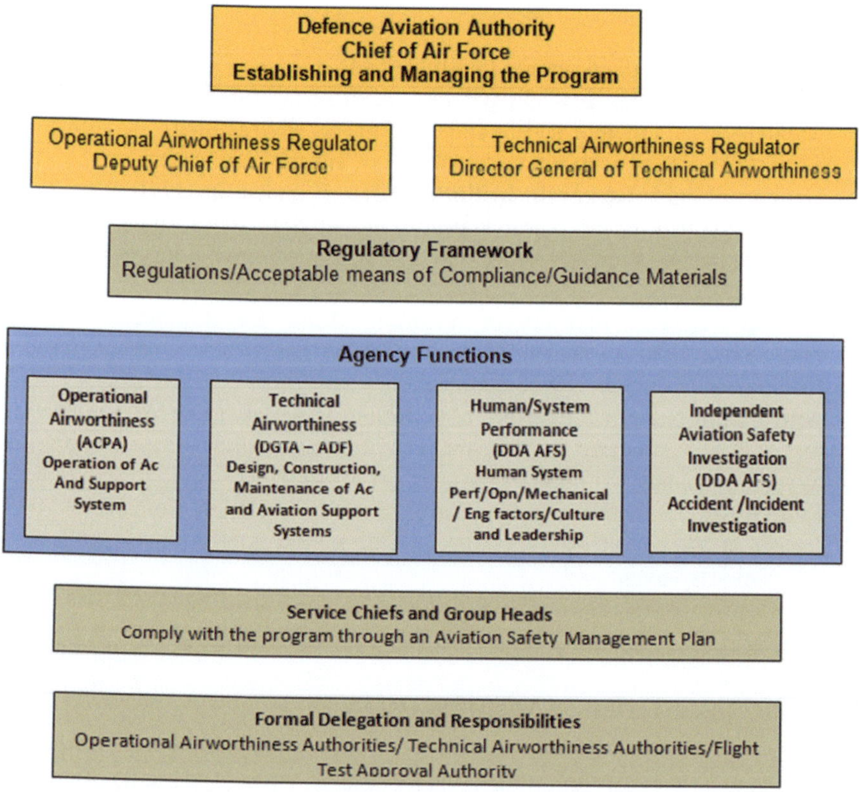

Fig. 5 Organisation structure of Australian airworthiness procedure [19]

9 Military Airworthiness in Russia

The military airworthiness procedure that was followed in the erstwhile Soviet Union is not very clearly available. There are various aviation *design bureaus and manufacturing complexes*. Operational requirements of the military stores are created by the air force and these design houses create designs to meet those military requirements. During the design and development process, the design bureaus take support from various state research institutes on aerodynamics and other aircraft systems. In fact the design bureau creates number of design alternatives. These competing designs are evaluated against each other and a winner is chosen which would then be assigned to one or more manufacturing complexes. Soviet Union had developed their own state standards GOST (*Russian: ГОСТ an acronym for gosudarstvennyy standart*). All Russian designs use these state standards. The production and quality system 'Oboronsertifika' followed in the defence industries is similar to international quality standards [20].

Among the various scientific institutions involved in aircraft design and certification a few important ones are:

(a) **Central Aero-Hydrodynamic Institute** (**TsAGI**). Founded in 1918, located in Zhukovsky near Moscow. TsAGI has expertise in both test and analysis in aerodynamics, structure and aircraft systems as well as aircraft certification. They are entrusted with the responsibility of aircraft certification.
(b) **Gromov Flight Research Institute (LII)**. Founded in 1941, located in Zhukovsky. LII is the research centre for all flight tests and evaluations. International Aviation and Space Salon (MAKS) is hosted at this airport every two years.
(c) **Baranov Central Institute of Aviation Motor Development (TsIAM)**. Founded in 1930, located in Moscow. TsIAM has all necessary engine test facilities and takes up engine certification.
(d) **All-Russian Research Institute of Aviation Materials** (**VIAM**). Founded in 1932, VIAM is located in Moscow. VIAM is responsible for qualification of aviation materials.
(e) **State Research Institute of Aviation Systems** (**GosNIIAS**). Founded in 1946, located in Moscow. GosNIAS is responsible for development and certification of aviation systems.

10 European Defence Agencies (EDA)

On the similar lines of European Aviation Safety Agency (EASA), European Defence Agency has been formed in November 2008 [21, 22]. Under the chairmanship of EDA, *Military Airworthiness Authorities (MAWA)* forum was established by Defence Ministers of Europe to harmonise the European *Military*

Airworthiness Regulations (EMAR) of participating Member States. MAWA Forum comprises representatives from the Military Airworthiness Authorities of *participating Member States (pMS)* and industry representatives. The driving factors and benefits as anticipated are enhanced military aviation safety and improve cooperation between military and civil regulatory bodies of the pMS as well as to reduce competition and enhance interoperability.

The aim of the MAWA Forum is to harmonise the national military airworthiness regulations of its pMS through a common set of harmonised EMARs, Acceptable Means of Compliance and Guidance Material that can be implemented into national airworthiness regulations by all pMS. The EMARs will ensure greater standardisation of each pMS's approach to military airworthiness across Europe by detailing the procedures and processes necessary to achieve this.

11 Military Airworthiness in India

Military Airworthiness certification process in India has been modelled after the earlier British system. The system is based on ***concurrent design and clearance*** leading to eventual certification. Milestones in the design and development process are agreed between the design groups and the certification authority and the design is reviewed from airworthiness and safety point of view along with the progress of the design. At appropriate stages, test procedures are examined, approved and tests are carried out. The test results are reviewed for acceptance, redesign or retest. This approach was adopted since the certification authority, viz the *Chief Resident Engineer (CRE)* and the inspection authority *Chief Resident Inspector (CRI)* were co-located with Defence Public Sector *HAL (Hindustan Aeronautics Limited)* who is responsible for design and development (D&D). The background for adopting this concept is that if redesign or retest has to be undertaken because of airworthiness and safety compulsions, when the entire aircraft design is complete, it will result in a major effort affecting the cost and time schedules [23]. Though both CRE and CRI organisations have been changed to CEMILAC and DGAQA, the inspection and design certification procedure has practically remained the same.

However, this process conceptually greatly differs from those followed in US, UK and many other countries where airworthiness functions are delegated to the contracting firm through approved designers and airworthiness groups within the firm. At mutually accepted major milestones, reviews are carried out as per the agreed documentation of the contract. The Government, through the certification provision, holds the authorised personnel within the firm responsible for the airworthiness certification of the aircraft. In contrast, in India, the regulatory authorities CEMILAC and DGAQA hold interaction with the D&D team on a day-to-day basis and carry out spot checks to identify design/production deficiencies during the prototype design and development stage. This becomes counter-productive and causes delays as unless and until one stage is cleared the designer cannot proceed to the next stage. Needless to say, the benefit claimed for adopting of this process is not realised till confirmatory tests are conducted with the participation of all agencies. It may therefore be advisable to consider a process of self-certification by the approved firms with regulatory monitoring and control of approved signatories for design and quality assurance.

11.1 Indian Military Certification Procedure

For Military Certification in India, MOD Document ***'Procedures for Design Development of Military Aircraft and Airborne Stores—2002' (DDPMAS—2002) guide lines are followed*** [24]. *Section III, Chapter-1 of DDPMAS—2002,* discusses the concurrent certification procedure to be followed for D&D of an ab–initio aircraft. Chapters 3 and 4 discuss the procedures to be followed during licence production and bought out aircraft. This document, however, does not support certification of an already developed system. Besides this procedural document, *there is no other Certification Procedure Document or Manual Available.* The certification and continued airworthiness functions are shown in Fig. 6. The certification procedure is shown in the block diagram in Fig. 7.

Some of the salient features of this procedure are:

(1) CEMILAC accords Design Approval (DA) to the development house and approves authorised signatories.
(2) As per DDPMAS, both CEMILAC and DGAQA are to participate in PDR, CDR.
(3) DGAQA approves quality management system of the production house and approves inspectors.

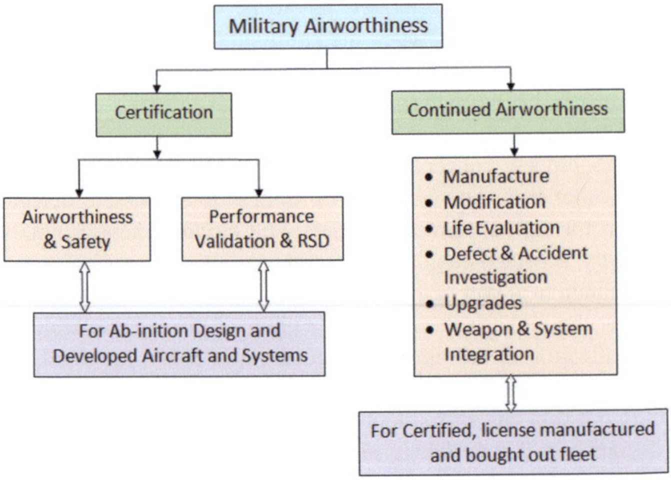

Fig. 6 Military airworthiness functions

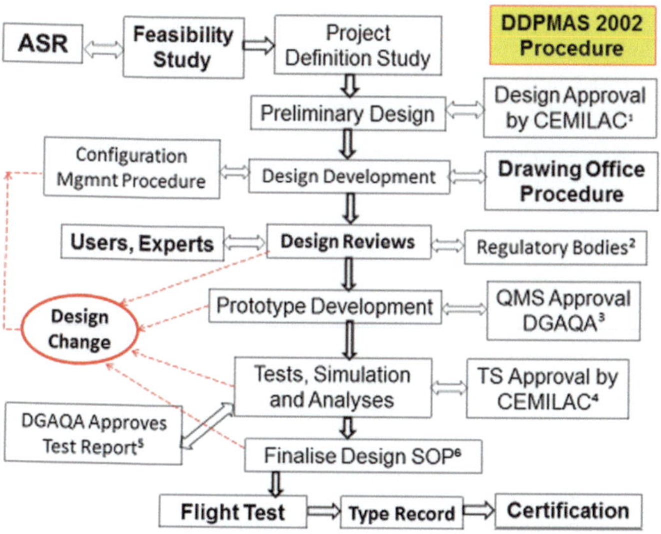

Fig. 7 Military certification procedure as per DDPMAS—2002, Section–III, Chapter-1

(4) Design SOP (Standard of Preparation) is to be finalised after all tests are completed, however, DGAQA carries out stage inspection during prototype fabrication and build up. To facilitate this activity, CEMILAC co-ordinate Design SOP in the detail design stage itself.

(5) Ground test schedule is approved by CEMILAC. Test rig is inspected and cleared by DGAQA. Test report is co-ordinated by DGAQA.

12 Discussion and Conclusion

Aviation regulatory activities for air systems are mainly separated into:

(a) Technical certification, ensuring that the aircraft meets the technical requirements set out in the regulations;

(b) Operational approval, which is the official declaration that the aircraft is compliant to intended operational exploitation.

While technical certification standards and procedures are well documented, there is no uniformity in the procedures followed in different countries as regards to the self-certification of the state aircraft. In fact, each state defines how and by whom their state certification will be carried out. Besides, the certification processes are insufficiently documented and for reasons of security not made available to others for scrutiny, though this does not mean that the regulatory processes are violated.

While most countries follow self-regularity control by their military airworthiness directorate, India follows a third-party certification route. Here again, the certification procedure and regulatory requirements are not well documented and available in public domain. In view of this, there is more of direct involvement by the regulatory bodies in day-to-day activities rather than relying on or holding the design and production agencies responsible for aviation safety. Besides, there is no apex body to guide and monitor for a unified approach towards design certification and product quality assurances in a development project.

References

1. FAA Order 8130.2H dated 02/04/2015, Airworthiness Certification of Products and Articles —National policy
2. Gratton G (2015) Initial airworthiness, determining the acceptability of new airborne systems. Springer International Publishing, Switzerland
3. De Florio F (2006) Airworthiness: an introduction to aircraft certification. Elsevier Publication, UK
4. Biswas K et al (2015) Comparison and unification of civil and military aircraft certification procedures. SP-CAD/2015/1000, Dec 2015, CSIR, NAL Bangalore
5. SAE ARP 4761 Guidelines and methods for conduction safety assessment process on civil airborne systems and equipment. SAE, 1996-12
6. Mil–Std–882D: Standard practice for system safety assessment. DoD, USA, 10 Feb 2000
7. PS Langford, The principles and practice of airworthiness control. Aeronautical Journal, January 1976
8. Airworthiness Manual: International Civil Aviation Organisation. Doc 9760, 3rd Ed 2013
9. Kundu AJ (2010) Aircraft design. Cambridge University Press
10. Roskam J (1985) Airplane design, vols I–VII. Roskam Aviation and Engineering Corporation, Ottawa, Kansas
11. MAA 01 (2012) Military Authority regulatory policy, military aviation authority. Govt of UK, Issue 2, 23 Feb 2012
12. RA 1500: Certification of UK military registered air system. Regulatory Article, Initial Issue, Military Aviation Authority, Govt of UK
13. DME 5000: Design, modification engineering regulatory articles. Military Aviation Authority, Govt of UK
14. Def-Stan–00 970: Design and airworthiness requirement for service aircraft. Military Aviation Authority, Govt of UK
15. Air Force Policy Directive 62—6 Developmental Engineering, USAF Airworthiness. DoD, USAF, 11 June 2010
16. Air Force Instruction 62—601 Developmental Engineering, USAF Airworthiness. DoD, USAF, 12 May 2011
17. Mil Std-516B. Airworthiness Certification Criteria. US DoD, 26 Sept 2005
18. FAA order no. 8100.101: Type Certification process for military commercial derivative Aircraft. US Department of Transportation National Policy, 09 July 2007
19. Australian Air Publication 7001.048. Defence Aviation Safety Program Manual. Department of Defence, Government of Australia, 04 Sept 2012
20. Aircraft Industry in Russia—Wikipedia, free encyclopaedia
21. www.eda.europa.eu: European Defence Agency, Military Airworthiness Recommendations (MAWA). Brussels, January 08, 2013

22. EASA and EDA strengthen civil and military cooperation in Aviation Safety, Press release, Paris 18 June 2013
23. Nagraj K et al (2005) Reference book on airworthiness and certification of military fixed wing aircraft. CEMILAC, Min of Defence, DRDO, Bangalore
24. DDPMAS—2002: Design, development and production of military aircraft and airborne stores. Department of Defence Production and Supplies, Ministry of Defence, Government of India
25. Raheja DG (2006) Assurance technologies: principles and practices. McGraw-Hill Engineering and Technology Management Series, Inc

Improving Assembly Line Efficiency and Output Using Assembly Line Balancing and Reducing Non-Value-Added Activities

Prateek Sharma and Meeta Sharma

1 Introduction

Assembly line balancing refers to properly assigning tasks to workstations such that time at each station is nearly the same. In line balancing, the leveling of workload across all the processes in a cell or value stream is completed to remove bottleneck from the flow, thus making it smoother. This task is carried out in the companies to achieve one-piece flow of product so that the wastes occurring in the system due to waiting can be eliminated. The assembly lines generally are of two parts as single-model assembly line and multi-model assembly line. A single-model assembly line is one which is designed for a single product or model across the line and a multi-model assembly line is one which is designed for different products. In multi-model, variety of products can be made on the same assembly line. The main objective of line balancing is to eliminate the wastes related to defects, overproduction, transportation, and waiting.

The major work done related to this field is summarized and presented ahead. Kumar and Mahto[1] have presented an actual case of industries where different components are manufactured to achieve productivity improvement, applied group technology for reducing production cost, and number of workstations.

Morshed and Palash [2] had focused on improving the efficiency of single-model assembly line by reducing non-value-added activities, cycle time, and distribution of workload at each station by line balancing.

P. Sharma (✉) · M. Sharma (✉)
Department of Mechanical and Automation Engineering,
Amity University, Noida, Uttar Pradesh, India
e-mail: prateek181292@gmail.com

M. Sharma
e-mail: contactmeeta@yahoo.com

S. Singh et al. (eds.), *Proceedings of the International Conference on Modern Research in Aerospace Engineering*, Lecture Notes in Mechanical Engineering,
https://doi.org/10.1007/978-981-10-5849-3_32

Kumar and Mahto [3] have reviewed works of different authors in the area of assembly line balancing and have found out the developments and trends in industries to reduce the workstations and the total equipment cost.

Kriengkorakot and Pianthong [4] presented an up-to-date review and have discussed the developments in classification of assembly line balancing problems. They have also presented suggestions and summaries for future in the field of assembly line balancing.

Kumar et al. [5] presented paper on a case study of wire harness assembly line and discussed about the improvement in the productivity and quality using kaizen and line balancing. They have calculated balance rate before and after improvement.

Amardeep et al. [6] presented the work on overall efficiency of a single-model assembly line.

Sathish and Lakshmanan [7] evaluated through a case study of an instrument cluster assembly line and showed that there is improvement in the efficiency by assembly line balancing. The efficiency has been improved from 46 to 86% by reducing the workstations and the number of operators.

Chandra [8] focuses on the crucial area of productivity improvement with the astute use of work study technique mixed with modern soft skills. It also concluded that the systematic application of method and time study improves the productivity and profit with reduction in costs.

Nordin et al. [9] explored the extent to which the lean manufacturing has been implemented in the Malaysian automotive industries. In the study, the barriers such the implementations of lean manufacturing, i.e., lack of understanding and the attitude of the employees as are also analyzed.

Talib Bon and Daim [10] worked on the time and motion study in order to increase the production and also identified the possible improvements that can be made by observing the manpower so that the production can be increased and simultaneously cost can be reduced.

Krishna Kumari et al. [11] presented a case study of small-scale industry to implement lean system to improve the productivity by eliminating the non-value-added activities. The literature survey provides the idea for designing the assembly line base.

Sharma and Sharma [12] in their paper presented a case study of wire harness assembly line and the productivity and efficiency were improved by using lean manufacturing concepts.

Balancing the assembly line keeping the takt time and cycle time in mind can help in increasing the efficiency as well as production of the assembly line. Non-value-added activities are a major source of waste occurring in any manufacturing system. These are the activities that do not add any value to the product, i.e., the activities that are not being paid by the customer to the company but the company has to pay for the same. Identifying these types of activities and eliminating them can help the companies save their resources.

2 Problem Modeling

Assembly line balancing is very important to achieve increased efficiency and productivity. Efficiency of the assembly line helps in determining the effectiveness of any system. In this problem, the required results have been obtained by balancing the takt time for improvement.

The takt time is the average time between two successive units and can be evaluated as,

$$T = \frac{T_a}{D}$$

where "T" is the takt time, "T_a" is the available time or shift time, and "D" is the demand of the customer per day.

Cycle time is the total time from start of process to end of process, i.e., maximum time allowed at each workstation. For a smooth operation, the cycle time should be less than takt time so that the demand from the customer end can be met.

Efficiency: It can be defined as the ratio of total minutes produced to the total minutes available

$$\eta = \frac{Total\ Minutes\ Produced}{Total\ Minutes\ Availabe}$$

$$\eta = \frac{Output * T_s * 100}{T_a * N}$$

where "*Output*" refers to the actual number of products produced in the shift, "T_s" is the standard time of the product. Standard time is the time required to make a complete product from start to end, "T_a" is the available time or shift time, and "N" is the total number of operators.

Line balancing basically refers to the balancing of any assembly line. It is the term commonly used for properly assigning tasks to workstations such that time at each station remains same. Line balancing means the leveling of workload across all the processes in a cell or value stream to remove bottleneck from the flow.

3 Results and Discussions

In the present case study the assembly line balancing of existing system and the proposed system have been analyzed by finding the takt time, cycle time, and efficiency.

Figure 1 shows the existing layout of the assembly line. In this layout, the process starts from the kit stations represented by KIT in the figure. The kits made by the operators are routed on the trolley and the same continues till the last workstation. Then the complete kit is routed on the board for tapping. The boards

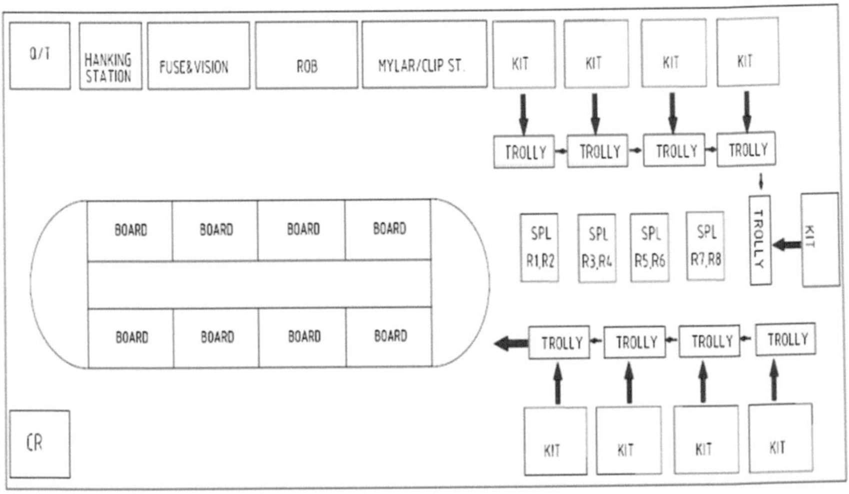

Fig. 1 Layout of existing assembly line

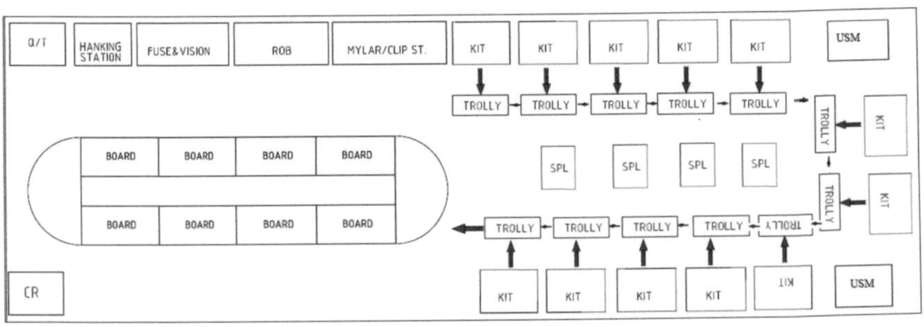

Fig. 2 Proposed layout of assembly line

are mounted on a conveyor. After tapping is completed, the harness is sent to myler where the dimensions of the harness are checked, then to ring outboard for checking the circuits, then to the vision for plugging the fuse, and finally to the hanging area for packing and quality check. In this process, the takt time is taken as 7 min.

3.1 Proposed Layout

Figure 2 shows the proposed layout. In this layout, 3 stations have been added considering takt time of 5.50 min. So the work has been divided to these stations accordingly. The blocks indicated by USM represent ultrasonic machine for the welding of splices. The splices can be made here and are kept in the racks represented by SPL in the figure. This will help in eliminating the transportation as well as reducing the inventory.

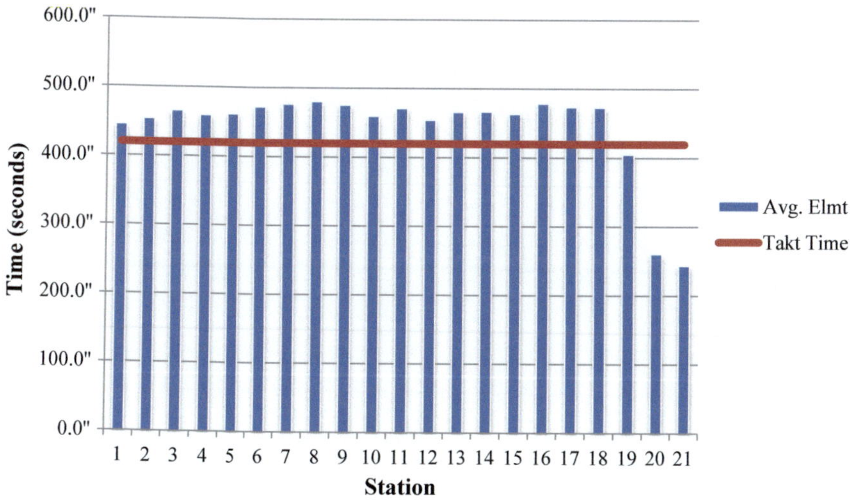

Fig. 3 Time of the stations for current layout

Table 1 Calculations for current layout

Available time (min)	Standard time (min)	Number of operators	Output	Efficiency
465	129	26	54	57%

The following results were obtained after studying the existing and the proposed layouts.

Figure 3 shows the average time for the stations depicted in blue and the line in red shows the takt time. From the graph, it is clear that the average time for almost all the stations is above the takt time, so the required output and efficiency cannot be achieved. Calculations for this layout are shown in Table 1.

3.2 *Proposed Layout*

Figure 4 shows the average time of stations in blue and the line in red shows the takt time. So all the stations have been balanced and are under limits, thus cycle time has also been balanced. The calculations for the proposed layout are shown in Table 2.

Table 3 shows the comparison of the two layouts. As it is clear that the efficiency and production have increased, thus the layout can be implemented.

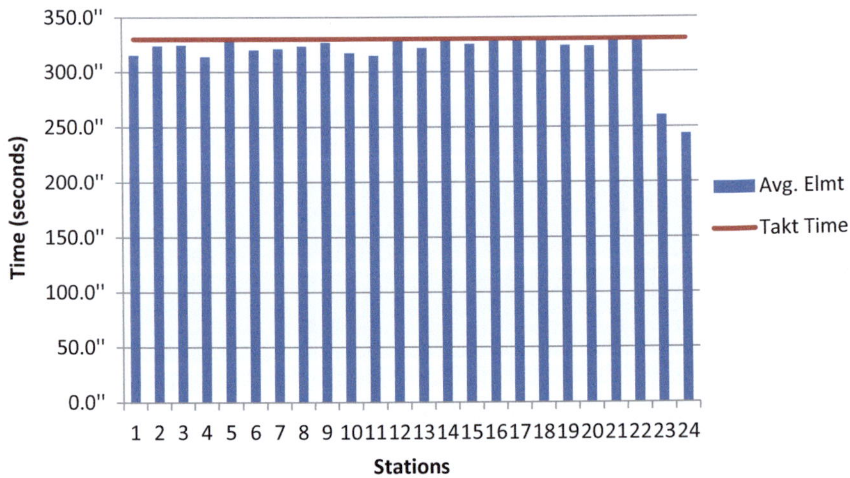

Fig. 4 Time of stations for proposed layout

Table 2 Calculations for proposed layout

Available time (min)	Standard time (min)	Number of operators	Output	Efficiency
465	129	30	86	80%

Table 3 Comparison of layouts

S. no.	Layout	Standard time (min)	Available time (min)	Takt time (min)	Number of operators	Output	Efficiency (%)
1.	Current layout	129	465	7	26	54	57
2	Proposed layout	129	465	5.50	30	86	80

4 Conclusions

The analysis of the assembly lines (existing and proposed layouts), provides the following conclusions:

i. The takt time has been reduced from 7 to 5.50 min.
ii. The per day output has increased from 54 units to 86 units.
iii. The efficiency of the assembly line has increased from 57 to 80%.

The proposed assembly line is more effective than existing one and therefore thus can be implemented for such conditions.

References

1. Kumar N, Mahto D (2013) Productivity improvement through process analysis for optimizing assembly line in packaging industries. Glob J Res Eng Ind Eng 13(3) (Version 1.0 Year 2013)
2. Morshed Md N, Palash KS (2014) Assembly line balancing to improve productivity using work sharing method in apparel industry. Glob J Res Eng Ind Eng 14(3) (Version 1.0 Year 2014)
3. Kumar N, Mahto D (2013) Assembly line balancing: a review of developments and trends in approach to industrial application. Glob J Res Eng Ind Eng 13(2) (Version 1.0 Year 2013)
4. Kriengkorakot N, Pianthong N (2007) The assembly line balancing problem: review articles. KKU Eng J 34(2):133–140
5. Kumar A, Chaudhary G, Kalra M, Jha BK (2014) Optimization of cycle time for wire harness assembly line balancing and Kaizen approach. Int J Innov Res Sci Eng Technol 3(8)
6. Amardeep, Rangaswamy TM, Gautham J (2013) Line balancing of single model assembly line. Int J Innov Res Sci Eng Technol 2(5)
7. Sathish S, Lakshmanan AR (2015) Assembly line efficiency measurement and line balancing—a case study on automobile cluster assembly line. Res J Appl Sci Eng Technol
8. Chandra PV (2013) An effort to apply work and time study techniques in a manufacturing unit for enhancing productivity. Int J Innov Res Sci Eng Technol 2(8)
9. Nordin N, Md Deros B, Abd Wahab D (2010) A survey on lean manufacturing implementation in Malaysian automotive industry. Int J Innov Manag Technol 1(4)
10. Talib Bon A, Daim D (2010) Time motion study in determination of time standard in manpower process. In: Proceedings from 3rd engineering conference on advancement in mechanical and manufacturing for sustainable environment, 14–16 April 2010, Kuching, Sarawak, Malaysia
11. Krishna Kumari S, Balaji AN, Sundar R (2014) Productivity improvement of an industry by implementing lean manufacturing principles. Int J Innov Res Sci Eng Technol 3(3)
12. Sharma P, Sharma M (2016) Improving efficiency of a wire harness assembly line using lean manufacturing concepts. In: National conference on mechanical engineering—ideas, innovations and initiatives (NCMEI3-2016), AMU Aligarh, 16–17 April 2016

Mechanical Behavior of Laminated Composite Curved Tubes

Hamidreza Yazdani Sarvestani and Mehdi Hojjati

1 Introduction

Composite curved tubes are used as primary structures in many engineering fields such as aerospace, offshore, and infrastructure industries. These structures usually are moderately thick, and they are subjected to certain loads such as tension, torsion, shear, and bending. The study on mechanical behavior of composite tubes subjected to mechanical loadings is required to figure out their design capability. Due to the anisotropic behavior, composite curved tubes are much more difficult in analysis, especially in an ultimate stress analysis than an isotropic tube. A large number of studies have been conducted to obtain stresses and perform failure analyses on composite straight and curved tubes.

Ting [1] and Chen et al. [2] investigated a cylindrical anisotropic circular tube subjected to pressure, shear, torsion, and extensive loads for axisymmetric deformation of a homogeneous tube assuming the stresses as a function of radial distance. Shearing and radial stresses in curved beams were developed based on satisfying both equilibrium equations and static boundary conditions on the surfaces of beams [3]. Dryden [4] obtained stress distributions across a functionally graded circular beam subjected to pure bending by using stress functions. The free vibration analysis was performed on functionally graded beams with curved axis by using the finite element method to discretize the motion equations [5]. A first-order shear deformation theory was used to study static and free vibration behavior of generally laminated curved beams [6]. Wang and Liu [7] presented elasticity

H. Yazdani Sarvestani (✉) · M. Hojjati
Department of Mechanical and Industrial Engineering, Concordia University, Montreal, Quebec H3G 1M8, Canada
e-mail: h_yazd@encs.concordia.ca

S. Singh et al. (eds.), *Proceedings of the International Conference on Modern Research in Aerospace Engineering*, Lecture Notes in Mechanical Engineering,
https://doi.org/10.1007/978-981-10-5849-3_33

solutions for curved beams with orthotropic functionally graded layers subjected to a uniform load on the outer surface by use of Airy stress function method. A mathematical model was developed to analyze the mechanical behavior of laminated glass curved beams [8].

The review shows that there is a lack of research on the stress analysis of thick composite curved tubes subjected to mechanical loadings. Although finite element methods can be used for analyzing such structures, it is necessary to do the meshing for each structure every time some dimensions or lay-up sequences are changed. In response to this need for an alternative, rapid and low computational cost method, the present work is devoted to develop a method that can provide stresses for thick composite curved tubes subjected to pure bending moment with simple inputs. Displacement approach of toroidal elasticity (TE) and layerwise method is used.

2 Formulation

By using the developed displacement field of single-layer composite curved tubes and layerwise method, a new displacement-based method is proposed to analyze thick laminated composite curved tubes. Note that the detailed derivations can be found in [9, 10]. Here, the necessary formulations are explained briefly allowing readers to understand the procedure.

2.1 Displacement Field of Single-Layer Composite Curved Tubes

2.1.1 Governing Equations in Toroidal Coordinates

A thick laminated composite curved tube with a bend radius R, mean radius R_I, and thickness h is subjected to a pure bending moment, M, as shown in Fig. 1. The cross section is bounded by radii a and b. Toroidal coordinate system (r, ϕ, θ) is placed at the mid-span of the composite curved tube, where r and ϕ are polar coordinates in the plane of the curved tube cross section and θ defines the position of the curved tube cross section.

A derivation of the governing equations for linear three-dimensional theory of elasticity in the toroidal coordinates has been given for isotropic materials by Zhu and Redekop [11]. The theory is extended here to cover the case for orthotropic materials. The toroidal governing equations are presented as [11]:

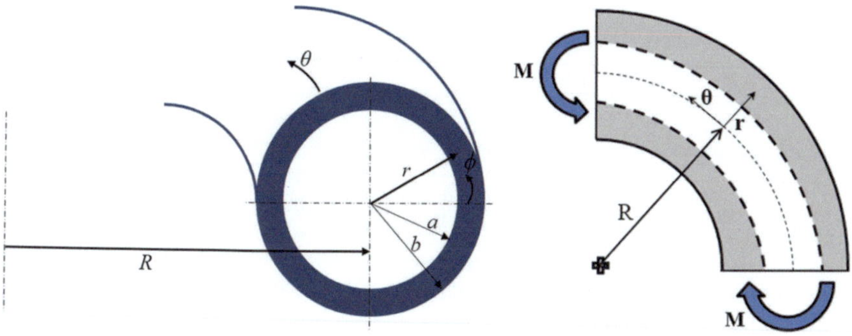

Fig. 1 Geometry and the coordinate system of a composite curved tube

$$\frac{\partial\sigma_{rr}}{\partial r}+\frac{1}{r}\frac{\partial\sigma_{r\phi}}{\partial\phi}+\frac{1}{r}(\sigma_{rr}-\sigma_{\phi\phi})+\frac{1}{\rho}\left(\frac{\partial\sigma_{r\theta}}{\partial\theta}+(\sigma_{rr}-\sigma_{\phi\phi})\cos\phi-\sigma_{r\phi}\sin\phi\right)=0$$
$$\frac{\partial\sigma_{r\phi}}{\partial r}+\frac{1}{r}\frac{\partial\sigma_{\phi\phi}}{\partial\phi}+\frac{2}{r}\sigma_{r\phi}+\frac{1}{\rho}\left(\frac{\partial\sigma_{\phi\theta}}{\partial\theta}+\sigma_{r\phi}\cos\phi-(\sigma_{\phi\phi}-\sigma_{\theta\theta})\sin\phi\right)=0$$
$$\frac{\partial\sigma_{r\theta}}{\partial r}+\frac{1}{r}\frac{\partial\sigma_{\phi\theta}}{\partial\phi}+\frac{1}{r}\sigma_{r\theta}+\frac{1}{\rho}\left(\frac{\partial\sigma_{\theta\theta}}{\partial\theta}+2\sigma_{r\theta}\cos\phi-2\sigma_{\phi\theta}\sin\phi\right)=0 \tag{1}$$

where $\rho=R+r\cos\phi$. The kinematics relations are:

$$\begin{aligned}&\varepsilon_{rr}=\frac{\partial u}{\partial r},\quad \varepsilon_{r\phi}=\frac{1}{2}\left(\frac{1}{r}\frac{\partial u}{\partial\phi}+\frac{\partial v}{\partial r}-\frac{v}{r}\right),\\&\varepsilon_{\phi\phi}=\frac{u}{r}+\frac{1}{r}\frac{\partial v}{\partial\phi},\quad \varepsilon_{r\theta}=\frac{1}{2}\left(\frac{\partial w}{\partial r}+\frac{1}{\rho}\frac{\partial u}{\partial\theta}-\frac{w}{\rho}\cos\phi\right),\\&\varepsilon_{\theta\theta}=\frac{1}{\rho}\left(u\cos\phi-v\sin\phi+\frac{\partial w}{\partial\theta}\right),\quad \varepsilon_{\phi\theta}=\frac{1}{2}\left(\frac{1}{r}\frac{\partial w}{\partial\phi}+\frac{1}{\rho}\frac{\partial v}{\partial\theta}+\frac{w}{\rho}\sin\phi\right)\end{aligned} \tag{2}$$

where u, v, and w represent the displacement components in the r, ϕ, and θ directions, respectively. Upon substitution of Eq. (3) in the constitutive equations for orthotropic materials, stress–displacement relations are obtained. Substituting stress–displacement relations in Eq. (1), Navier equations in toroidal coordinates are obtained as:

$$
\begin{aligned}
&\left(\frac{1}{\rho}\right)^0 U + \left(\frac{1}{\rho}\right)^1 \bar{U} + \left(\frac{1}{\rho}\right)^2 \hat{U} = 0 \\
&\left(\frac{1}{\rho}\right)^0 V + \left(\frac{1}{\rho}\right)^1 \bar{V} + \left(\frac{1}{\rho}\right)^2 \hat{V} = 0 \\
&\left(\frac{1}{\rho}\right)^0 W + \left(\frac{1}{\rho}\right)^1 \bar{W} + \left(\frac{1}{\rho}\right)^2 \hat{W} = 0
\end{aligned} \tag{3}
$$

where the coefficients in Eq. (3) are defined in [9]. The three Navier equations serve as the fundamental equations for the displacement-based toroidal elasticity. As it is impossible to find an exact solution for the Navier equations, the method of successive approximation is used to obtain an approximate solution.

2.1.2 General Solution for In-Plane Pure Bending

A direct solution is formulated to avoid the complications which arise from the use of stress function. The solution for the displacement is thus sought in the following form:

$$
\begin{aligned}
u &= ar^m \cos(n\phi)\cos(\theta) \\
v &= br^m \sin(n\phi)\cos(\theta) \\
w &= cr^{\bar{m}} \cos(n\phi)\sin(\theta)
\end{aligned} \tag{4}
$$

The quantities a, b, c, m, $\bar{m}$, and n are constants and parameters which are determined in the reference [9]. Eventually, the general form displacement field of single-layer composite curved tubes of the kth plane and up to the nth order is presented as (the detailed derivation can be found in [9]):

$$
\begin{aligned}
u^{(k)}(r,\phi,\theta) &= \varepsilon^n B_n^{(k)}(r)\cos(n\phi)\cos(\theta) \\
v^{(k)}(r,\phi,\theta) &= \varepsilon^n A_n^{(k)}(r)\sin(n\phi)\cos(\theta) \\
w^{(k)}(r,\phi,\theta) &= \varepsilon^n C_n^{(k)}(r)\cos((n-1)\phi)\sin(\theta)
\end{aligned} \tag{5}
$$

where

$$
\begin{aligned}
B_n^{(k)}(r) &= a_n^{(k)} B_{1,m_n^{(k)}} r^{m_n^{(k)}} + b_n^{(k)} B_{1,m_n^{\prime(k)}} r^{m_n^{\prime(k)}} + c_n^{(k)} B_{1,-m_n^{(k)}} r^{-m_n^{(k)}} + d_n^{(k)} B_{1,-m_n^{\prime(k)}} r^{-m_n^{\prime(k)}} \\
A_n^{(k)}(r) &= -a_n^{(k)} A_{2,m_n^{(k)}} r^{m_n^{(k)}} - b_n^{(k)} A_{2,m_n^{\prime(k)}} r^{m_n^{\prime(k)}} - c_n^{(k)} A_{2,-m_n^{(k)}} r^{-m_n^{(k)}} - d_n^{(k)} A_{2,-m_n^{\prime(k)}} r^{-m_n^{\prime(k)}} \\
C_n^{(k)}(r) &= e_n^{(k)} r^{\bar{m}_n^{(k)}} + f_n^{(k)} r^{-\bar{m}_n^{(k)}}
\end{aligned} \tag{6a}
$$

$$\bar{m}_n^{(k)} = \pm n \left(\frac{\bar{C}_{66}^{(k)}}{\bar{C}_{55}^{(k)}} \right)^{\frac{1}{2}}$$

$$B_{1,m_n^{(k)}} = -\left(\bar{C}_{22}^{(k)} + \frac{1}{2}\bar{C}_{44}^{(k)} \right) + \frac{n}{(n+1)} m_n^{(k)} \left(\frac{1}{2}\bar{C}_{44}^{(k)} + \bar{C}_{12}^{(k)} \right) \tag{6b}$$

$$A_{2,m_n^{(k)}} = -\left(\bar{C}_{22}^{(k)} + \frac{1}{2}\bar{C}_{44}^{(k)} \right) - \frac{n}{(n+1)} m_n^{(k)} \left(\frac{1}{2}\bar{C}_{44}^{(k)} + \bar{C}_{12}^{(k)} \right)$$

and $m_n^{(k)}$, $-m_n^{(k)}, m_n^{'(k)}$ and $-m_n^{'(k)}$ are the 4 roots of the following equation:

$$\left(\frac{1}{2}\bar{C}_{44}^{(k)}\bar{C}_{11}^{(k)} \right) m_n^{(k)4} + \left(-n^2\bar{C}_{11}^{(k)}\bar{C}_{22}^{(k)} - \frac{1}{2}\bar{C}_{11}^{(k)}\bar{C}_{44}^{(k)} - \frac{1}{2}\bar{C}_{44}^{(k)}\bar{C}_{22}^{(k)} + n^2\bar{C}_{12}^{(k)}\bar{C}_{44}^{(k)} + n^2\bar{C}_{12}^{(k)2} \right) m_n^{(k)2} + \frac{n^4 - 2n^2 + 1}{2}\bar{C}_{22}^{(k)}\bar{C}_{44}^{(k)} = 0 \tag{6c}$$

with n, here and in what follows, being a dummy index implying summation of orders from $n = 0$ to the nth order. $\bar{C}_{ij}^{(k)}$ represents the off-axis stiffness. Also, n and k present the order number (i.e., $n = 0, 1, 2, 3, 4, \ldots$) and plane number (i.e., $k = 1, 2, \ldots, N + 1$), respectively.

2.2 *Displacement Field of Thick Laminated Composite Curved Tubes*

2.2.1 Layerwise Theory (LWT)

In LWT, the displacement components of a generic point in the laminate are assumed as:

$$\begin{aligned} U(z,\phi,\theta) &= u_k(\phi,\theta)\Phi_k(z) \\ V(z,\phi,\theta) &= v_k(\phi,\theta)\Phi_k(z) \\ W(z,\phi,\theta) &= w_k(\phi,\theta)\Phi_k(z) \ (k=1,2,\ldots,N+1) \end{aligned} \tag{7}$$

with k, here and in what follows, being a dummy index implying summation of terms from $k = 1$ to $k = N+1$. Note that z is the local direction starting from the mid-thickness of the curved tube cross section. The variable N corresponds to the total number of numerical layers within the laminated composite curved tube. In Eq. (7), U, V, and W represent the total displacement components in the z, ϕ, and θ directions, respectively. Moreover, $u_k(\phi, \theta)$, $v_k(\phi, \theta)$, and $w_k(\phi, \theta)$ represent the displacements of the points initially located at the kth plane within the laminated composite curved tube in the z, ϕ, and θ directions, respectively. $\Phi_k(z)$ is the global Lagrangian interpolation function associated with the kth plane. The procedure for solving a specific problem can now be outlined. The displacement components of single-layer composite curved tubes are taken based on Eq. (5). By applying layerwise method, Eq. (7), the displacement components of thick laminated composite curved tubes are obtained. Then, by employing the displacement components and their corresponding strains in the principle of minimum total potential energy and using the fundamental lemma of calculus of variations, the equilibrium equations of composite curved tubes under pure bending moment are derived. By applying LWT displacement field (Eq. (7)) on general displacement field of single-layer composite curved tubes in Eq. (5), the elasticity displacement field of thick laminated composite curved tubes is rewritten up to the nth order in the following form as:

$$
\begin{aligned}
U(z,\phi,\theta) &= \varepsilon^n \begin{pmatrix} a_n B_{1,m_n^{(k)}} (R_1+z)^{m_n^{(k)}} + b_n B_{1,m_n^{'(k)}} (R_1+z)^{m_n^{'(k)}} \\ + c_n B_{1,-m_n^{(k)}} (R_1+z)^{-m_n^{(k)}} + d_n B_{1,-m_n^{'(k)}} (R_1+z)^{-m_n^{'(k)}} \end{pmatrix} \cos(n\phi)\cos(\theta) + U_k(\phi)\Phi_k(z) \\
V(z,\phi,\theta) &= -\varepsilon^n \begin{pmatrix} a_n A_{2,m_n^{(k)}} (R_1+z)^{m_n^{(k)}} + b_n A_{2,m_n^{'(k)}} (R_1+z)^{m_n^{'(k)}} \\ + c_n A_{2,-m_n^{(k)}} (R_1+z)^{-m_n^{(k)}} + d_n A_{2,-m_n^{'(k)}} (R_1+z)^{-m_n^{'(k)}} \end{pmatrix} \sin(n\phi)\cos(\theta) + V_k(\phi)\Phi_k(z) \\
W(z,\phi,\theta) &= \varepsilon^n \left(e_n (R_1+z)^{\bar{m}_n^{(k)}} + f_n (R_1+z)^{-\bar{m}_n^{(k)}} \right) \cos((n-1)\phi)\sin(\theta) + W_k(\phi)\Phi_k(z)
\end{aligned}
\tag{8}
$$

In order to satisfy the interfacial continuities of the displacement components, it is necessary that the constants appearing in Eq. (8) (i.e., a_n, b_n, c_n, d_n, e_n, and f_n) to be the same for all layers. The equilibrium equations of a thick composite curved tube with N numerical layers are obtained by employing the strain–displacement relations in the principle of minimum total potential energy [12], carrying out necessary integrations and employing the fundamental lemma of calculus of variations. The results are, in general, $3(N+1)$ local equilibrium equations corresponding to $3(N+1)$ unknown functions U_k, V_k, and W_k and, in general, six global equilibrium equations for every order ($n = 0, 1, 2, 3, \ldots$) up to the nth order associated with the six parameters a_n, b_n, c_n, d_n, e_n, and f_n of the same order. The equilibrium equations of thick laminated composite curved tubes under bending moments up to the nth order are obtained as:

$$
\begin{aligned}
&\delta U_k: N_z^k + M_\phi^k + M_\theta^k \cos\phi - \frac{dR_{z\phi}^k}{d\phi} = 0 \\
&\delta V_k: Q_\phi^k - R_{z\phi}^k + M_\theta^k \sin\phi \frac{dM_\phi^k}{d\phi} = 0 \\
&\delta W_k: Q_\theta^k - R_\theta^k \cos\phi + R_{\phi\theta}^k \sin\phi - \frac{dM_{\phi\theta}^k}{d\phi} = \frac{4}{\pi}\frac{M}{R_1+z}\Phi_k \quad (k = 1, 2, \ldots, N+1)
\end{aligned}
\tag{9}
$$

$$\delta a_n: \int_{-\pi/4}^{\pi/4}\int_{-h/2}^{+h/2}\int_{-\pi}^{\pi}(\Theta 1)\,\mathrm{d}\phi \mathrm{d}z \mathrm{d}\theta = 0 \text{ for } n = 0,\ 1,\ 2,\ 3,\ \ldots \tag{10a}$$

$$\delta b_n: \int_{-\pi/4}^{\pi/4}\int_{-h/2}^{+h/2}\int_{-\pi}^{\pi}(\Theta 2)\,\mathrm{d}\phi \mathrm{d}z \mathrm{d}\theta = 0 \text{ for } n = 1, 2, 3, \ldots \tag{10b}$$

$$\delta c_n: \int_{-\pi/4}^{\pi/4}\int_{-h/2}^{+h/2}\int_{-\pi}^{\pi}(\Theta 3)\,\mathrm{d}\phi \mathrm{d}z \mathrm{d}\theta = 0 \text{ for } n = 1, 2, 3, \ldots \tag{10c}$$

$$\delta d_n: \int_{-\pi/4}^{\pi/4}\int_{-h/2}^{+h/2}\int_{-\pi}^{\pi}(\Theta 4)\,\mathrm{d}\phi \mathrm{d}z \mathrm{d}\theta = 0 \text{ for } n = 2, 3, \ldots \tag{10d}$$

$$\delta e_n: \int_{-\pi/4}^{\pi/4}\int_{-h/2}^{+h/2}\int_{-\pi}^{\pi}(\Theta 5)\,\mathrm{d}\phi \mathrm{d}z \mathrm{d}\theta = -\sqrt{2}\varepsilon^n M (R_1 + z)^{\bar{m}_n^{(k)} - 1}\cos((n-1)\phi) \quad \text{for } n = 1, 2, 3, \ldots \tag{10e}$$

$$\delta f_n: \int_{-\pi/4}^{\pi/4}\int_{-h/2}^{+h/2}\int_{-\pi}^{\pi}(\Theta 6)\,\mathrm{d}\phi \mathrm{d}z \mathrm{d}\theta = -\sqrt{2}\varepsilon^n M (R_1 + z)^{-\bar{m}_n^{(k)} - 1}\cos((n-1)\phi) \quad \text{for } n = 1, 2, 3, \ldots \tag{10f}$$

where the functions $\Theta 1$, $\Theta 2$, $\Theta 3$, $\Theta 4$, $\Theta 5$ and, $\Theta 6$ in Eqs. (10a–10f) are defined in [10]. Note that the numbers of the global equilibrium equations, Eqs. (10a–10f), depending on the numbers of unknown constants are employed in Eq. (8) to develop the displacement field of composite curved tubes. Based on the detailed derivation in [9], there are one unknown constant for the zeroth order (i.e., a_0), five unknown constants for the 1st order (i.e., a_1, b_1, c_1, e_1, and f_1), and six unknown constants for the nth order (n = 2, 3, ...) (i.e., a_n, b_n, c_n, d_n, e_n and f_n). The following boundary conditions must be satisfied on the free inner and outer curved surfaces:

$$R_\theta^k = Q_x^k = N_z^k = 0 \text{ at } (z = \pm h/2)\ (k = 1, 2, \ \ldots, \ N+1) \tag{11}$$

Note that in Eq. (11) the superscript k refers to the kth interface in the laminated composite curved tube. By substituting the strain–displacement relations in three-dimensional constitutive law and the subsequent results in Eq. (9), the stress resultants based on displacement components are obtained which are presented in [10]. The local displacement-based equilibrium equations are obtained by substituting Eq. (9), the stress resultants based on displacement components, in Eq. (9). In addition, the global equilibrium equations of composite curved tubes are expressed in terms of displacement functions by substituting the strain-displacement relations in three-dimensional constitutive law and the subsequent results in Eqs. (10a–10f).

3 Analytical Solution

The system of local displacement equilibrium equations (Eq. (9)) shows *3(N + 1)* coupled ordinary differential equations with constant coefficients which may be displayed in a matrix form as:

$$[M]\{\eta''\}+[K_1]\{\eta'\}+[K_2]\{\eta\}=\{F\} \tag{12}$$

where

$$\begin{aligned}&\{\eta\}=\left\{\{U\}^T,\{V\}^T,\{W\}^T\right\}^T\\&\{U\}=\{U_1,U_2,\ldots,U_{N+1}\}^T,\ \{V\}=\{V_1,V_2,\ldots,V_{N+1}\}^T,\ \{W\}=\{W_1,W_2,\ldots,W_{N+1}\}^T\end{aligned} \tag{13}$$

The coefficient matrices [M], [K_1], [K_2], and {F} in Eq. (12) are defined in [10]. It is confirmed that the general solution of Eq. (12) is presented as [13]:

$$\begin{aligned}\{\eta\}&=[X_0][\exp(\phi T_0)]\{k_1\}+\\&[X_0]\int_0^{\phi}\left(\begin{array}{c}[\exp(T_0(\phi-s))][X_0]^{-1}\left[\exp\left(s\left([M]^{-1}[K_1]+[X_0][T_0][X_0]^{-1}\right)\right)\right]\\ \left(\{k_1\}+\int_0^{s}\left(\left[\exp\left(u\left([M]^{-1}[K_1]+[X_0][T_0][X_0]^{-1}\right)\right)\right][M]^{-1}\{F(u)\}\right)du\right)\end{array}\right)ds\end{aligned} \tag{14}$$

where

$$[T_0]_{(N+1)\times(N+1)}=diag(T_1,T_2,\ldots,T_{N+1}) \tag{15}$$

And T_i *(i = 1, 2,..., N + 1)* are Jordan blocks associated to the eigenvalues of matrix [Q] where

$$[Q]_{6(N+1)\times 6(N+1)}=\begin{bmatrix}[0]_{3(N+1)\times 3(N+1)} & [I]_{3(N+1)\times 3(N+1)}\\ -[M]^{-1}[K_2] & -[M]^{-1}[K_1]\end{bmatrix} \tag{16}$$

In addition,

$$[X_0]_{3(N+1)\times 3(N+1)} = [X_1, X_2, \ldots, X_{N+1}] \tag{17}$$

where X_i $(i = 1, 2, \ldots, N + 1)$ are co-solutions of $[X_i][T_i]^2 + [M]^{-1}[K_1][X_i][T_i] + [M]^{-1}[K_2][X_i] = [0]$.

Vector $\{k_1\}$ in Eq. (14) is unknown vector representing $3(N + 1)$ integration constants. The constants a_n, b_n, c_n, d_n, e_n, and f_n must be calculated within the analysis. The boundary conditions in Eq. (11) are first imposed to calculate the vector $\{k_1\}$ in terms of the unknown constants a_n, b_n, c_n, d_n, e_n, and f_n. These constants are then obtained by the satisfaction of the global equilibrium conditions in Eqs. (10a–10f).

4 Results and Discussion

All physical plies are assumed to have equal thickness (= *0.1* mm) and are modeled as being made up of *p* numerical layers. In all the subsequent calculations, *p* is set equal to 12 [14, 15]. The mechanical properties of the composite curved tube are given in Table 1 [16]. The stress components are normalized as $\bar{\sigma}_{ij} = \sigma_{ij}/\sigma_0$, where $\sigma_0 = (M \cdot r)/(\pi/64^*(OD^4 - ID^4))$, where the outer diameter and the inner diameter of the composite curved tube are presented as *OD* and *ID*, respectively. In addition, a thick laminated composite curved tube spanning a curved segment of 90° with $R/a = 10$ is considered (*see Fig. 1*). In the present cases, the composite curved tube sections with the $[90°_{20}/0°_{20}]$ lay-up sequence have an internal radius of *28* mm and an external radius of *32* mm, i.e., a wall thickness of *4* mm. The results are presented based on the proposed method at θ = 0°. Since the radial stress affects delamination and the hoop stress affects buckling, interlaminar radial and hoop stress distributions are investigated in this section.

Here, the results obtained for a laminated composite curved tube using the developed method are compared with results obtained using FEM (ANSYS). To ensure the verifications are done in all layers of the laminated composite curved tube, the comparison must be performed by using the developed method and FEM (ANSYS) along tube thickness. Figure 2 presents the comparison of the

Table 1 Mechanical properties of the composite curved tube

Properties	E_1 (GPa)	$E_2 = E_3$ (GPa)	$G_{12} = G_{13} = G_{23}$ (GPa)	$\upsilon_{12} = \upsilon_{13}$	υ_{23}
Carbon AS4/PEKK	140	10	5.56	0.31	0.33

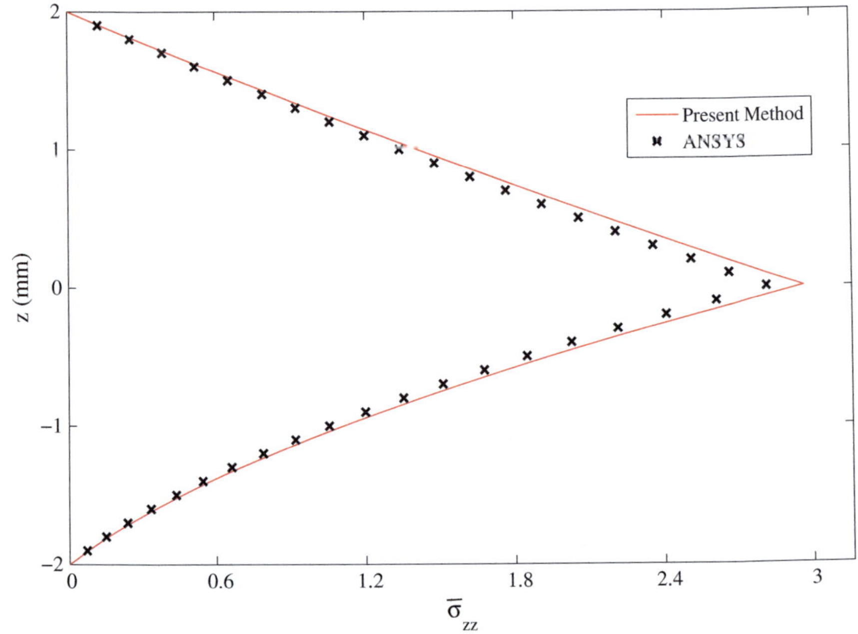

Fig. 2 Comparison of the interlaminar radial stress, $\bar{\sigma}_{zz}$, obtained using the present method and ANSYS of the $[90°_{20}/0°_{20}]$ laminated curved tubes at $\phi = 270°$ along tube thickness

interlaminar radial stress, $\bar{\sigma}_{zz}$, of the $[90°_{20}/0°_{20}]$ laminated composite curved tube at $\phi = 270°$ along the tube thickness based on the present method and ANSYS. There is a good agreement seen along tube thickness between the theoretical analysis and FEM results. Comparison of the hoop stress, $\bar{\sigma}_{\phi\phi}$, of the thick $[90°_{20}/0°_{20}]$ laminated composite curved tube at $\phi = 90°$ along tube thickness using the developed method and FEM is shown in Fig. 3. FEM results make a good agreement with the results obtained by the present method along the tube thickness. Note that the 0° layer group experiences a compressive stress, while the hoop stress of the 90° layer group (in the inner surface of the curved tube) is tensile.

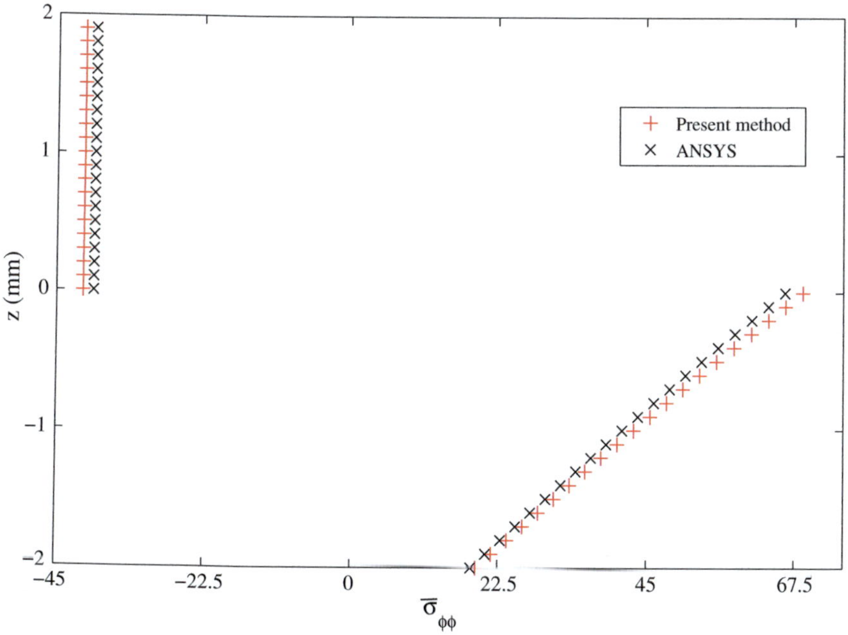

Fig. 3 Comparison of the hoop stresses, $\bar{\sigma}_{\phi\phi}$, obtained using the present method and ANSYS of the $[90°_{20}/0°_{20}]$ laminated curved tube at $\phi = 90°$ along tube thickness

5 Conclusion

The displacement-based approaches were used to study stresses in thick laminated composite curved tubes subjected to pure bending moment. The most general form of the displacement field of thick laminated composite curved tubes was derived using toroidal elasticity (TE) and layerwise method. The accuracy of the results was examined by comparing the FEM results with those of the proposed method. The key advantage of the developed method is its suitability for parametric study, enabling simple inputs and being fast to run. The method we propose in this paper does not require meshing. It simplifies greatly inputs that the user has to do, once the program for solution is available. This presents a clear advantage over FEM.

Acknowledgements The financial contributions from the Natural Sciences and Engineering Research Council of Canada (NSERC), Bell Helicopter Textron Canada Ltd., and Concordia University are appreciated.

References

1. Ting TCT (1999) New solution to pressuring, shearing, torsion and extension of cylindrically anisotropic elastic circular tube or bar. Proc R Soc Lond A 455:3527–3542
2. Chen T, Chung CT, Lin WL (2000) A revisit of a cylindrically anisotropic tube subjected to pressuring, shearing, torsion and extension and a uniform temperature change. Int J Solids Struct 37:5143–5159
3. Yu AM, Nie GH (2005) Explicit solutions for shearing and radial stresses in curved beams. Mech Res Commun 32(3):323–331
4. Dryden J (2007) Bending of inhomogeneous curved bars. Int J Solids Struct 44(11–12): 4158–4166
5. Piovan MT, Domini S, Ramirez JM (2012) In-plane and out-of-plane dynamics and buckling of functionally graded circular curved beams. Compos Struct 94(11):3194–3206
6. Hajianmaleki M, Qatu MS (2012) Static and vibration analyses of thick, generally laminated deep curved beams with different boundary conditions. Compos B Eng 43(4):1767–1775
7. Wang M, Liu Y (2013) Elasticity solutions for orthotropic functionally graded curved beams. Eur J Mech A Solids 37:8–16
8. Aşık MZ, Dural E, Yetmez M, Uzhan T (2014) A mathematical model for the behavior of laminated uniformly curved glass beams. Compos B Eng 58:593–604
9. Yazdani Sarvestani H, Hoa SV, Hojjati M (2016) Three-dimensional stress analysis of orthotropic curved tubes-part 1: single-layer solution. Eur J Mech A/Solids
10. Yazdani Sarvestani H, Hojjati M (2016) Three-dimensional stress analysis of orthotropic curved tubes-part 2: laminate solution. Eur J Mech A/Solids
11. Zhu Y, Redekop D (1994) An out-of-plane displacement solution in toroidal elasticity. Int J Press Vessel Pip 58:309–319
12. Fung YC, Tong P (2001) Classical and computational solid mechanics. World Scientific, New Jersey
13. Jódar L, Navarro E (1992) Solving coupled systems of linear second-order differential equations knowing a part of the spectrum of the companion matrix. J Comput Appl Math 39 (1):115–119
14. Yazdani Sarvestani H, Hoa SV, Hojjati M (2016) Stress analysis of thick orthotropic cantilever tubes under transverse loading. Adv Compos Mater 1:28
15. Yazdani Sarvestani H, Hoa SV, Hojjati M (2016) Effects of shear loading on stress distributions at sections in thick composite tubes. Compos Struct 140:433–445
16. Derisi B (2008) Development of thermoplastic composite tubes for large deformation. In: PhD thesis, Concordia University, Montreal, Canada

Cruise Missile Mission Planning Using Genetic Algorithm

G. Naresh Kumar, Shweta Dadarya, Akshay Verandani, A.K. Sarkar and S.E. Talole

Nomenclature

α	Angle of attack
β	Side-slip angle
Γ_i	Threat intensity
τ_{a_n}	Time constant of normal accel^n
τ_{Φ}	Time constant of roll
Φ	Roll angle
Φ_c	Roll angle command
Ψ	Yaw angle
Θ	Pitch angle

G. Naresh Kumar (✉) · S. Dadarya · A.K. Sarkar
Defence Research and Development Laboratory, Hyderabad, India
e-mail: gaddam.nareshkumar@gmail.com

A. Verandani
National Institute of Technology, Goa, India

S.E. Talole
Defence Institute of Advanced Technology, Pune, India

S. Singh et al. (eds.), *Proceedings of the International Conference on Modern Research in Aerospace Engineering*, Lecture Notes in Mechanical Engineering,
https://doi.org/10.1007/978-981-10-5849-3_34

a_n	Normal acceleration
a_{nc}	Normal acceleration command
a_x, a_y, a_z	Accelerations in earth frame
r_i	Radius of *i*th way-point
E_i	Expected fitness of a trajectory
F_i	Trajectory fitness
g	Acceleration due to gravity
Z_{nom}	Vehicle altitude
J_i	Trajectory risk
k	Bit position
l	String length
p_c	Probability of crossover
p_m	Probability of mutation
s	Distance along the trajectory
t	Time
Δt	Time interval
t_i	Initial time of segment
t_f	Final time of segment
$T(x, y)$	Threat function
V	Velocity vector
V_{nom}	Nominal velocity
v_x, v_y, v_z	Velocities in earth frame

x, y, z	Positions in earth frame
X_A, Y_A, Z_A	Vehicle frame axes
X_E, Y_E, Z_E	Earth frame axes

1 Introduction

Mission planning system (MPS) is a primary system to execute mission of cruise vehicles. The typical functions of MPS are navigation, guidance, situation assessment, and online path and tactical planning. The need for mission planning emerged in very early days and more focus is generated after development of unmanned air vehicles like missiles and UAVs. To this end, vast literature exists on both non-real-time (just before launch) and real-time (after launch) algorithms for mission planning. It got evolved initially based on heuristics approach [1] and transformed into optimal control problem and further in evolutionary algorithms. A three-dimensional A* mission planning algorithm [2], an hybrid approach based on geometrical route and graph-based planning algorithms [3], a multidisciplinary approach [4], a hybrid optimal control theory-based formulation [5, 6], evolutionary flight path planning algorithm [7], a two-layer controller design [8], mission planning and dynamic reassignment algorithms [9], a three-level mission planner [10], quantum immune clone algorithm [11], a multi-agent system [12] are some references to cite that dealt with the mission planning approaches.

The purpose of the present research is to investigate how a genetic algorithm can be implemented in the MPS as a search procedure to achieve better performance and to make it more realizable. Genetic algorithms are well suited to the control commands optimization problem because of robustness and global convergence. This study addresses the implementation of a genetic algorithm in the MPS to achieve robust performance characteristics and computational efficiency. In current scenario, the CV navigation is performed with the basic sensors like INS, GPS, radio altimeter (RA), and air data system (ADS). The CV is designed to navigate according to the planned flight plan. The guidance parameters are computed by the MPS using basic sensor parameters [13].

The remaining paper is organized as follows. The problem formulation and flight trajectory planning using GA are explained in Sect. 2. Flight control system and vehicle dynamical equations as well as objective function in terms of minimizing threat and risk functions are described in Sect. 3. Fitness function and search space definition are presented in Sect. 4. The results are presented in Sect. 5, and Sect. 6 concludes this work.

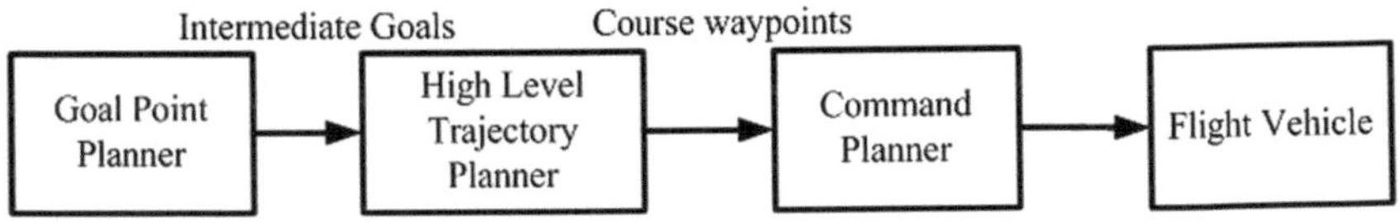

Fig. 1 Structure of the mission planning system

2 Problem Formulation

Flight vehicle mission planning system is designed to formulate optimal strategies to accomplish specific mission objectives. Detailed database of both objective and threat information is the main input required at battle field. Figure 1 indicates the structure of MPS, and it consists of mainly three parts as enumerated below:

1. Goal-point Planner (GP): GP is the highest level of the MPS. Its function is to generate a sequence of intermediate goals and geographic locations along with associated time and energy constraints to accomplish given mission objectives.
2. High-level Trajectory Planner (HLTP): Using knowledge of operational facilities and major threat concentrations (can be man-made or natural), the HLTP identifies a maximum survivability flight path such that the intermediate goals, associated time, and energy constraints are met while avoiding major threat concentrations. The flight path is in general characterized by way-points of a nominal separation determined by the vehicle operating condition, such as velocity, altitude, and turning radius.
3. Command Planner (CP): The purpose of the CP is to generate flight control commands such that the trajectory will minimize the risk associated with the low-level threat database while satisfying the way-point restrictions. Since the nonlinear dynamics of the flight vehicle and its control system are embedded in the evaluation of the risk function, it provides a platform on which the flight vehicle limitations are recognized from the outset at the expense of having to optimize complex, constrained, and nonlinear equations.

The mission objectives, i.e., the way-point, threat information, and time constraints are available to the HLTP before the launch. In a real-time scenario, it is required that once the mission objectives are available, the HLTP converges to an optimum trajectory as quickly as possible. Based on the solution, the normal acceleration and roll angle commands are fed to the command planner before the liftoff.

3 Flight Control System and Vehicle Dynamics

Flight vehicle dynamics and control system of a typical cruise vehicle are considered here for modeling and simulation. The flight environment is chosen based on the characteristics of a cruise vehicle operating at low level in a terrain following (TF)/terrain avoidance (TA) mode [14] which are:

1. Nominal velocity: $V_{nom} = 250$ m/s
2. Nominal altitude: $z_{nom} = -200$ m
3. Maximum normal acceleration: $|a_{nc}|_{max} = 40\text{m/s}^2$

With the cruise vehicle characteristics defined as above, the minimum turning radius is approximately 1500 m (since, $R_{min} = (V_{nom}^2)/|a_{nc}|_{max}$). The flight control inputs are taken to be normal acceleration a_{nc} and roll angle Φ. The flight control system (FCS) is modeled as a first-order lag between each of the command variables and the corresponding physical quantities as follows:

$$a_n = \frac{a_{nc}}{1 + s\tau_{a_n}} \qquad \text{and} \qquad \Phi = \frac{\Phi_c}{1 + s\tau_{\Phi}} \tag{1}$$

where τ_{a_n} and τ_{Φ} are the time constants associated with the normal acceleration and roll angle control channels of the FCS, respectively. It is justified in [15], the use of Eq. 1 by showing that the errors resulting from these simple models were of the same order as those due to air turbulence and could, therefore, be ignored in the presence of velocity and position feedback.

The standard body axis system and inertial axis system (Earth reference frame) are used in this work. For the purpose of this research, the angle of attack α and side-slip angle β are neglected so that the velocity vector V of the cruise vehicle coincides with the X_A-axis in the cruise vehicle body axis system. The Z_E-axis of the inertial axis system points vertically downward, while the X_E- and Y_E-axes lie in the horizontal plane. The acceleration of the cruise vehicle in the inertial axis system referred to the normal acceleration in the cruise vehicle body axis system is given in Eq. 2. Level flight at a nominal altitude without deviation can be obtained by reducing the trajectory dimensionality as in Eq. 6. The maximum roll angle is constrained to be $\pm\Phi \leq 75^o$.

$$\begin{bmatrix} a_x \\ a_y \\ a_z \end{bmatrix} = a_n \begin{bmatrix} \cos\Phi\sin\Theta\cos\Psi + \sin\Phi\sin\Psi \\ \cos\Phi\sin\Theta\sin\Psi - \sin\Phi\cos\Psi \\ \cos\Phi\cos\Theta \end{bmatrix} + \begin{bmatrix} 0 \\ 0 \\ g \end{bmatrix} \tag{2}$$

$$\left[\dot{x}, \dot{y}, \dot{z}\right] = \left[v_x, v_y, v_z\right] \tag{3}$$

$$\dot{a}_n = \frac{a_{nc} - a_n}{\tau_{a_n}} \quad \text{where,} \quad a_n = \frac{-g}{\cos\Phi} \tag{4}$$

$$\dot{\Phi} = \frac{\Phi_c - \Phi}{\tau_{\Phi}} \tag{5}$$

$$\left[z(0), v_z, \dot{v}_z\right] = \left[Z_{nom}, 0, 0\right] \tag{6}$$

$$\text{where,} \quad \Psi = \tan^{-1}\left[\frac{v_y}{v_x}\right] \quad \text{and,} \quad \Theta = \sin^{-1}\left[\frac{-v_z}{\sqrt{v_x^2 + v_y^2 + v_z^2}}\right] \tag{7}$$

3.1 Threat Function and Trajectory Risk

The threat function $T(x, y)$ is a measure of the threat to which the cruise vehicle is exposed when its position is (x, y). The nature of the threat function is defined by the actual threat that it represents. A radar installation can be shielded by a mountain and will result in a threat that is a function of the distance as well as direction of the cruise vehicle from the threat. The threat data may also be available as tabulated numerical values in the MPS. The following analytical form of the threat function is adopted [16]:

$$T(x, y) = \sum_{i=1}^{n} \frac{\Gamma_i}{\sqrt{(x - x_i)^2 + (y - y_i)^2}} > 0 \tag{8}$$

where (x_i, y_i) and Γ_i are the position and intensity of the ith threat respectively. The risk J associated with an arbitrary trajectory ζ is defined as the integral of the threat function with respect to distance along the trajectory:

$$J = \int_{\zeta} T(x, y)ds \tag{9}$$

The optimal trajectory has the property that it minimizes the associated risk subject to the constraints. For a particular trajectory segment over the interval $[t_i, t_f]$, the risk can be written as:

$$J = \int_{t_i}^{t_f} T[x(t), y(t)]\frac{ds}{dt}dt \tag{10}$$

where, $$\frac{ds}{dt} = \sqrt{\left(\frac{dx}{dt}\right)^2 + \left(\frac{dy}{dt}\right)^2} = V_{nom} \tag{11}$$

It is assumed that the cruise vehicle is flying at constant speed. Hence, Eq. 10 can be simplified to:

$$J = V_{nom} \int_{t_i}^{t_f} T[x(t), y(t)]dt \tag{12}$$

The risk is used in the genetic algorithm to evaluate and compare the fitness of different population members.

4 Fitness Function and Search Space

For the control commands optimization problem, the flight trajectories are evaluated according to the risk associated with each trajectory where the risk is defined by Eq. 12. The following risk-to-fitness transformation is used for maximizing fitness function.

$$F_i = \frac{1}{1 + J_i} \tag{13}$$

Each way-point has an associated nonzero capture radius which is an indication of how close to the way-point the flight trajectory must pass. This constraint would seem to pose no problem, as trajectories which pass outside of these capture radii can be assigned zero fitness and will not be selected to the mating pool. However, trajectories which pass very close outside the capture radii usually contain valuable genetic information that will be lost if this method is employed. To preserve the genetic information in the population, a penalty method is used which degrades the fitness in relation to the degree of constraint violation. The penalty function transforms the constrained problem to an unconstrained problem by associating a penalty with all constraint violations which is included in the objective function (fitness) evaluation. The degree of constraint violation for the ith way-point is defined to be:

$$h_i = \frac{r_i}{d_i} \tag{14}$$

where d_i is the closest distance between trajectory and ith way-point and r_i is capture radius of ith way-point. The penalty function is included in the objective function defined in Eq. 13 as follows:

$$F_i = \frac{1}{1 + (J_i + \Omega \sum_{i=1}^{n} P(h_i))} \tag{15}$$

where P is penalty function and Ω is penalty coefficient. A number of alternatives exist for the penalty function P. In the research done by [1], the penalty function is defined as the square of the degree of constraint violation h_i as follows:

$$P(h_i) = \begin{cases} h_i^2, & h_i > 1 \\ 0, & \text{otherwise} \end{cases} \tag{16}$$

The optimal solution at any time is, therefore, the population member that is within bounds and with the highest fitness. This optimal solution is kept in memory and updated each time that a solution within constraints and a higher fitness is generated. This ensures that there are flight control commands available to the FCS for the next trajectory segment at any time after the first legitimate solution has been generated (Table 1).

5 Simulation Studies

Different way-point and threat distributions were taken to represent a set of diverse realistic scenarios as tabulated in Table 2. GA code developed on C-language environment was downloaded from IIT, Kanpur Web site [17–19] for current use.

Table 1 GA control variables

Sl. no.	Variable	Value
1	Crossover probability p_c	0.7
2	Mutation probability p_m	0.02
3	Population size	250
4	No. of generations	250

Table 2 Various test scenarios

Case no.	Way-points (x[m], y[m]), r[m]	Threats (x[m], y[m])	$\|a_{nc}\|_{max}[m/s^2]$
1	(4000, 1000), 200	(2000, 3000)	40
	(5000, 4800), 200	(4000, 3000)	
2	(4000, 1000), 200	(2000, 3000)	30
	(5000, 4800), 200	(4000, 3000)	
3	(4000, 1000), 200	(2000, 3000)	50
	(5000, 4800), 200	(4000, 3000)	
4	(5000, 0), 200	(2000, 2500)	40
	(8000, 1500), 200	(2000, 2500)	
		(6500, 1500)	
5	(3000, 2750), 250	(2000, 4000)	40
	(7000, 2100), 700	(6000, 4500)	

The same can be used for real-time applications in future. The results obtained for different test segments are shown in Fig. 2. Solid circles represent the way-points and × symbols indicate the threats/hurdles. Figure 2a and b indicates the final optimum normal acceleration and roll angle after 250 generations. Every generation executes 250 trajectories (i.e., population size = 250), and best solutions are propagated to next generation. Figure 2c shows the optimum trajectory ensuring all constraints for case 1, 4, and 5 ($|a_{nc}|_{max} = 40\,\mathrm{m/s^2}$). The performance of last generation (i.e., 250th generation) is presented in Fig. 2e, i.e., all 250 trajectories. Figure 2d indicates the final optimum trajectory for case 2–3 with different lateral acceleration capabilities. It is clear from the trajectories depicted in Fig. 2d that case 3 trajectories are taking sharp turn behind 5 km down range point compared to case 2 due to higher lateral acceleration. Best fitness of each generation and average fitness are indicated in Fig. 2f typically for case 1.

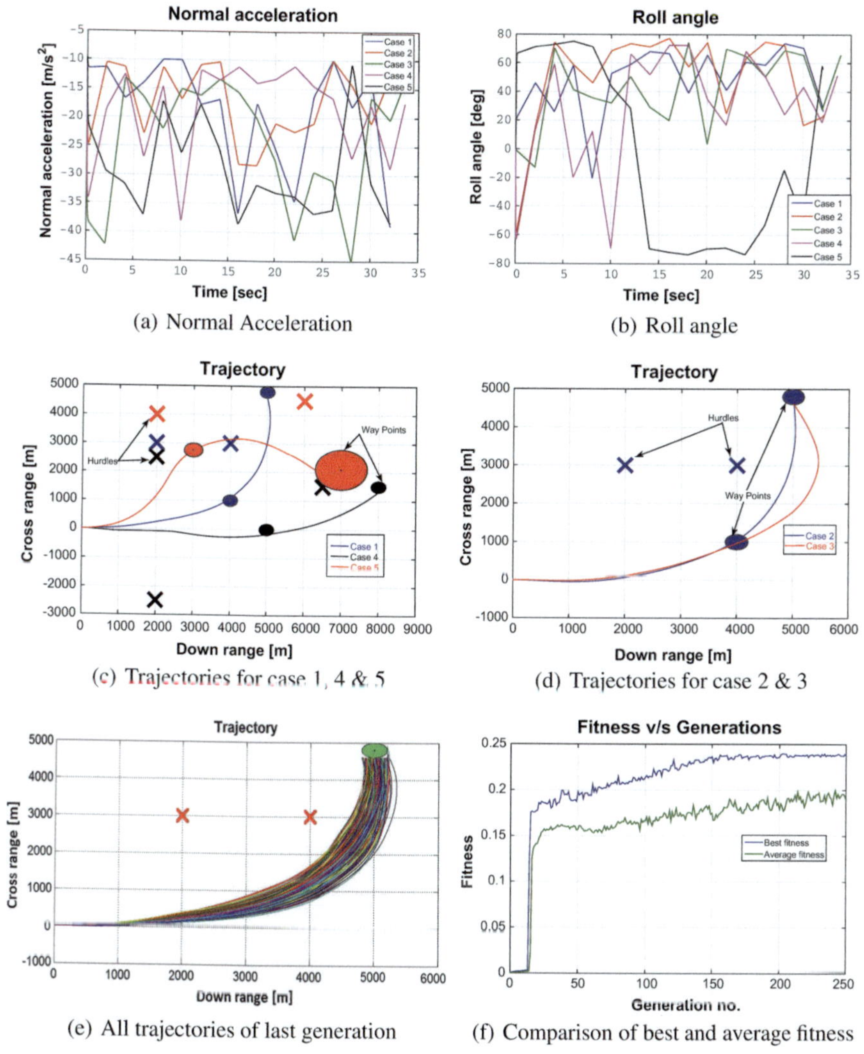

(a) Normal Acceleration

(b) Roll angle

(c) Trajectories for case 1, 4 & 5

(d) Trajectories for case 2 & 3

(e) All trajectories of last generation

(f) Comparison of best and average fitness

Fig. 2 Results

6 Conclusion

An attempt is made to generate an optimum flight path at critical stage of combat scenario against highly defended targets. Mission planning software was developed and optimum flight path generated using genetic algorithm. C-language-based GA code used is here for easy implementation on real-time environment. In addition to air defense, terrain features are also considered as constraint to ensure collision free trajectory. The flight control commands and the flight path are, respectively, constrained by the cruise vehicle dynamics and trajectory objectives. Results are quite satisfactory in terms of specified way-points, threats, and impact accuracy.

Acknowledgements The authors thank competent authorities of Defence Research and Development Laboratory (DRDL), Hyderabad, India for granting permission to publish this piece of work.

References

1. Helgason RV, Kennington JL, Lewis KR (2001) Cruise missile mission planning: a heuristic algorithm for automatic path generation. J Heuristics 7:473–494
2. Dicheva Svetlana, Bestaoui Yasmina (2014) Three-dimensional A* dynamic mission planning for an airborne launch vehicle. J Aerosp Info Syst 11:98–105
3. Szczerba RJ, Galkowski P, Glickstein IS, Ternullo N (2000) Robust algorithm for real-time route planning. IEEE Trans Aerosp Electron Syst 36:869–878
4. McManus IA, Walker RA (2006) Multidisciplinary approach to intelligent unmanned-airborne-vehicles mission planning. J Aircr 43:318–335
5. Ross IM, D'Souza CN (2005) Hybrid optimal control framework for mission planning. J Guidance, Control, Dyn 28:686–697
6. Englander JA, Conway BA, Williams T (2012) Automated mission planning via evolutionary algorithms. J Guidance, Control, Dyn 35:1878–1887
7. Vaidyanathan R, Hocaoglu C, Prince TS, Quinn RD (2002) Evolutionary path planning for autonomous air vehicles using multiresolution path representation. IEEE International Conference on Intelligent Robots and Systems (IROS)
8. Yadav V, Wang X, Balakrishnan S (2006) Neural network approach for obstacle avoidance in 3-D environments for UAVs. Proceedings of the 2006 American control conference, pp 3667–3672
9. Myers D, Batta R, Karwan M (2016) A real-time network approach for including obstacles and flight dynamics in uav route planning. J Defense Model Simul: Appl Method Technol
10. Vachtsevanos G, Tang L, Drozeski G, Gutierrez L (2005) From mission planning to flight control of unmanned aerial vehicles: strategies and implementation tools. Ann Rev Control 29:101–115
11. Xian Y, He B, Liu G, Lei G (2012) Cruise missile route planning based on quantum immune clone algorithm. J Info Comput Sci 9:2097–2105
12. Giovanini L, Baldeurd J, Katebi R (2007) Autonomous and decentralized mission planning for clusters of UUVs. Intern J Control 80:1169–1179
13. Yanushevsky R (2011) Guidance of unmanned aerial vehicles. CRC Press, Taylor & Francis Group
14. Lee D, Periaux J, Gonzalez LF (2010) UAS mission path planning system (MPPS) using hybrid-game coupled to multi-objective optimizer. J Dyn Syst, Meas Control 132
15. El-Dirani T (1991) Fidelity of flight control systems in a real-time optimal trajectory planner. Technical Report Master's Thesis, Massachusetts Institute of Technology
16. van Deventer PG (1993) Real-time trajectory optimization. Technical Report Master's Thesis, Massachusetts Institute of Technology
17. Deb K (2014) Kanpur genetic algorithms laboratory. Technical Report. https://www.iitk.ac.in/kangal/index.html
18. Deb K (2004) Optimization for engineering design. Algorithms and examples. Prentice-Hall of India Pvt, Ltd
19. Samanta S (2014) Genetic algorithm: an approach for optimization (Using MATLAB). Intern J Latest Trends Eng Technol 3:261–267

Design of Power Efficient SRAM on FPGA

Tarun Agrawal

1 Introduction

Now small and lightweight electronic gadgets, including cellular phones, digital watches, processors, incorporate ASICs along with digital processors and/or memories such as CAM, SRAM and DRAM [1]. It takes a long time to design such application-specific integrated circuits. In VLSI industry for design such IC's the methodologies, includes standard cells, gate arrays, PLDs and CPLDs. Now most of the VLSI industries use FPGA for testing and analysis circuit [2]. As the density of active/passive component increases, placement and routing for FPGAs require more attention to achieve successful design. Once the placement in FPGA is fixed, the time spent on routing in FPGA may be enormous. Interconnect estimation in FPGA at the time of design stage is the process of predicting the routing-resource requirement [3]. The design stage at which the estimation is completed, it determines the reliability and accuracy that is needed at the time of the estimation time interval. In FPGA placement, design stage wants the most reliable estimation just before routing. For designing SRAM, a new technique is proposed [4], memristor-based nonvolatile SRAM cell, that uses a combination of memristor and MOSFET. For designing single cell, it uses three MOS transistors and two memristors. Here read is very faster and also very less power consumption. Apart from SRAM and DRAM another memory is proposed in [5], Time-based access memory (TAM), totally different from SRAM and DRAM. It doesn't make use of decoder, so it is simple and fast. It makes all of its available data to the processor in a system, so that processor doesn't need to access the shared memory in LIFO or FIFO serial fashion. The DDV increases as their size reduces [6]. Static RAM is especially vulnerable to device-to-device variability, since SRAM has a higher component packing density.

T. Agrawal (✉)
Department of Electronics and Communication, G.L.A. University, Mathura, India
e-mail: tarun.ag.1506@gmail.com

S. Singh et al. (eds.), *Proceedings of the International Conference on Modern Research in Aerospace Engineering*, Lecture Notes in Mechanical Engineering,
https://doi.org/10.1007/978-981-10-5849-3_35

2 SRAM Cell

A low power Static RAM cell may be designed by using cross-coupled connected inverters. Here, states are held by cross-coupled inverters (M1, M2, M5, and M6). The advantage of this arrangement is that the leakage power dissipation is very small due to very small leakage current, and another advantage is great immune to noise and it can be operated at low supply voltage (Fig. 1).

3 IO Standard

3.1 LVTTL

For seven series FPGA, high range IO bank is available. LVTTL is a general purpose IO standard used for 3.3 V operation voltage circuit applications [7]. This IO buffer uses CMOS single-ended input buffer and push-pull output buffer which results in high-speed operation and less delay. This standard uses zero-volt reference voltage/termination voltage. LVTTL is available for both unidirectional and bidirectional termination (Figs. 2, 3, 4, 5 and 6).

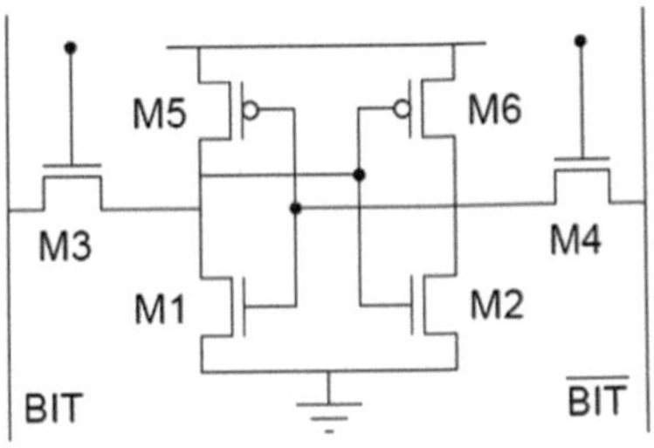

Fig. 1 CMOS SRAM 6T cell

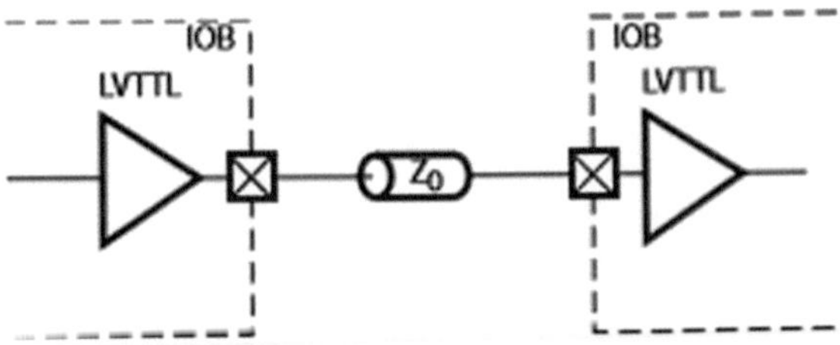

Fig. 2 LVTTL unidirectional termination

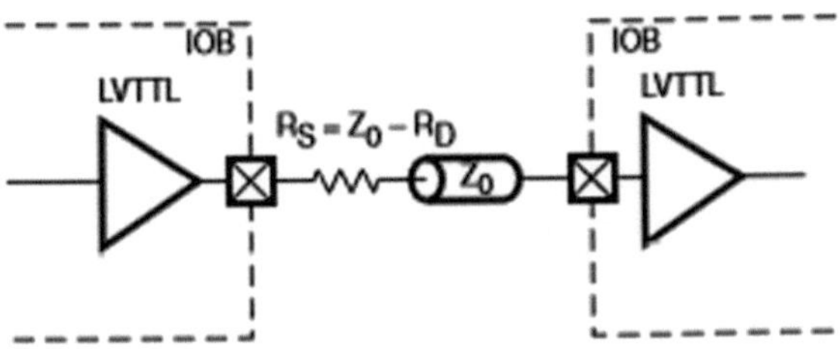

Fig. 3 LVTTL unidirectional termination for impedance matching

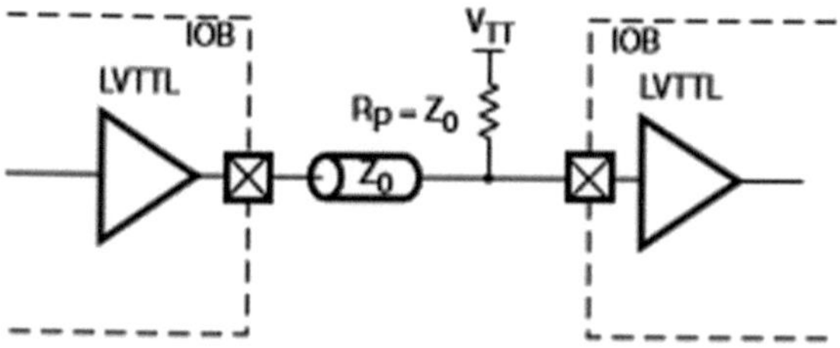

Fig. 4 LVTTL unidirectional termination for avoid transmission line effect

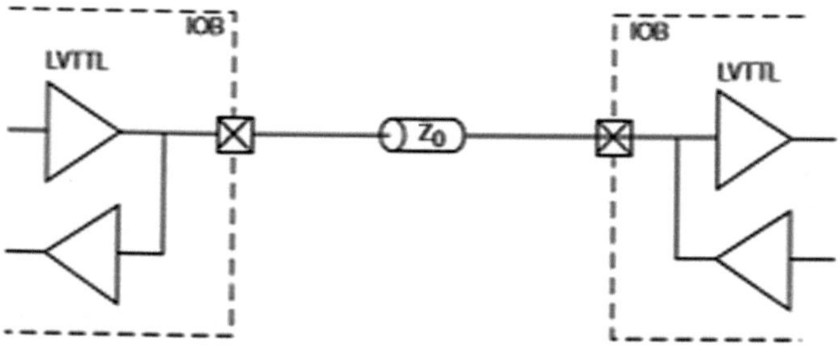

Fig. 5 LVTTL bidirectional termination

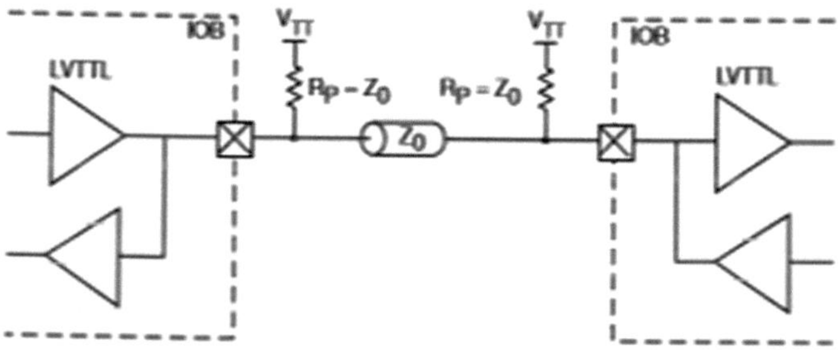

Fig. 6 LVTTL bidirectional termination for avoiding transmission line effect

3.2 *Mobile-DDR*

It is used for low power applications, so it is also known as "LP-DDR." It can be used up-to 1.8 V applications. It uses single-ended CMOS input buffer and differential output buffer (Figs. 7, 8 and 9).

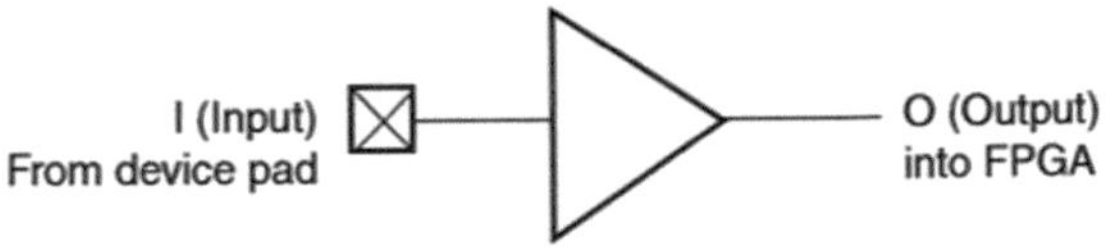

Fig. 7 Single-ended input buffer

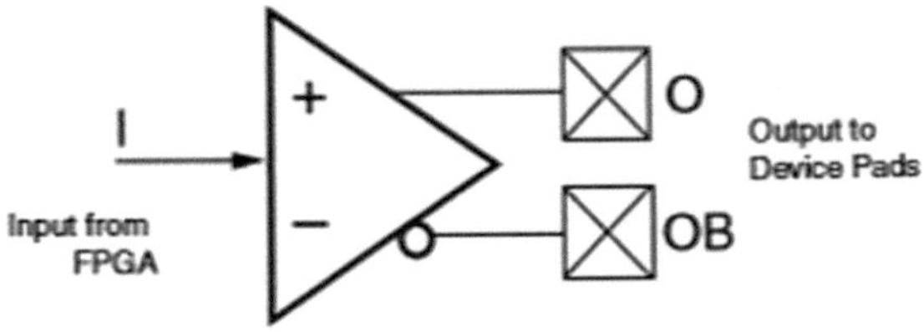

Fig. 8 Differential output buffer

Fig. 9 Artix-7, XC7A100T device

4 Proposed Architecture of Memory

In this work, SRAM is designed using HDL. The size of memory is 64 kb, where the input bit is 64. Figure 10 shows the RTL diagram of memory (Figs. 11 and 12).

5 Power Analysis

5.1 Power Consumption at 1.0 GHz Operating Frequency

In this work, power dissipation is calculated by using two different IO Standards. Table 1 shows the various powers by using LVTTL and Mobile-DDR IO standards. Figure 13 shows the variation in total power dissipation by using LVTTL and Mobile-DDR IO Standards. By comparing both IO Standards, Mobile-DDR dissipates less power as compared with LVTTL IO Standard. By using Mobile-DDR in place of LVTTL, there is saving of 43.45% IO power and 31.26% total power.

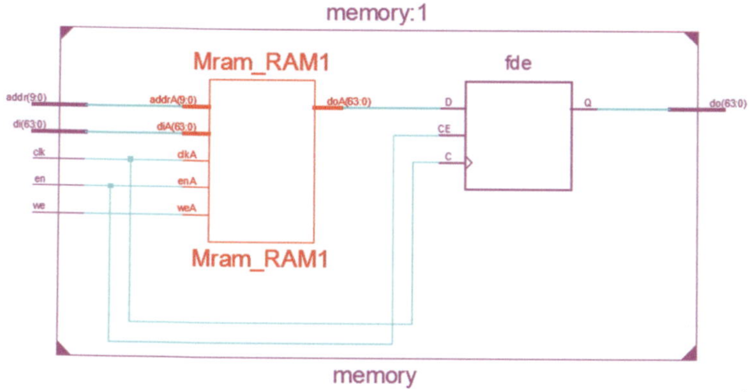

Fig. 10 RTL diagram of 64 kb memory

Fig. 11 Package view of memory

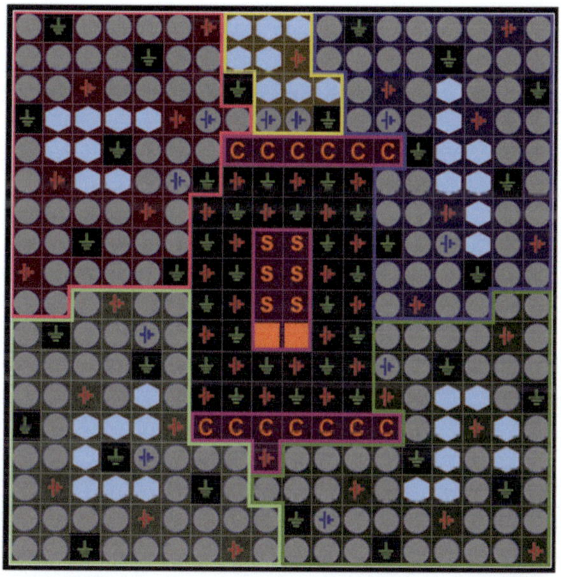

Fig. 12 Power categorization

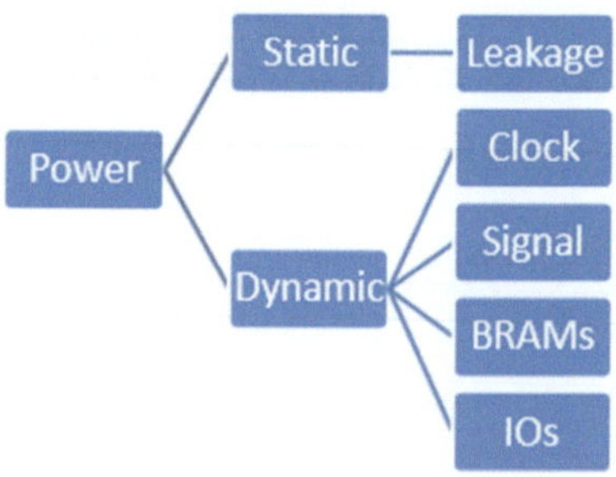

Table 1 Power consumption at 1 GHz operating frequency

IO standard	Clock (w)	Signal (w)	BRAMs (w)	IOs (w)	Leakage (w)	Total (w)
LVTTL	0.012	0.029	0.125	0.695	0.122	0.982
Mobile-DDR	0.012	0.029	0.125	0.393	0.116	0.675

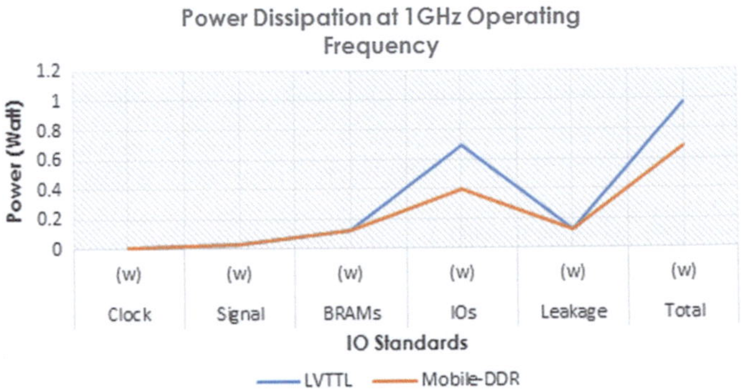

Fig. 13 Power dissipation at 1 GHz operating frequency

Table 2 Power consumption at 1.5 GHz operating frequency

IO standard	Clock (w)	Signal (w)	BRAMs (w)	IOs (w)	Leakage (w)	Total (w)
LVTTL	0.018	0.043	0.187	1.042	0.121	1.418
Mobile-DDR	0.018	0.043	0.187	0.497	0.119	0.865

5.2 *Power Consumption at 1.5 GHz Operating Frequency*

In this work, power dissipation is calculated for 64 kb SRAM memory by using LVTTL and Mobile-DDR IO Standards. Table 2 illustrates the amount of power dissipation by using Mobile-DDR and LVTTL IO standards. As a result, between LVTTL and Mobile-DDR IO standards, LVTTL dissipates more power as compared with Mobile-DDR. Figure 14 shows the variation in total power dissipation by using both IO Standards. If Mobile-DDR is used in place of LVTTL IO Standard, we can save 52.30% IO Power and 39% total power at 1.5 GHz operating frequency.

5.3 *Power Consumption at 2 GHz Operating Frequency*

In this work, clock, signal, BRAMs, Ios, and leakage power dissipation are calculated by using Mobile-DDR and LVTTL IO Standard. Table 3 illustrates the

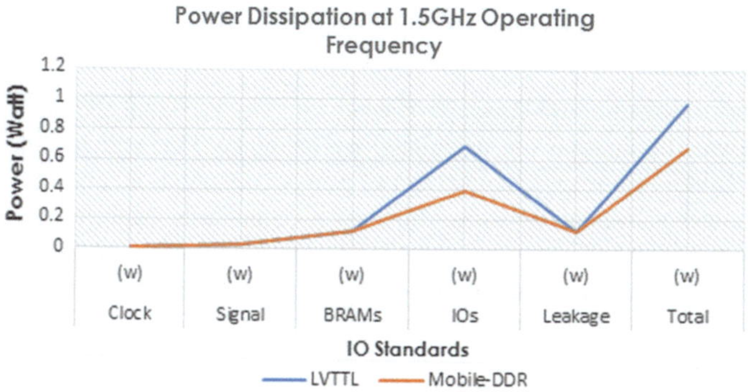

Fig. 14 Power dissipation at 1.5 GHz operating frequency

Table 3 Power consumption at 2 GHz operating frequency

IO standard	Clock (w)	Signal (w)	BRAMs (w)	IOs (w)	Leakage (w)	Total (w)
LVTTL	0.025	0.058	0.249	1.389	0.133	1.854
Mobile-DDR	0.025	0.058	0.249	0.601	0.121	1.054

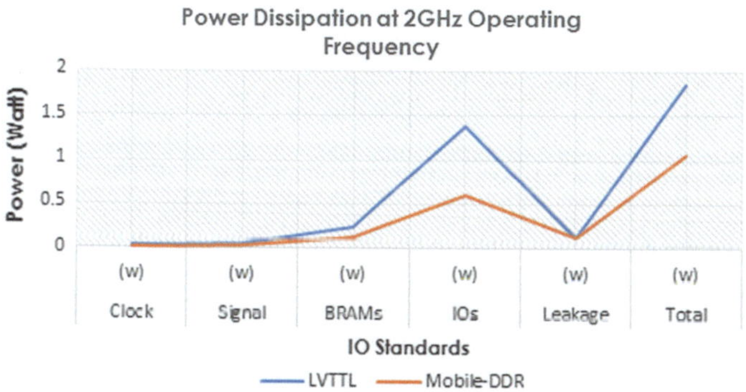

Fig. 15 Power dissipation at 2 GHz operating frequency

amount of power consumed by 64 kb SRAM by using both IO standards, and Fig. 15 shows the variation in total power dissipation by using different IO Standards. And find that between these IO standards Mobile-DDR IO Standard is more power efficient one and LVTTL dissipates more power. At 2 GHz operating frequency when Mobile-DDR is used in place of LVTTL, 43.14% total power and 56.73% IO power can be saved.

5.4 Power Consumption at 2.5 GHz Operating Frequency

In this work, all static and dynamic power, i.e., clock, signal, BRAMs, Ios, and leakage power dissipation are calculated by using LVTTL and Mobile-DDR IO Standard. Table 4 illustrates the amount of power consumption, and Fig. 16 shows the total power variations by using different IO Standards. Mobile-DDR IO Standard is more power efficient one, and LVTTL dissipates more power. At 2.5 GHz operating frequency when Mobile-DDR is used in place of LVTTL, 45.8% total power and 59.30% IO power can be saved.

5.5 Power Consumption at 3 GHz Operating Frequency

Clock, signal, BRAMs, Ios, and leakage power dissipation are calculated by using different IO Standards. Table 5 illustrates the amount of power consumed by 64 kb SRAM by using LVTTL and Mobile-DDR IO standards, and Fig. 17 shows the variation in total power dissipation by using different IO Standards. Between both IO standards, Mobile-DDR IO Standard is more power efficient. At 3 GHz operating frequency when Mobile-DDR is used in place of LVTTL, 47.45% total power and 61.14% IO power can be saved.

Table 4 Power consumption at 2.5 GHz operating frequency

IO standard	Clock (w)	Signal (w)	BRAMs (w)	IOs (w)	Leakage (w)	Total (w)
LVTTL	0.031	0.072	0.311	1.737	0.139	2.291
Mobile-DDR	0.031	0.072	0.311	0.706	0.124	1.244

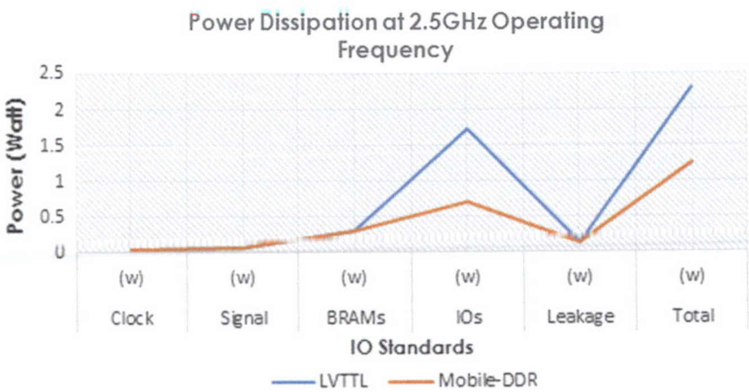

Fig. 16 Power dissipation at 2.5 GHz operating frequency

Table 5 Power consumption at 3 GHz operating frequency

IO standard	Clock (w)	Signal (w)	BRAMs (w)	IOs (w)	Leakage (w)	Total (w)
LVTTL	0.037	0.087	0.374	2.084	0.146	2.727
Mobile-DDR	0.037	0.087	0.374	0.810	0.126	1.433

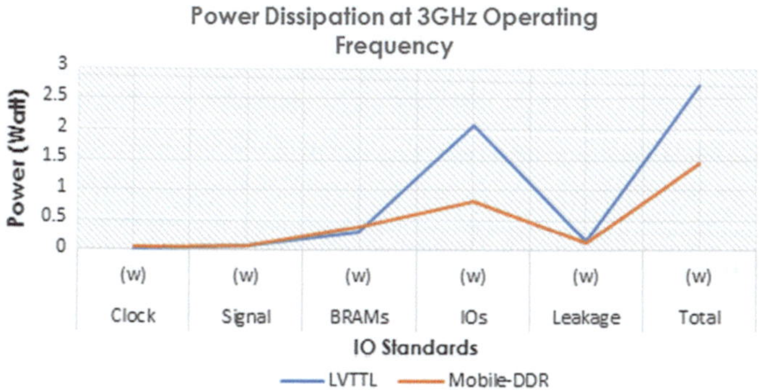

Fig. 17 Power dissipation at 3 GHz operating frequency

Table 6 Power consumption at 3.5 GHz operating frequency

IO standard	Clock (w)	Signal (w)	BRAMs (w)	IOs (w)	Leakage (w)	Total (w)
LVTTL	0.043	0.101	0.436	2.431	0.153	3.165
Mobile-DDR	0.043	0.101	0.436	0.914	0.129	1.623

5.6 *Power Consumption at 3.5 GHz Operating Frequency*

In this section, clock, signal, BRAMs, Ios, and leakage power dissipation are calculated by using different IO Standards. Table 6 illustrates the amount of power consumed by 64 kb SRAM by using Mobile-DDR and LVTTL IO standard, and Fig. 18 shows the variation in total power dissipation by using different IO Standards. Mobile-DDR IO Standard is more power efficient. At 3.5 GHz operating frequency when Mobile-DDR is used in place of LVTTL then 15.6% leakage power, 48.72% total power, and 62.40% IO power can be saved.

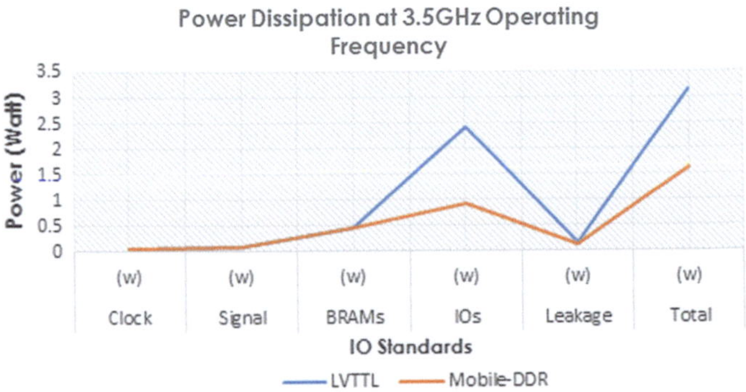

Fig. 18 Power dissipation at 3.5 GHz operating frequency

6 Conclusion

In this work, 64 kb SRAM is designed on 28 nm Artix-7 FPGA and LVTTL and Mobile-DDR IO standard. At different operating frequency, power dissipation is calculated and compared with each other and find that Mobile-DDR is more power efficient among all, at 1, 2, and 3 GHz operating frequency if Mobile-DDR is used in place of LVTTL, 31.26, 43.14, and 47.45% total power can be saved and 43.45, 56.73, and 61.14% IO power can be saved.

7 Future Scopes

In this work, we are using LVTTL and Mobile-DDR IO standards in energy efficient design of RAM. There is also open scope to design memory using different IO standards like LVCMOS, HSUL, HSTL, SSTL, and PCI. Now, our target design is to design memory on 18 nm technology and using FinFet. We can migrate to 3-D IC, system on chips (SoCs), and other programmable logic devices and re-implement same memory on that. In future, we shall design high-speed memory that will be able to operate with frequency of 1 THz and beyond.

References

1. Yu KC, Fan ML, SU P (Member, IEEE), Chuang CT (2016) Evaluation of monolithic 3-D logic circuits and 6T SRAMs with InGaAs-n/Ge-p ultra-thin-body MOSFETs. J Electron Device Soc
2. Wilkerson C, Alameldeen AR, Chishti Z, Wu W, Somasekhar D, Lu S-L (2010) Reducing cache power with low-cost, multi-bit error-correcting codes. In: Proceedings of the 37th Annual ISCA, pp 92–93

3. Ahn J-H, Hong S-H, Kim S-J, Ko J-B, Shin S-W, Lee S-D et al (2004) An experimental 256 Mb non-volatile DRAM with cell plate boosted programming technique. In: Proceedings, IEEE ISSCC, vol 1, pp 42–512
4. Agrawal T, Srivastava V, Kumar A (2015) Designing of power efficient ROM using LVTTL and mobile-DDR IO standard on 28nm FPGA. In: International conference on computational intelligence and communication networks
5. Pandey B, Pattanaik M (2013) Energy efficient VLSI design and implementation on 28 nm FPGA, Lambert Academic Publisher, Germany, ISBN: 978-3-659-47759-1, EAN: 9783659477591
6. Calson I, Andersson S, Natarajan S, Alvandpour A (2004) Solid-State Circuits Conference. p 215
7. Lee D (2013) Area efficient ROM-embedded SRAM cache. IEEE Trans Very Large Scale Integr VLSI Sys 21(9)

FSC
www.fsc.org
MIX
Papier aus verantwortungsvollen Quellen
Paper from responsible sources
FSC® C105338

If you have any concerns about our products,
you can contact us on
ProductSafety@springernature.com

In case Publisher is established outside the EU,
the EU authorized representative is:
Springer Nature Customer Service Center GmbH
Europaplatz 3, 69115 Heidelberg, Germany

Printed by Libri Plureos GmbH
in Hamburg, Germany